Code of Federal Regulations

CODE OF FEDERAL REGULATIONS

Title 7
Agriculture

Parts 1000 to 1199

Revised as of January 1, 2020

Containing a codification of documents
of general applicability and future effect

As of January 1, 2020

Published by the Office of the Federal Register
National Archives and Records Administration
as a Special Edition of the Federal Register

T0175237

Table of Contents

Cite this Code: **CFR**

To cite the regulations in this volume use title, part and section number. Thus, 7 CFR 1000.1 refers to title 7, part 1000, section 1.

Explanation

The Code of Federal Regulations is a codification of the general and permanent rules published in the Federal Register by the Executive departments and agencies of the Federal Government. The Code is divided into 50 titles which represent broad areas subject to Federal regulation. Each title is divided into chapters which usually bear the name of the issuing agency. Each chapter is further subdivided into parts covering specific regulatory areas.

Each volume of the Code is revised at least once each calendar year and issued on a quarterly basis approximately as follows:

Title 1 through Title 16..as of January 1
Title 17 through Title 27..as of April 1
Title 28 through Title 41..as of July 1
Title 42 through Title 50..as of October 1

The appropriate revision date is printed on the cover of each volume.

LEGAL STATUS

The contents of the Federal Register are required to be judicially noticed (44 U.S.C. 1507). The Code of Federal Regulations is prima facie evidence of the text of the original documents (44 U.S.C. 1510).

HOW TO USE THE CODE OF FEDERAL REGULATIONS

The Code of Federal Regulations is kept up to date by the individual issues of the Federal Register. These two publications must be used together to determine the latest version of any given rule.

To determine whether a Code volume has been amended since its revision date (in this case, January 1, 2020), consult the "List of CFR Sections Affected (LSA)," which is issued monthly, and the "Cumulative List of Parts Affected," which appears in the Reader Aids section of the daily Federal Register. These two lists will identify the Federal Register page number of the latest amendment of any given rule.

EFFECTIVE AND EXPIRATION DATES

Each volume of the Code contains amendments published in the Federal Register since the last revision of that volume of the Code. Source citations for the regulations are referred to by volume number and page number of the Federal Register and date of publication. Publication dates and effective dates are usually not the same and care must be exercised by the user in determining the actual effective date. In instances where the effective date is beyond the cut-off date for the Code a note has been inserted to reflect the future effective date. In those instances where a regulation published in the Federal Register states a date certain for expiration, an appropriate note will be inserted following the text.

OMB CONTROL NUMBERS

The Paperwork Reduction Act of 1980 (Pub. L. 96–511) requires Federal agencies to display an OMB control number with their information collection request.

Many agencies have begun publishing numerous OMB control numbers as amendments to existing regulations in the CFR. These OMB numbers are placed as close as possible to the applicable recordkeeping or reporting requirements.

PAST PROVISIONS OF THE CODE

Provisions of the Code that are no longer in force and effect as of the revision date stated on the cover of each volume are not carried. Code users may find the text of provisions in effect on any given date in the past by using the appropriate List of CFR Sections Affected (LSA). For the convenience of the reader, a "List of CFR Sections Affected" is published at the end of each CFR volume. For changes to the Code prior to the LSA listings at the end of the volume, consult previous annual editions of the LSA. For changes to the Code prior to 2001, consult the List of CFR Sections Affected compilations, published for 1949-1963, 1964-1972, 1973-1985, and 1986-2000.

"[RESERVED]" TERMINOLOGY

The term "[Reserved]" is used as a place holder within the Code of Federal Regulations. An agency may add regulatory information at a "[Reserved]" location at any time. Occasionally "[Reserved]" is used editorially to indicate that a portion of the CFR was left vacant and not dropped in error.

INCORPORATION BY REFERENCE

What is incorporation by reference? Incorporation by reference was established by statute and allows Federal agencies to meet the requirement to publish regulations in the Federal Register by referring to materials already published elsewhere. For an incorporation to be valid, the Director of the Federal Register must approve it. The legal effect of incorporation by reference is that the material is treated as if it were published in full in the Federal Register (5 U.S.C. 552(a)). This material, like any other properly issued regulation, has the force of law.

What is a proper incorporation by reference? The Director of the Federal Register will approve an incorporation by reference only when the requirements of 1 CFR part 51 are met. Some of the elements on which approval is based are:

(a) The incorporation will substantially reduce the volume of material published in the Federal Register.

(b) The matter incorporated is in fact available to the extent necessary to afford fairness and uniformity in the administrative process.

(c) The incorporating document is drafted and submitted for publication in accordance with 1 CFR part 51.

What if the material incorporated by reference cannot be found? If you have any problem locating or obtaining a copy of material listed as an approved incorporation by reference, please contact the agency that issued the regulation containing that incorporation. If, after contacting the agency, you find the material is not available, please notify the Director of the Federal Register, National Archives and Records Administration, 8601 Adelphi Road, College Park, MD 20740-6001, or call 202-741-6010.

CFR INDEXES AND TABULAR GUIDES

A subject index to the Code of Federal Regulations is contained in a separate volume, revised annually as of January 1, entitled CFR INDEX AND FINDING AIDS. This volume contains the Parallel Table of Authorities and Rules. A list of CFR titles, chapters, subchapters, and parts and an alphabetical list of agencies publishing in the CFR are also included in this volume.

An index to the text of "Title 3—The President" is carried within that volume.

The Federal Register Index is issued monthly in cumulative form. This index is based on a consolidation of the "Contents" entries in the daily Federal Register.

A List of CFR Sections Affected (LSA) is published monthly, keyed to the revision dates of the 50 CFR titles.

REPUBLICATION OF MATERIAL

There are no restrictions on the republication of material appearing in the Code of Federal Regulations.

INQUIRIES

For a legal interpretation or explanation of any regulation in this volume, contact the issuing agency. The issuing agency's name appears at the top of odd-numbered pages.

For inquiries concerning CFR reference assistance, call 202–741–6000 or write to the Director, Office of the Federal Register, National Archives and Records Administration, 8601 Adelphi Road, College Park, MD 20740-6001 or e-mail *fedreg.info@nara.gov*.

THIS TITLE

Title 7—AGRICULTURE is composed of fifteen volumes. The parts in these volumes are arranged in the following order: Parts 1–26, 27–52, 53–209, 210–299, 300–399, 400–699, 700–899, 900–999, 1000–1199, 1200–1599, 1600–1759, 1760–1939, 1940–1949, 1950–1999, and part 2000 to end. The contents of these volumes represent all current regulations codified under this title of the CFR as of January 1, 2020.

The Food and Nutrition Service current regulations in the volume containing parts 210–299 include the Child Nutrition Programs and the Food Stamp Program. The regulations of the Federal Crop Insurance Corporation are found in the volume containing parts 400–699.

All marketing agreements and orders for fruits, vegetables and nuts appear in the one volume containing parts 900–999. All marketing agreements and orders for milk appear in the volume containing parts 1000–1199.

For this volume, Susannah C. Hurley was Chief Editor. The Code of Federal Regulations publication program is under the direction of John Hyrum Martinez, assisted by Stephen J. Frattini.

Title 7—Agriculture

(This book contains parts 1000 to 1199)

SUBTITLE B—REGULATIONS OF THE DEPARTMENT OF AGRICULTURE (CONTINUED)

Subtitle B—Regulations of the Department of Agriculture (Continued)

CHAPTER X—AGRICULTURAL MARKETING SERVICE (MARKETING AGREEMENTS AND ORDERS; MILK), DEPARTMENT OF AGRICULTURE

PART 1000—GENERAL PROVISIONS OF FEDERAL MILK MARKETING ORDERS

AUTHORITY: 7 U.S.C. 601–674, and 7253.

SOURCE: 64 FR 47899, Sept. 1, 1999, unless otherwise noted.

Subpart A—Scope and Purpose

§ 1000.1 Scope and purpose of this part 1000.

This part sets forth certain terms, definitions, and provisions which shall be common to and apply to Federal milk marketing order in 7 CFR, chapter X, except as specifically defined otherwise, or modified, or otherwise provided, in an individual order in 7 CFR, chapter X.

Subpart B—Definitions

§ 1000.2 General definitions.

(a) *Act* means Public Act No. 10, 73d Congress, as amended and as reenacted and amended by the Agricultural Marketing Agreement Act of 1937, as amended (7 U.S.C. 601 *et seq.*).

(b) *Order* or *Federal milk order* means the applicable part of 7 CFR, chapter X, issued pursuant to Section 8c of the Act as a Federal milk marketing order (as amended).

(c) *Department* means the U.S. Department of Agriculture.

(d) *Secretary* means the Secretary of Agriculture of the United States or any officer or employee of the Department to whom authority has heretofore been delegated, or to whom authority may hereafter be delegated, to act in his stead.

(e) *Person* means any individual, partnership, corporation, association, or other business unit.

§ 1000.3 Route disposition.

Route disposition means a delivery to a retail or wholesale outlet (except a plant), either directly or through any distribution facility (including disposition from a plant store, vendor, or vending machine) of a fluid milk product in consumer-type packages or dispenser units classified as Class I milk.

§ 1000.4 Plant.

(a) Except as provided in paragraph (b) of this section, *plant* means the land, buildings, facilities, and equipment constituting a single operating unit or establishment at which milk or milk products are received, processed, or packaged, including a facility described in paragraph (b)(2) of this section if the facility receives the milk of more than one dairy farmer.

(b) Plant shall not include:

(1) A separate building without stationary storage tanks that is used only as a reload point for transferring bulk milk from one tank truck to another or a separate building used only as a distribution point for storing packaged fluid milk products in transit for route disposition; or

(2) An on-farm facility operated as part of a single dairy farm entity for the separation of cream and skim or the removal of water from milk.

§ 1000.5 Distributing plant.

Distributing plant means a plant that is approved by a duly constituted regulatory agency for the handling of Grade A milk at which fluid milk products are processed or packaged and from which there is route disposition or transfers of packaged fluid milk products to other plants.

§ 1000.6 Supply plant.

Supply plant means a plant approved by a duly constituted regulatory agency for the handling of Grade A milk that receives milk directly from dairy farmers and transfers or diverts fluid milk products to other plants or manufactures dairy products on its premises.

§ 1000.8 Nonpool plant.

Nonpool plant means any milk receiving, manufacturing, or processing plant other than a pool plant. The following categories of nonpool plants are further defined as follows:

(a) *A plant fully regulated under another Federal order* means a plant that is fully subject to the pricing and pooling provisions of another Federal order.

(b) *Producer-handler plant* means a plant operated by a producer-handler as defined under any Federal order.

(c) *Partially regulated distributing plant* means a nonpool plant that is not a plant fully regulated under another Federal order, a producer-handler plant, or an exempt plant, from which there is route disposition in the marketing area during the month.

(d) *Unregulated supply plant* means a supply plant that does not qualify as a pool supply plant and is not a plant fully regulated under another Federal order, a producer-handler plant, or an exempt plant.

(e) *An exempt plant* means a plant described in this paragraph that is exempt from the pricing and pooling provisions of any order provided that the operator of the plant files reports as prescribed by the market administrator of any marketing area in which the plant distributes packaged fluid milk products to enable determination of the handler's exempt status:

(1) A plant that is operated by a governmental agency that has no route disposition in commercial channels;

(2) A plant that is operated by a duly accredited college or university disposing of fluid milk products only through the operation of its own facilities with no route disposition in commercial channels;

(3) A plant from which the total route disposition is for individuals or institutions for charitable purposes without remuneration; or

(4) A plant that has route disposition and packaged sales of fluid milk products to other plants of 150,000 pounds or less during the month.

§ 1000.9 Handler.

Handler means:

(a) Any person who operates a pool plant or a nonpool plant.

(b) Any person who receives packaged fluid milk products from a plant for resale and distribution to retail or wholesale outlets, any person who as a broker negotiates a purchase or sale of fluid milk products or fluid cream products from or to any pool or nonpool plant, and any person who by purchase or direction causes milk of producers to be picked up at the farm and/or moved to a plant. Persons who qualify as handlers only under this paragraph under any Federal milk order are not subject to the payment provisions of §§_____.70, _____.71, _____.72, _____.73, _____.76, and _____.85 of that order.

(c) Any cooperative association with respect to milk that it receives for its account from the farm of a producer and delivers to pool plants or diverts to nonpool plants pursuant to § _____.13 of the order. The operator of a pool plant receiving milk from a cooperative association may be the handler for such milk if both parties notify the market administrator of this agreement prior to the time that the milk is delivered to the pool plant and the plant operator purchases the milk on the basis of farm bulk tank weights and samples.

§ 1000.14 Other source milk.

Other source milk means all skim milk and butterfat contained in or represented by:

(a) Receipts of fluid milk products and bulk fluid cream products from any source other than producers, handlers described in § 1000.9(c) and § 1135.11, or pool plants;

(b) Products (other than fluid milk products, fluid cream products, and products produced at the plant during the same month) from any source which are reprocessed, converted into, or combined with another product in the plant during the month; and

(c) Receipts of any milk product (other than a fluid milk product or a fluid cream product) for which the handler fails to establish a disposition.

§ 1000.15 Fluid milk product.

(a) Except as provided in paragraph (b) of this section, *fluid milk product* shall mean any milk products in fluid or frozen form that are intended to be used as beverages containing less than 9 percent butterfat and 6.5 percent or more nonfat solids or 2.25 percent or more true milk protein. Sources of such nonfat solids/protein include but are not limited to: Casein, whey protein concentrate, milk protein concentrate, dry whey, caseinates, lactose, and any similar dairy derived ingredient. Such products include, but are not limited to: Milk, fat-free milk, lowfat milk, light milk, reduced fat milk, milk drinks, eggnog and cultured buttermilk, including any such beverage products that are flavored, cultured, modified with added or reduced nonfat solids, sterilized, concentrated, or reconstituted. As used in this part, the term concentrated milk means milk that contains not less than 25.5 percent, and not more than 50 percent, total milk solids.

(b) The term fluid milk product shall not include:

(1) Any product that contains less than 6.5 percent nonfat milk solids and contains less than 2.25 percent true milk protein; whey; plain or sweetened evaporated milk/skim milk; sweetened condensed milk/skim milk; yogurt containing beverages with 20 or more percent yogurt by weight and kefir; products especially prepared for infant feeding or dietary use (meal replacement) that are packaged in hermetically sealed containers; and products that meet the compositional standards specified in paragraph (a) of this section but contain no fluid milk products included in paragraph (a) of this section.

(2) The quantity of skim milk equivalent in any modified product specified in paragraph (a) of this section that is greater than an equal volume of an unmodified product of the same nature and butterfat content.

[64 FR 47899, Sept. 1, 1999, as amended at 75 FR 51931, Aug. 24, 2010]

§ 1000.16 Fluid cream product.

Fluid cream product means cream (other than plastic cream or frozen cream), including sterilized cream, or a mixture of cream and milk or skim milk containing 9 percent or more butterfat, with or without the addition of other ingredients.

§ 1000.17 [Reserved]

§ 1000.18 Cooperative association.

Cooperative association means any cooperative marketing association of producers which the Secretary determines is qualified under the provisions of the Capper-Volstead Act, has full authority in the sale of milk of its members, and is engaged in marketing milk or milk products for its members. A federation of 2 or more cooperatives incorporated under the laws of any state will be considered a cooperative association under any Federal milk order if all member cooperatives meet the requirements of this section.

§ 1000.19 Commercial food processing establishment.

Commercial food processing establishment means any facility, other than a milk plant, to which fluid milk products and fluid cream products are disposed of, or producer milk is diverted, that uses such receipts as ingredients in food products and has no other disposition of fluid milk products other than those received in consumer-type packages (1 gallon or less). Producer milk diverted to commercial food processing establishments shall be subject to the same provisions relating to diversions to plants, including, but not limited to, §§ _____.13 and _____.52 of each Federal milk order.

Subpart C—Rules of Practice and Procedure Governing Market Administrators

§ 1000.25 Market administrator.

(a) *Designation.* The agency for the administration of the order shall be a market administrator selected by the Secretary and subject to removal at the Secretary's discretion. The market administrator shall be entitled to compensation determined by the Secretary.

(b) *Powers.* The market administrator shall have the following powers with respect to each order under his/her administration:

(1) Administer the order in accordance with its terms and provisions;

(2) Maintain and invest funds outside of the United States Department of the Treasury for the purpose of administering the order;

(3) Make rules and regulations to effectuate the terms and provisions of the order;

(4) Receive, investigate, and report complaints of violations to the Secretary; and

(5) Recommend amendments to the Secretary.

(c) *Duties.* The market administrator shall perform all the duties necessary to administer the terms and provisions of each order under his/her administration, including, but not limited to, the following:

(1) Employ and fix the compensation of persons necessary to enable him/her to exercise the powers and perform the duties of the office;

(2) Pay out of funds provided by the administrative assessment, except expenses associated with functions for which the order provides a separate charge, all expenses necessarily incurred in the maintenance and functioning of the office and in the performance of the duties of the office, including the market administrator's compensation;

(3) Keep records which will clearly reflect the transactions provided for in the order and upon request by the Secretary, surrender the records to a successor or such other person as the Secretary may designate;

(4) Furnish information and reports requested by the Secretary and submit office records for examination by the Secretary;

(5) Announce publicly at his/her discretion, unless otherwise directed by the Secretary, by such means as he/she deems appropriate, the name of any handler who, after the date upon which the handler is required to perform such act, has not:

(i) Made reports required by the order;

(ii) Made payments required by the order; or

(iii) Made available records and facilities as required pursuant to § 1000.27;

(6) Prescribe reports required of each handler under the order. Verify such reports and the payments required by the order by examining records (including such papers as copies of income tax

reports, fiscal and product accounts, correspondence, contracts, documents or memoranda of the handler, and the records of any other persons that are relevant to the handler's obligation under the order), by examining such handler's milk handling facilities, and by such other investigation as the market administrator deems necessary for the purpose of ascertaining the correctness of any report or any obligation under the order. Reclassify skim milk and butterfat received by any handler if such examination and investigation discloses that the original classification was incorrect; ·

(7) Furnish each regulated handler a written statement of such handler's accounts with the market administrator promptly each month. Furnish a corrected statement to such handler if verification discloses that the original statement was incorrect; and

(8) Prepare and disseminate publicly for the benefit of producers, handlers, and consumers such statistics and other information concerning operation of the order and facts relevant to the provisions thereof (or proposed provisions) as do not reveal confidential information.

Subpart D—Rules Governing Order Provisions

§ 1000.26 Continuity and separability of provisions.

(a) *Effective time.* The provisions of the order or any amendment to the order shall become effective at such time as the Secretary may declare and shall continue in force until suspended or terminated.

(b) *Suspension or termination.* The Secretary shall suspend or terminate any or all of the provisions of the order whenever he/she finds that such provision(s) obstructs or does not tend to effectuate the declared policy of the Act. The order shall terminate whenever the provisions of the Act authorizing it cease to be in effect.

(c) *Continuing obligations.* If upon the suspension or termination of any or all of the provisions of the order there are any obligations arising under the order, the final accrual or ascertainment of which requires acts by any handler, by the market administrator or by any other person, the power and duty to perform such further acts shall continue notwithstanding such suspension or termination.

(d) *Liquidation.* (1) Upon the suspension or termination of any or all provisions of the order the market administrator, or such other liquidating agent designated by the Secretary, shall, if so directed by the Secretary, liquidate the business of the market administrator's office, dispose of all property in his/her possession or control, including accounts receivable, and execute and deliver all assignments or other instruments necessary or appropriate to effectuate any such disposition; and

(2) If a liquidating agent is so designated, all assets and records of the market administrator shall be transferred promptly to such liquidating agent. If, upon such liquidation, the funds on hand exceed the amounts required to pay outstanding obligations of the office of the market administrator and to pay necessary expenses of liquidation and distribution, such excess shall be distributed to contributing handlers and producers in an equitable manner.

(e) *Separability of provisions.* If any provision of the order or its application to any person or circumstances is held invalid, the application of such provision and of the remaining provisions of the order to other persons or circumstances shall not be affected thereby.

Subpart E—Rules of Practice and Procedure Governing Handlers

§ 1000.27 Handler responsibility for records and facilities.

Each handler shall maintain and retain records of its operations and make such records and its facilities available to the market administrator. If adequate records of a handler, or of any other persons, that are relevant to the obligation of such handler are not maintained and made available, any skim milk and butterfat required to be reported by such handler for which adequate records are not available shall be considered as used in the highest-priced class.

(a) *Records to be maintained.* (1) Each handler shall maintain records of its

11

operations (including, but not limited to, records of purchases, sales, processing, packaging, and disposition) as are necessary to verify whether such handler has any obligation under the order and if so, the amount of such obligation. Such records shall be such as to establish for each plant or other receiving point for each month:

(i) The quantities of skim milk and butterfat contained in, or represented by, products received in any form, including inventories on hand at the beginning of the month, according to form, time, and source of each receipt;

(ii) The utilization of all skim milk and butterfat showing the respective quantities of such skim milk and butterfat in each form disposed of or on hand at the end of the month; and

(iii) Payments to producers, dairy farmers, and cooperative associations, including the amount and nature of any deductions and the disbursement of money so deducted.

(2) Each handler shall keep such other specific records as the market administrator deems necessary to verify or establish such handler's obligation under the order.

(b) *Availability of records and facilities.* Each handler shall make available all records pertaining to such handler's operations and all facilities the market administrator finds are necessary to verify the information required to be reported by the order and/or to ascertain such handler's reporting, monetary, or other obligation under the order. Each handler shall permit the market administrator to weigh, sample, and test milk and milk products and observe plant operations and equipment and make available to the market administrator such facilities as are necessary to carry out his/her duties.

(c) *Retention of records.* All records required under the order to be made available to the market administrator shall be retained by the handler for a period of 3 years to begin at the end of the month to which such records pertain. If, within such 3-year period, the market administrator notifies the handler in writing that the retention of such records, or of specified records, is necessary in connection with a proceeding under section 8c(15)(A) of the

Act or a court action specified in such notice, the handler shall retain such records, or specified records, until further written notification from the market administrator. The market administrator shall give further written notification to the handler promptly upon the termination of the litigation or when the records are no longer necessary in connection therewith.

§ 1000.28   Termination of obligations.

(a) Except as provided in paragraphs (b) and (c) of this section, the obligation of any handler to pay money required to be paid under the terms of the order shall terminate 2 years after the last day of the month during which the market administrator receives the handler's report of receipts and utilization on which such obligation is based, unless within such 2-year period, the market administrator notifies the handler in writing that such money is due and payable. Service of such written notice shall be complete upon mailing to the handler's last known address and it shall contain, but need not be limited to, the following information:

(1) The amount of the obligation;

(2) The month(s) on which such obligation is based; and

(3) If the obligation is payable to one or more producers or to a cooperative association, the name of such producer(s) or such cooperative association, or if the obligation is payable to the market administrator, the account for which it is to be paid.

(b) If a handler fails or refuses, with respect to any obligation under the order, to make available to the market administrator all records required by the order to be made available, the market administrator may notify the handler in writing, within the 2-year period provided for in paragraph (a) of this section, of such failure or refusal. If the market administrator so notifies a handler, the said 2-year period with respect to such obligation shall not begin to run until the first day of the month following the month during which all such records pertaining to such obligation are made available to the market administrator.

(c) Notwithstanding the provisions of paragraphs (a) and (b) of this section, a handler's obligation under the order to pay money shall not be terminated with respect to any transaction involving fraud or willful concealment of a fact, material to the obligation, on the part of the handler against whom the obligation is sought to be imposed.

(d) Unless the handler files a petition pursuant to section 8c(15)(A) of the Act and the applicable rules and regulations (7 CFR 900.50 through 900.71) within the applicable 2-year period indicated below, the obligation of the market administrator:

(1) To pay a handler any money which such handler claims is due under the terms of the order shall terminate 2 years after the end of the month during which the skim milk and butterfat involved in the claim were received; or

(2) To refund any payment made by a handler (including a deduction or offset by the market administrator) shall terminate 2 years after the end of the month during which payment was made by the handler.

Subpart F—Classification of Milk

§ 1000.40   Classes of utilization.

Except as provided in § 1000.42, all skim milk and butterfat required to be reported pursuant to § ——.30 of each Federal milk order shall be classified as follows:

(a) *Class I milk* shall be all skim milk and butterfat:

(1) Disposed of in the form of fluid milk products, except as otherwise provided in this section;

(2) In packaged fluid milk products in inventory at the end of the month; and

(3) In shrinkage assigned pursuant to § 1000.43(b).

(b) *Class II milk* shall be all skim milk and butterfat:

(1) In fluid milk products in containers larger than 1 gallon and fluid cream products disposed of or diverted to a commercial food processing establishment if the market administrator is permitted to audit the records of the commercial food processing establishment for the purpose of verification. Otherwise, such uses shall be Class I;

(2) Used to produce:

(i) Cottage cheese, lowfat cottage cheese, dry curd cottage cheese, ricotta cheese, pot cheese, Creole cheese, and any similar soft, high-moisture cheese resembling cottage cheese in form or use;

(ii) Milkshake and ice milk mixes (or bases), frozen desserts, and frozen dessert mixes distributed in half-gallon containers or larger and intended to be used in soft or semi-solid form;

(iii) Aerated cream, frozen cream, sour cream, sour half-and-half, sour cream mixtures containing non-milk items; yogurt, including yogurt containing beverages with 20 percent or more yogurt by weight and kefir, and any other semi-solid product resembling a Class II product;

(iv) Custards, puddings, pancake mixes, coatings, batter, and similar products;

(v) Buttermilk biscuit mixes and other buttermilk for baking that contain food starch in excess of 2% of the total solids, provided that the product is labeled to indicate the food starch content;

(vi) Products especially prepared for infant feeding or dietary use (meal replacements) that are packaged in hermetically sealed containers and products that meet the compositional standards of § 1000.15(a) but contain no fluid milk products included in § 1000.15(a).

(vii) Candy, soup, bakery products and other prepared foods which are processed for general distribution to the public, and intermediate products, including sweetened condensed milk, to be used in processing such prepared food products;

(viii) A fluid cream product or any product containing artificial fat or fat substitutes that resembles a fluid cream product, except as otherwise provided in paragraph (c) of this section; and

(ix) Any product not otherwise specified in this section; and

(3) In shrinkage assigned pursuant to § 1000.43(b).

(c) *Class III milk* shall be all skim milk and butterfat:

(1) Used to produce:

(i) Cream cheese and other spreadable cheeses, and hard cheese of types that may be shredded, grated, or crumbled;

(ii) Plastic cream, anhydrous milkfat, and butteroil; and

(2) In shrinkage assigned pursuant to § 1000.43(b).

(d) *Class IV milk* shall be all skim milk and butterfat:

(1) Used to produce:

(i) Butter; and

(ii) Evaporated or sweetened condensed milk in a consumer-type package; and

(iii) Any milk product in dried form;

(2) In inventory at the end of the month of fluid milk products and fluid cream products in bulk form;

(3) In the skim milk equivalent of nonfat milk solids used to modify a fluid milk product that has not been accounted for in Class I; and

(4) In shrinkage assigned pursuant to § 1000.43(b).

(e) *Other uses.* Other uses include skim milk and butterfat used in any product described in this section that is dumped, used for animal feed, destroyed, or lost by a handler in a vehicular accident, flood, fire, or similar occurrence beyond the handler's control. Such uses of skim milk and butterfat shall be assigned to the lowest priced class for the month to the extent that the quantities destroyed or lost can be verified from records satisfactory to the market administrator.

[64 FR 47899, Sept. 1, 1999, as amended at 65 FR 82833, Dec. 28, 2000; 68 FR 7064, Feb. 12, 2003; 69 FR 21952, Apr. 23, 2004; 75 FR 51931, Aug. 24, 2010]

§ 1000.41 [Reserved]

§ 1000.42 Classification of transfers and diversions.

(a) *Transfers and diversions to pool plants.* Skim milk or butterfat transferred or diverted in the form of a fluid milk product or transferred in the form of a bulk fluid cream product from a pool plant or a handler described in § 1135.11 of this chapter to another pool plant shall be classified as Class I milk unless the handlers both request the same classification in another class. In either case, the classification shall be subject to the following conditions:

(1) The skim milk and butterfat classified in each class shall be limited to the amount of skim milk and butterfat, respectively, remaining in such class at the receiving plant after the computations pursuant to § 1000.44(a)(9)

and the corresponding step of § 1000.44(b);

(2) If the transferring plant received during the month other source milk to be allocated pursuant to § 1000.44(a)(3) or the corresponding step of § 1000.44(b), the skim milk or butterfat so transferred shall be classified so as to allocate the least possible Class I utilization to such other source milk; and

(3) If the transferring handler received during the month other source milk to be allocated pursuant to § 1000.44(a)(8) or (9) or the corresponding steps of § 1000.44(b), the skim milk or butterfat so transferred, up to the total of the skim milk and butterfat, respectively, in such receipts of other source milk, shall not be classified as Class I milk to a greater extent than would be the case if the other source milk had been received at the receiving plant.

(b) *Transfers and diversions to a plant regulated under another Federal order.* Skim milk or butterfat transferred or diverted in the form of a fluid milk product or transferred in the form of a bulk fluid cream product from a pool plant to a plant regulated under another Federal order shall be classified in the following manner. Such classification shall apply only to the skim milk or butterfat that is in excess of any receipts at the pool plant from a plant regulated under another Federal order of skim milk and butterfat, respectively, in fluid milk products and bulk fluid cream products, respectively, that are in the same category as described in paragraph (b)(1) or (2) of this section:

(1) As Class I milk, if transferred as packaged fluid milk products;

(2) If transferred or diverted in bulk form, classification shall be in the classes to which allocated under the other order:

(i) If the operators of both plants so request in their reports of receipts and utilization filed with their respective market administrators, transfers in bulk form shall be classified as other than Class I to the extent that such utilization is available for such classification pursuant to the allocation provisions of the other order;

(ii) If diverted, the diverting handler must request a classification other than Class I. If the plant receiving the diverted milk does not have sufficient utilization available for the requested classification and some of the diverted milk is consequently assigned to Class I use, the diverting handler shall be given the option of designating the entire load of diverted milk as producer milk at the plant physically receiving the milk. Alternatively, if the diverting handler so chooses, it may designate which dairy farmers whose milk was diverted during the month will be designated as producers under the order physically receiving the milk. If the diverting handler declines to accept either of these options, the market administrator will prorate the portion of diverted milk in excess of Class II, III, and IV use among all the dairy farmers whose milk was received from the diverting handler on the last day of the month, then the second-to-last day, and continuing in that fashion until the excess diverted milk has been assigned as producer milk under the receiving order; and

(iii) If information concerning the classes to which such transfers or diversions were allocated under the other order is not available to the market administrator for the purpose of establishing classification under this paragraph, classification shall be Class I, subject to adjustment when such information is available.

(c) *Transfers and diversions to producer-handlers and to exempt plants.* Skim milk or butterfat that is transferred or diverted from a pool plant to a producer-handler under any Federal order or to an exempt plant shall be classified:

(1) As Class I milk if transferred or diverted to a producer-handler;

(2) As Class I milk if transferred to an exempt plant in the form of a packaged fluid milk product; and

(3) In accordance with the utilization assigned to it by the market administrator if transferred or diverted in the form of a bulk fluid milk product or transferred in the form of a bulk fluid cream product to an exempt plant. For this purpose, the receiving handler's utilization of skim milk and butterfat in each class, in series beginning with Class IV, shall be assigned to the extent possible to its receipts of skim milk and butterfat, in bulk fluid cream products, and bulk fluid milk products, respectively, pro rata to each source.

(d) *Transfers and diversions to other nonpool plants.* Skim milk or butterfat transferred or diverted in the following forms from a pool plant to a nonpool plant that is not a plant regulated under another order, an exempt plant, or a producer-handler plant shall be classified:

(1) As Class I milk, if transferred in the form of a packaged fluid milk product; and

(2) As Class I milk, if transferred or diverted in the form of a bulk fluid milk product or transferred in the form of a bulk fluid cream product, unless the following conditions apply:

(i) If the conditions described in paragraphs (d)(2)(i)(A) and (B) of this section are met, transfers or diversions in bulk form shall be classified on the basis of the assignment of the nonpool plant's utilization, excluding the milk equivalent of both nonfat milk solids and concentrated milk used in the plant during the month, to its receipts as set forth in paragraphs (d)(2)(ii) through (viii) of this section:

(A) The transferring handler or diverting handler claims such classification in such handler's report of receipts and utilization filed pursuant to §____.30 of each Federal milk order for the month within which such transaction occurred; and

(B) The nonpool plant operator maintains books and records showing the utilization of all skim milk and butterfat received at such plant which are made available for verification purposes if requested by the market administrator;

(ii) Route disposition in the marketing area of each Federal milk order from the nonpool plant and transfers of packaged fluid milk products from such nonpool plant to plants fully regulated thereunder shall be assigned to the extent possible in the following sequence:

(A) Pro rata to receipts of packaged fluid milk products at such nonpool plant from pool plants;

(B) Pro rata to any remaining unassigned receipts of packaged fluid milk products at such nonpool plant from plants regulated under other Federal orders;

(C) Pro rata to receipts of bulk fluid milk products at such nonpool plant from pool plants; and

(D) Pro rata to any remaining unassigned receipts of bulk fluid milk products at such nonpool plant from plants regulated under other Federal orders;

(iii) Any remaining Class I disposition of packaged fluid milk products from the nonpool plant shall be assigned to the extent possible pro rata to any remaining unassigned receipts of packaged fluid milk products at such nonpool plant from pool plants and plants regulated under other Federal orders;

(iv) Transfers of bulk fluid milk products from the nonpool plant to a plant regulated under any Federal order, to the extent that such transfers to the regulated plant exceed receipts of fluid milk products from such plant and are allocated to Class I at the receiving plant, shall be assigned to the extent possible in the following sequence:

(A) Pro rata to receipts of fluid milk products at such nonpool plant from pool plants; and

(B) Pro rata to any remaining unassigned receipts of fluid milk products at such nonpool plant from plants regulated under other Federal orders;

(v) Any remaining unassigned Class I disposition from the nonpool plant shall be assigned to the extent possible in the following sequence:

(A) To such nonpool plant's receipts from dairy farmers who the market administrator determines constitute regular sources of Grade A milk for such nonpool plant; and

(B) To such nonpool plant's receipts of Grade A milk from plants not fully regulated under any Federal order which the market administrator determines constitute regular sources of Grade A milk for such nonpool plant;

(vi) Any remaining unassigned receipts of bulk fluid milk products at the nonpool plant from pool plants and plants regulated under other Federal orders shall be assigned, pro rata among such plants, to the extent possible first to any remaining Class I utilization and then to all other utilization, in sequence beginning with Class IV at such nonpool plant;

(vii) Receipts of bulk fluid cream products at the nonpool plant from pool plants and plants regulated under other Federal orders shall be assigned, pro rata among such plants, to the extent possible to any remaining utilization, in sequence beginning with Class IV at such nonpool plant; and

(viii) In determining the nonpool plant's utilization for purposes of this paragraph, any fluid milk products and bulk fluid cream products transferred from such nonpool plant to a plant not fully regulated under any Federal order shall be classified on the basis of the second plant's utilization using the same assignment priorities at the second plant that are set forth in this paragraph.

§ 1000.43   General classification rules.

In determining the classification of producer milk pursuant to § 1000.44, the following rules shall apply:

(a) Each month the market administrator shall correct for mathematical and other obvious errors all reports filed pursuant to § ___.30 of each Federal milk order and shall compute separately for each pool plant, for each handler described in § 1000.9(c) and § 1135.11 of this chapter, the pounds of skim milk and butterfat, respectively, in each class in accordance with §§ 1000.40 and 1000.42, and paragraph (b) of this section.

(b) *Shrinkage and Overage.* For purposes of classifying all milk reported by a handler pursuant to § ___.30 of each Federal milk order the market administrator shall determine the shrinkage or overage of skim milk and butterfat for each pool plant and each handler described in § 1000.9(c) and § 1135.11 of this chapter by subtracting total utilization from total receipts. Any positive difference shall be shrinkage, and any negative difference shall be overage.

(1) Shrinkage incurred by pool plants qualified pursuant to § ___.7 of any Federal milk order shall be assigned to the lowest-priced class to the extent that such shrinkage does not exceed:

(i) Two percent of the total quantity of milk physically received at the plant directly from producers' farms on the basis of farm weights and tests;

(ii) Plus 1.5 percent of the quantity of bulk milk physically received on a basis other than farm weights and tests, excluding concentrated milk received by agreement for other than Class I use;

(iii) Plus .5 percent of the quantity of milk diverted by the plant operator to another plant on a basis other than farm weights and tests; and

(iv) Minus 1.5 percent of the quantity of bulk milk transferred to other plants, excluding concentrated milk transferred by agreement for other than Class I use.

(2) A handler described in § 1000.9(c) or § 1135.11 of this chapter that delivers milk to plants on a basis other than farm weights and tests shall receive a lowest-priced-class shrinkage allowance of .5 percent of the total quantity of such milk picked up at producers' farms.

(3) Shrinkage in excess of the amounts provided in paragraphs (b)(1) and (2) of this section shall be assigned to existing utilization in series starting with Class I. The shrinkage assigned pursuant to this paragraph shall be added to the handler's reported utilization and the result shall be known as the *gross utilization in each class.*

(c) If any of the water but none of the nonfat solids contained in the milk from which a product is made is removed before the product is utilized or disposed of by the handler, the pounds of skim milk in such product that are to be considered under this part as used or disposed of by the handler shall be an amount equivalent to the nonfat milk solids contained in such product plus all of the water originally associated with such solids. If any of the nonfat solids contained in the milk from which a product is made are removed before the product is utilized or disposed of by the handler, the pounds of skim milk in such product that are to be considered under this part as used or disposed of by the handler shall be an amount equivalent to the nonfat milk solids contained in such product plus all of the water and nonfat solids originally associated with such solids determined on a protein equivalent basis.

(d) Skim milk and butterfat contained in receipts of bulk concentrated fluid milk and nonfluid milk products that are reconstituted for fluid use shall be assigned to Class I use, up to the reconstituted portion of labeled reconstituted fluid milk products, on a pro rata basis (except for any Class I use of specific concentrated receipts that is established by the handler) prior to any assignments under § 1000.44. Any remaining skim milk and butterfat in concentrated receipts shall be assigned to uses under § 1000.44 on a pro rata basis, unless a specific use of such receipts is established by the handler.

[64 FR 47899, Sept. 1, 1999, as amended at 75 FR 51931, Aug. 24, 2010]

§ 1000.44 Classification of producer milk.

For each month the market administrator shall determine for each handler described in § 1000.9(a) for each pool plant of the handler separately and for each handler described in § 1000.9(c) and § 1135.11 of this chapter the classification of producer milk by allocating the handler's receipts of skim milk and butterfat to the handler's gross utilization of such receipts pursuant to § 1000.43(b)(3) as follows:

(a) Skim milk shall be allocated in the following manner:

(1) Subtract from the pounds of skim milk in Class I the pounds of skim milk in:

(i) Receipts of packaged fluid milk products from an unregulated supply plant to the extent that an equivalent amount of skim milk disposed of to such plant by handlers fully regulated under any Federal order is classified and priced as Class I milk and is not used as an offset for any other payment obligation under any order;

(ii) Packaged fluid milk products in inventory at the beginning of the month. This paragraph shall apply only if the pool plant was subject to the provisions of this paragraph or comparable provisions of another Federal order in the immediately preceding month;

(iii) Fluid milk products received in packaged form from plants regulated under other Federal orders; and

(iv) To the extent that the receipts described in paragraphs (a)(1)(i) through (iii) of this section exceed the gross Class I utilization of skim milk, the excess receipts shall be subtracted pursuant to paragraph (a)(3)(vi) of this section.

(2) Subtract from the pounds of skim milk in Class II the pounds of skim milk in the receipts of skim milk in bulk concentrated fluid milk products and in other source milk (except other source milk received in the form of an unconcentrated fluid milk product or a fluid cream product) that is used to produce, or added to, any product in Class II (excluding the quantity of such skim milk that was classified as Class IV milk pursuant to § 1000.40(d)(3)). To the extent that the receipts described in this paragraph exceed the gross Class II utilization of skim milk, the excess receipts shall be subtracted pursuant to paragraph (a)(3)(vi) of this section.

(3) Subtract from the pounds of skim milk remaining in each class, in series beginning with Class IV, the pounds of skim milk in:

(i) Receipts of bulk concentrated fluid milk products and other source milk (except other source milk received in the form of an unconcentrated fluid milk product);

(ii) Receipts of fluid milk products and bulk fluid cream products for which appropriate health approval is not established and from unidentified sources;

(iii) Receipts of fluid milk products and bulk fluid cream products from an exempt plant;

(iv) Fluid milk products and bulk fluid cream products received from a producer-handler as defined under the order in this part, or any other Federal order;

(v) Receipts of fluid milk products from dairy farmers for other markets; and

(vi) The excess receipts specified in paragraphs (a)(1)(iv) and (a)(2) of this section.

(4) Subtract from the pounds of skim milk remaining in all classes other than Class I, in sequence beginning with Class IV, the receipts of fluid milk products from an unregulated supply plant that were not previously subtracted in this section for which the handler requests classification other than Class I, but not in excess of the pounds of skim milk remaining in these other classes combined.

(5) Subtract from the pounds of skim milk remaining in all classes other than Class I, in sequence beginning with Class IV, receipts of fluid milk products from an unregulated supply plant that were not previously subtracted in this section, and which are in excess of the pounds of skim milk determined pursuant to paragraphs (a)(5)(i) and (ii) of this section;

(i) Multiply by 1.25 the pounds of skim milk remaining in Class I at this allocation step; and

(ii) Subtract from the result in paragraph (a)(5)(i) the pounds of skim milk in receipts of producer milk and fluid milk products from other pool plants.

(6) Subtract from the pounds of skim milk remaining in all classes other than Class I, in sequence beginning with Class IV, the pounds of skim milk in receipts of bulk fluid milk products from a handler regulated under another Federal order that are in excess of bulk fluid milk products transferred or diverted to such handler, if other than Class I classification is requested, but not in excess of the pounds of skim milk remaining in these classes combined.

(7) Subtract from the pounds of skim milk remaining in each class, in series beginning with Class IV, the pounds of skim milk in fluid milk products and bulk fluid cream products in inventory at the beginning of the month that were not previously subtracted in this section.

(8) Subtract from the pounds of skim milk remaining in each class at the plant receipts of skim milk in fluid milk products from an unregulated supply plant that were not previously subtracted in this section and that were not offset by transfers or diversions of fluid milk products to the unregulated supply plant from which fluid milk products to be allocated at this step were received. Such subtraction shall be pro rata to the pounds of skim milk in Class I and in Classes II, III, and IV combined, with the quantity prorated to Classes II, III, and IV combined being subtracted in sequence beginning with Class IV.

(9) Subtract from the pounds of skim milk remaining in each class the pounds of skim milk in receipts of bulk fluid milk products from a handler regulated under another Federal order that are in excess of bulk fluid milk

products transferred or diverted to such handler that were not subtracted in paragraph (a)(6) of this section. Such subtraction shall be pro rata to the pounds of skim milk in Class I and in Classes II, III, and IV combined, with the quantity prorated to Classes II, III, and IV combined being subtracted in sequence beginning with Class IV, with respect to whichever of the following quantities represents the lower proportion of Class I milk:

(i) The estimated utilization of skim milk of all handlers in each class as announced for the month pursuant to § 1000.45(a); or

(ii) The total pounds of skim milk remaining in each class at this allocation step.

(10) Subtract from the pounds of skim milk remaining in each class the pounds of skim milk in receipts of fluid milk products and bulk fluid cream products from another pool plant and from a handler described in § 1135.11 of this chapter according to the classification of such products pursuant to § 1000.42(a).

(11) If the total pounds of skim milk remaining in all classes exceed the pounds of skim milk in producer milk, subtract such excess from the pounds of skim milk remaining in each class in series beginning with Class IV.

(b) Butterfat shall be allocated in accordance with the procedure outlined for skim milk in paragraph (a) of this section.

(c) The quantity of producer milk in each class shall be the combined pounds of skim milk and butterfat remaining in each class after the computations pursuant to paragraphs (a) and (b) of this section.

§ 1000.45 Market administrator's reports and announcements concerning classification.

(a) Whenever required for the purpose of allocating receipts from plants regulated under other Federal orders pursuant to § 1000.44(a)(9) and the corresponding step of § 1000.44(b), the market administrator shall estimate and publicly announce the utilization (to the nearest whole percentage) in Class I during the month of skim milk and butterfat, respectively, in producer milk of all handlers. The estimate

shall be based upon the most current available data and shall be final for such purpose.

(b) The market administrator shall report to the market administrators of other Federal orders as soon as possible after the handlers' reports of receipts and utilization are received, the class to which receipts from plants regulated under other Federal orders are allocated pursuant to §§ 1000.43(d) and 1000.44 (including any reclassification of inventories of bulk concentrated fluid milk products), and thereafter any change in allocation required to correct errors disclosed on the verification of such report.

(c) The market administrator shall furnish each handler operating a pool plant and each handler described in § 1135.11 of this chapter who has shipped fluid milk products or bulk fluid cream products to a plant fully regulated under another Federal order the class to which the shipments were allocated by the market administrator of the other Federal order on the basis of the report by the receiving handler and, as necessary, any changes in the allocation arising from the verification of such report.

(d) The market administrator shall report to each cooperative association which so requests, the percentage of producer milk delivered by members of the association that was used in each class by each handler receiving the milk. For the purpose of this report, the milk so received shall be prorated to each class in accordance with the total utilization of producer milk by the handler.

Subpart G—Class Prices

§ 1000.50 Class prices, component prices, and advanced pricing factors.

Class prices per hundredweight of milk containing 3.5 percent butterfat, component prices, and advanced pricing factors shall be as follows. The prices and pricing factors described in paragraphs (a), (b), (c), (e), (f), and (q) of this section shall be based on a weighted average of the most recent 2 weekly prices announced by the National Agricultural Statistical Service (NASS) before the 24th day of the month. These prices shall be announced on or before the 23rd day of the month and shall apply to milk received during the following month. The prices described in paragraphs (g) through (p) of this section shall be based on a weighted average for the preceding month of weekly prices announced by NASS on or before the 5th day of the month and shall apply to milk received during the preceding month. The price described in paragraph (d) of this section shall be derived from the Class II skim milk price announced on or before the 23rd day of the month preceding the month to which it applies and the butterfat price announced on or before the 5th day of the month following the month to which it applies.

(a) *Class I price.* The Class I price per hundredweight, rounded to the nearest cent, shall be 0.965 times the Class I skim milk price plus 3.5 times the Class I butterfat price.

(b) *Class I skim milk price.* The Class I skim milk price per hundredweight shall be the adjusted Class I differential specified in §1000.52, plus the adjustment to Class I prices specified in §§1005.51(b), 1006.51(b) and 1007.51(b) of this chapter, plus the simple average of the advanced pricing factors computed in paragraph (q)(1) and (2) of this section rounded to the nearest cent, plus $0.74 per hundredweight.

(c) *Class I butterfat price.* The Class I butterfat price per pound shall be the adjusted Class I differential specified in §1000.52 divided by 100, plus the adjustments to Class I prices specified in §§1005.51(b), 1006.51(b) and 1007.51(b) divided by 100, plus the advanced butterfat price computed in paragraph (q)(3) of this section.

(d) *The Class II price* per hundredweight, rounded to the nearest cent, shall be .965 times the Class II skim milk price plus 3.5 times the Class II butterfat price.

(e) *Class II skim milk price.* The Class II skim milk price per hundredweight shall be the advanced Class IV skim milk price computed in paragraph (q)(2) of this section plus 70 cents.

(f) *Class II nonfat solids price.* The Class II nonfat solids price per pound, rounded to the nearest one-hundredth

cent, shall be the Class II skim milk price divided by 9.

(g) *Class II butterfat price.* The Class II butterfat price per pound shall be the butterfat price plus $0.007.

(h) *Class III price.* The Class III price per hundredweight, rounded to the nearest cent, shall be 0.965 times the Class III skim milk price plus 3.5 times the butterfat price.

(i) *Class III skim milk price.* The Class III skim milk price per hundredweight, rounded to the nearest cent, shall be the protein price per pound times 3.1 plus the other solids price per pound times 5.9.

(j) *Class IV price.* The Class IV price per hundredweight, rounded to the nearest cent, shall be 0.965 times the Class IV skim milk price plus 3.5 times the butterfat price.

(k) *Class IV skim milk price.* The Class IV skim milk price per hundredweight, rounded to the nearest cent, shall be the nonfat solids price per pound times 9.

(l) *Butterfat price.* The butterfat price per pound, rounded to the nearest one-hundredth cent, shall be the U.S. average NASS AA Butter survey price reported by the Department for the month, less 17.15 cents, with the result multiplied by 1.211.

(m) *Nonfat solids price.* The nonfat solids price per pound, rounded to the nearest one-hundredth cent, shall be the U.S. average NASS nonfat dry milk survey price reported by the Department for the month, less 16.78 cents and multiplying the result by 0.99.

(n) *Protein price.* The protein price per pound, rounded to the nearest one-hundredth cent, shall be computed as follows:

(1) Compute a weighted average of the amounts described in paragraphs (n)(1)(i) and (ii) of this section:

(i) The U.S. average NASS survey price for 40-lb. block cheese reported by the Department for the month; and

(ii) The U.S. average NASS survey price for 500-pound barrel cheddar cheese (38 percent moisture) reported by the Department for the month plus 3 cents;

(2) Subtract 20.03 cents from the price computed pursuant to paragraph (n)(1) of this section and multiply the result by 1.383;

(3) Add to the amount computed pursuant to paragraph (n)(2) of this section an amount computed as follows:

(i) Subtract 20.03 cents from the price computed pursuant to paragraph (n)(1) of this section and multiply the result by 1.572; and

(ii) Subtract 0.9 times the butterfat price computed pursuant to paragraph (l) of this section from the amount computed pursuant to paragraph (n)(3)(i) of this section; and

(iii) Multiply the amount computed pursuant to paragraph (n)(3)(ii) of this section by 1.17.

(o) *Other solids price.* The other solids price per pound, rounded to the nearest one-hundredth cent, shall be the U.S. average NASS dry whey survey price reported by the Department for the month minus 19.91 cents, with the result multiplied by 1.03.

(p) *Somatic cell adjustment.* The somatic cell adjustment per hundredweight of milk shall be determined as follows:

(1) Multiply 0.0005 by the weighted average price computed pursuant to paragraph (n)(1) of this section and round to the 5th decimal place;

(2) Subtract the somatic cell count of the milk (reported in thousands) from 350; and

(3) Multiply the amount computed in paragraph (p)(1) of this section by the amount computed in paragraph (p)(2) of this section and round to the nearest full cent.

(q) *Advanced pricing factors.* For the purpose of computing the Class I skim milk price, the Class II skim milk price, the Class II nonfat solids price, and the Class I butterfat price for the following month, the following pricing factors shall be computed using the weighted average of the 2 most recent NASS U.S. average weekly survey prices announced before the 24th day of the month:

(1) An advanced Class III skim milk price per hundredweight, rounded to the nearest cent, shall be computed as follows:

(i) Following the procedure set forth in paragraphs (n) and (o) of this section, but using the weighted average of the 2 most recent NASS U.S. average weekly survey prices announced before the 24th day of the month, compute a protein price and an other solids price;

(ii) Multiply the protein price computed in paragraph (q)(1)(i) of this section by 3.1;

(iii) Multiply the other solids price per pound computed in paragraph (q)(1)(i) of this section by 5.9; and

(iv) Add the amounts computed in paragraphs (q)(1)(ii) and (iii) of this section.

(2) An advanced Class IV skim milk price per hundredweight, rounded to the nearest cent, shall be computed as follows:

(i) Following the procedure set forth in paragraph (m) of this section, but using the weighted average of the 2 most recent NASS U.S. average weekly survey prices announced before the 24th day of the month, compute a nonfat solids price; and

(ii) Multiply the nonfat solids price computed in paragraph (q)(2)(i) of this section by 9.

(3) An advanced butterfat price per pound rounded to the nearest one-hundredth cent, shall be calculated by computing a weighted average of the 2 most recent U.S. average NASS AA Butter survey prices announced before the 24th day of the month, subtracting 17.15 cents from this average, and multiplying the result by 1.211.

[64 FR 47899, Sept. 1, 1999, as amended at 65 FR 82833, Dec. 28, 2000; 68 FR 7064, Feb. 12, 2003; 71 FR 78334, Dec. 29, 2006; 73 FR 14155, Mar. 17, 2008; 73 FR 44619, July 31, 2008; 84 FR 8591, Mar. 11, 2019; 84 FR 12483, Apr. 2, 2019]

§ 1000.51 [Reserved]

§ 1000.52 Adjusted Class I differentials.

The Class I differential adjusted for location to be used in § 1000.50(b) and (c) shall be as follows:

County/parish/city	State	FIPS code	Class I differential adjusted for location
AUTAUGA	AL	01001	3.30
BALDWIN	AL	01003	3.50
BARBOUR	AL	01005	3.45
BIBB	AL	01007	3.10
BLOUNT	AL	01009	3.10
BULLOCK	AL	01011	3.30
BUTLER	AL	01013	3.45
CALHOUN	AL	01015	3.10
CHAMBERS	AL	01017	3.10
CHEROKEE	AL	01019	3.10
CHILTON	AL	01021	3.10
CHOCTAW	AL	01023	3.30

County/parish/city	State	FIPS code	Class I differential adjusted for location	County/parish/city	State	FIPS code	Class I differential adjusted for location
CLARKE	AL	01025	3.45	CRITTENDEN	AR	05035	2.80
CLAY	AL	01027	3.10	CROSS	AR	05037	2.80
CLEBURNE	AL	01029	3.10	DALLAS	AR	05039	2.90
COFFEE	AL	01031	3.45	DESHA	AR	05041	2.90
COLBERT	AL	01033	2.90	DREW	AR	05043	2.90
CONECUH	AL	01035	3.45	FAULKNER	AR	05045	2.80
COOSA	AL	01037	3.10	FRANKLIN	AR	05047	2.80
COVINGTON	AL	01039	3.45	FULTON	AR	05049	2.60
CRENSHAW	AL	01041	3.45	GARLAND	AR	05051	2.80
CULLMAN	AL	01043	3.10	GRANT	AR	05053	2.90
DALE	AL	01045	3.45	GREENE	AR	05055	2.60
DALLAS	AL	01047	3.30	HEMPSTEAD	AR	05057	2.90
DE KALB	AL	01049	2.90	HOT SPRING	AR	05059	2.90
ELMORE	AL	01051	3.30	HOWARD	AR	05061	2.90
ESCAMBIA	AL	01053	3.45	INDEPENDENCE	AR	05063	2.60
ETOWAH	AL	01055	3.10	IZARD	AR	05065	2.60
FAYETTE	AL	01057	3.10	JACKSON	AR	05067	2.60
FRANKLIN	AL	01059	2.90	JEFFERSON	AR	05069	2.90
GENEVA	AL	01061	3.45	JOHNSON	AR	05071	2.80
GREENE	AL	01063	3.10	LAFAYETTE	AR	05073	3.10
HALE	AL	01065	3.10	LAWRENCE	AR	05075	2.60
HENRY	AL	01067	3.45	LEE	AR	05077	2.80
HOUSTON	AL	01069	3.45	LINCOLN	AR	05079	2.90
JACKSON	AL	01071	2.90	LITTLE RIVER	AR	05081	2.90
JEFFERSON	AL	01073	3.10	LOGAN	AR	05083	2.80
LAMAR	AL	01075	3.10	LONOKE	AR	05085	2.80
LAUDERDALE	AL	01077	2.90	MADISON	AR	05087	2.60
LAWRENCE	AL	01079	2.90	MARION	AR	05089	2.60
LEE	AL	01081	3.30	MILLER	AR	05091	3.10
LIMESTONE	AL	01083	2.90	MISSISSIPPI	AR	05093	2.60
LOWNDES	AL	01085	3.30	MONROE	AR	05095	2.80
MACON	AL	01087	3.30	MONTGOMERY	AR	05097	2.80
MADISON	AL	01089	2.90	NEVADA	AR	05099	2.90
MARENGO	AL	01091	3.30	NEWTON	AR	05101	2.60
MARION	AL	01093	3.10	OUACHITA	AR	05103	2.90
MARSHALL	AL	01095	2.90	PERRY	AR	05105	2.80
MOBILE	AL	01097	3.50	PHILLIPS	AR	05107	2.90
MONROE	AL	01099	3.45	PIKE	AR	05109	2.90
MONTGOMERY	AL	01101	3.30	POINSETT	AR	05111	2.60
MORGAN	AL	01103	2.90	POLK	AR	05113	2.80
PERRY	AL	01105	3.10	POPE	AR	05115	2.80
PICKENS	AL	01107	3.10	PRAIRIE	AR	05117	2.80
PIKE	AL	01109	3.45	PULASKI	AR	05119	2.80
RANDOLPH	AL	01111	3.10	RANDOLPH	AR	05121	2.60
RUSSELL	AL	01113	3.30	SALINE	AR	05125	2.80
SHELBY	AL	01117	3.10	SCOTT	AR	05127	2.80
ST. CLAIR	AL	01115	3.10	SEARCY	AR	05129	2.60
SUMTER	AL	01119	3.10	SEBASTIAN	AR	05131	2.80
TALLADEGA	AL	01121	3.10	SEVIER	AR	05133	2.90
TALLAPOOSA	AL	01123	3.10	SHARP	AR	05135	2.60
TUSCALOOSA	AL	01125	3.10	ST. FRANCIS	AR	05123	2.80
WALKER	AL	01127	3.10	STONE	AR	05137	2.60
WASHINGTON	AL	01129	3.45	UNION	AR	05139	3.10
WILCOX	AL	01131	3.30	VAN BUREN	AR	05141	2.80
WINSTON	AL	01133	3.10	WASHINGTON	AR	05143	2.60
ARKANSAS	AR	05001	2.90	WHITE	AR	05145	2.80
ASHLEY	AR	05003	3.10	WOODRUFF	AR	05147	2.80
BAXTER	AR	05005	2.60	YELL	AR	05149	2.80
BENTON	AR	05007	2.60	APACHE	AZ	04001	1.90
BOONE	AR	05009	2.60	COCHISE	AZ	04003	2.10
BRADLEY	AR	05011	2.90	COCONINO	AZ	04005	1.90
CALHOUN	AR	05013	2.90	GILA	AZ	04007	2.10
CARROLL	AR	05015	2.60	GRAHAM	AZ	04009	2.10
CHICOT	AR	05017	3.10	GREENLEE	AZ	04011	2.10
CLARK	AR	05019	2.90	LA PAZ	AZ	04012	2.10
CLAY	AR	05021	2.60	MARICOPA	AZ	04013	2.35
CLEBURNE	AR	05023	2.80	MOHAVE	AZ	04015	1.90
CLEVELAND	AR	05025	2.90	NAVAJO	AZ	04017	1.90
COLUMBIA	AR	05027	3.10	PIMA	AZ	04019	2.35
CONWAY	AR	05029	2.80	PINAL	AZ	04021	2.35
CRAIGHEAD	AR	05031	2.60	SANTA CRUZ	AZ	04023	2.10
CRAWFORD	AR	05033	2.80	YAVAPAI	AZ	04025	1.90

County/parish/city	State	FIPS code	Class I differential adjusted for location	County/parish/city	State	FIPS code	Class I differential adjusted for location
YUMA	AZ	04027	2.10	CROWLEY	CO ...	08025	2.45
ALAMEDA	CA	06001	1.80	CUSTER	CO ...	08027	2.45
ALPINE	CA	06003	1.70	DELTA	CO ...	08029	2.00
AMADOR	CA	06005	1.70	DENVER	CO ...	08031	2.55
BUTTE	CA	06007	1.70	DOLORES	CO ...	08033	1.90
CALAVERAS	CA	06009	1.70	DOUGLAS	CO ...	08035	2.55
COLUSA	CA	06011	1.70	EAGLE	CO ...	08037	1.90
CONTRA COSTA	CA	06013	1.80	EL PASO	CO ...	08041	2.55
DEL NORTE	CA	06015	1.80	ELBERT	CO ...	08039	2.45
EL DORADO	CA	06017	1.70	FREMONT	CO ...	08043	2.45
FRESNO	CA	06019	1.60	GARFIELD	CO ...	08045	2.00
GLENN	CA	06021	1.70	GILPIN	CO ...	08047	2.45
HUMBOLDT	CA	06023	1.80	GRAND	CO ...	08049	1.90
IMPERIAL	CA	06025	2.00	GUNNISON	CO ...	08051	1.90
INYO	CA	06027	1.60	HINSDALE	CO ...	08053	1.90
KERN	CA	06029	1.80	HUERFANO	CO ...	08055	2.45
KINGS	CA	06031	1.60	JACKSON	CO ...	08057	1.90
LAKE	CA	06033	1.80	JEFFERSON	CO ...	08059	2.55
LASSEN	CA	06035	1.70	KIOWA	CO ...	08061	2.35
LOS ANGELES	CA	06037	2.10	KIT CARSON	CO ...	08063	2.35
MADERA	CA	06039	1.60	LA PLATA	CO ...	08067	1.90
MARIN	CA	06041	1.80	LAKE	CO ...	08065	1.90
MARIPOSA	CA	06043	1.70	LARIMER	CO ...	08069	2.45
MENDOCINO	CA	06045	1.80	LAS ANIMAS	CO ...	08071	2.35
MERCED	CA	06047	1.70	LINCOLN	CO ...	08073	2.45
MODOC	CA	06049	1.70	LOGAN	CO ...	08075	2.35
MONO	CA	06051	1.60	MESA	CO ...	08077	2.00
MONTEREY	CA	06053	1.80	MINERAL	CO ...	08079	1.90
NAPA	CA	06055	1.80	MOFFAT	CO ...	08081	1.90
NEVADA	CA	06057	1.70	MONTEZUMA	CO ...	08083	1.90
ORANGE	CA	06059	2.10	MONTROSE	CO ...	08085	2.00
PLACER	CA	06061	1.70	MORGAN	CO ...	08087	2.35
PLUMAS	CA	06063	1.70	OTERO	CO ...	08089	2.45
RIVERSIDE	CA	06065	2.00	OURAY	CO ...	08091	1.90
SACRAMENTO	CA	06067	1.70	PARK	CO ...	08093	2.45
SAN BENITO	CA	06069	1.80	PHILLIPS	CO ...	08095	2.35
SAN BERNARDINO	CA	06071	1.80	PITKIN	CO ...	08097	1.90
SAN DIEGO	CA	06073	2.10	PROWERS	CO ...	08099	2.35
SAN FRANCISCO	CA	06075	1.80	PUEBLO	CO ...	08101	2.45
SAN JOAQUIN	CA	06077	1.70	RIO BLANCO	CO ...	08103	1.90
SAN LUIS OBISPO	CA	06079	1.80	RIO GRANDE	CO ...	08105	1.90
SAN MATEO	CA	06081	1.80	ROUTT	CO ...	08107	1.90
SANTA BARBARA	CA	06083	1.80	SAGUACHE	CO ...	08109	1.90
SANTA CLARA	CA	06085	1.80	SAN JUAN	CO ...	08111	1.90
SANTA CRUZ	CA	06087	1.80	SAN MIGUEL	CO ...	08113	1.90
SHASTA	CA	06089	1.70	SEDGWICK	CO ...	08115	2.35
SIERRA	CA	06091	1.70	SUMMIT	CO ...	08117	1.90
SISKIYOU	CA	06093	1.80	TELLER	CO ...	08119	2.45
SOLANO	CA	06095	1.80	WASHINGTON	CO ...	08121	2.35
SONOMA	CA	06097	1.80	WELD	CO ...	08123	2.45
STANISLAUS	CA	06099	1.70	YUMA	CO ...	08125	2.35
SUTTER	CA	06101	1.70	FAIRFIELD	CT	09001	3.15
TEHAMA	CA	06103	1.70	HARTFORD	CT	09003	3.15
TRINITY	CA	06105	1.80	LITCHFIELD	CT	09005	3.00
TULARE	CA	06107	1.60	MIDDLESEX	CT	09007	3.15
TUOLUMNE	CA	06109	1.70	NEW HAVEN	CT	09009	3.15
VENTURA	CA	06111	1.80	NEW LONDON	CT	09011	3.15
YOLO	CA	06113	1.70	TOLLAND	CT	09013	3.15
YUBA	CA	06115	1.70	WINDHAM	CT	09015	3.15
ADAMS	CO ...	08001	2.55	DISTRICT OF COLUMBIA	DC	11001	3.00
ALAMOSA	CO ...	08003	1.90	KENT	DE	10001	3.05
ARAPAHOE	CO ...	08005	2.55	NEW CASTLE	DE	10003	3.05
ARCHULETA	CO ...	08007	1.90	SUSSEX	DE	10005	3.05
BACA	CO ...	08009	2.35	ALACHUA	FL	12001	3.70
BENT	CO ...	08011	2.35	BAKER	FL	12003	3.70
BOULDER	CO ...	08013	2.45	BAY	FL	12005	3.70
BROOMFIELD	CO ...	08014	2.45	BRADFORD	FL	12007	3.70
CHAFFEE	CO ...	08015	1.90	BREVARD	FL	12009	4.00
CHEYENNE	CO ...	08017	2.35	BROWARD	FL	12011	4.30
CLEAR CREEK	CO ...	08019	2.45	CALHOUN	FL	12013	3.70
CONEJOS	CO ...	08021	1.90	CHARLOTTE	FL	12015	4.30
COSTILLA	CO ...	08023	1.90	CITRUS	FL	12017	4.00

County/parish/city	State	FIPS code	Class I differential adjusted for location	County/parish/city	State	FIPS code	Class I differential adjusted for location
CLAY	FL	12019	3.70	BRYAN	GA	13029	3.45
COLLIER	FL	12021	4.30	BULLOCH	GA	13031	3.30
COLUMBIA	FL	12023	3.70	BURKE	GA	13033	3.30
DADE	FL	12025	4.30	BUTTS	GA	13035	3.10
DE SOTO	FL	12027	4.00	CALHOUN	GA	13037	3.45
DIXIE	FL	12029	3.70	CAMDEN	GA	13039	3.45
DUVAL	FL	12031	3.70	CANDLER	GA	13043	3.30
ESCAMBIA	FL	12033	3.45	CARROLL	GA	13045	3.10
FLAGLER	FL	12035	4.00	CATOOSA	GA	13047	2.80
FRANKLIN	FL	12037	3.70	CHARLTON	GA	13049	3.45
GADSDEN	FL	12039	3.70	CHATHAM	GA	13051	3.45
GILCHRIST	FL	12041	3.70	CHATTAHOOCHEE	GA	13053	3.30
GLADES	FL	12043	4.30	CHATTOOGA	GA	13055	2.80
GULF	FL	12045	3.70	CHEROKEE	GA	13057	3.10
HAMILTON	FL	12047	3.70	CLARKE	GA	13059	3.10
HARDEE	FL	12049	4.00	CLAY	GA	13061	3.45
HENDRY	FL	12051	4.30	CLAYTON	GA	13063	3.10
HERNANDO	FL	12053	4.00	CLINCH	GA	13065	3.45
HIGHLANDS	FL	12055	4.00	COBB	GA	13067	3.10
HILLSBOROUGH	FL	12057	4.00	COFFEE	GA	13069	3.45
HOLMES	FL	12059	3.70	COLQUITT	GA	13071	3.45
INDIAN RIVER	FL	12061	4.00	COLUMBIA	GA	13073	3.10
JACKSON	FL	12063	3.70	COOK	GA	13075	3.45
JEFFERSON	FL	12065	3.70	COWETA	GA	13077	3.10
LAFAYETTE	FL	12067	3.70	CRAWFORD	GA	13079	3.30
LAKE	FL	12069	4.00	CRISP	GA	13081	3.45
LEE	FL	12071	4.30	DADE	GA	13083	2.80
LEON	FL	12073	3.70	DAWSON	GA	13085	3.10
LEVY	FL	12075	4.00	DE KALB	GA	13089	3.10
LIBERTY	FL	12077	3.70	DECATUR	GA	13087	3.45
MADISON	FL	12079	3.70	DODGE	GA	13091	3.45
MANATEE	FL	12081	4.00	DOOLY	GA	13093	3.45
MARION	FL	12083	4.00	DOUGHERTY	GA	13095	3.45
MARTIN	FL	12085	4.30	DOUGLAS	GA	13097	3.10
MONROE	FL	12087	4.30	EARLY	GA	13099	3.45
NASSAU	FL	12089	3.70	ECHOLS	GA	13101	3.45
OKALOOSA	FL	12091	3.45	EFFINGHAM	GA	13103	3.30
OKEECHOBEE	FL	12093	4.00	ELBERT	GA	13105	3.10
ORANGE	FL	12095	4.00	EMANUEL	GA	13107	3.30
OSCEOLA	FL	12097	4.00	EVANS	GA	13109	3.45
PALM BEACH	FL	12099	4.30	FANNIN	GA	13111	2.80
PASCO	FL	12101	4.00	FAYETTE	GA	13113	3.10
PINELLAS	FL	12103	4.00	FLOYD	GA	13115	3.10
POLK	FL	12105	4.00	FORSYTH	GA	13117	3.10
PUTNAM	FL	12107	3.70	FRANKLIN	GA	13119	3.10
SANTA ROSA	FL	12113	3.45	FULTON	GA	13121	3.10
SARASOTA	FL	12115	4.00	GILMER	GA	13123	3.10
SEMINOLE	FL	12117	4.00	GLASCOCK	GA	13125	3.10
ST. JOHNS	FL	12109	3.70	GLYNN	GA	13127	3.45
ST. LUCIE	FL	12111	4.00	GORDON	GA	13129	3.10
SUMTER	FL	12119	4.00	GRADY	GA	13131	3.45
SUWANNEE	FL	12121	3.70	GREENE	GA	13133	3.10
TAYLOR	FL	12123	3.70	GWINNETT	GA	13135	3.10
UNION	FL	12125	3.70	HABERSHAM	GA	13137	3.10
VOLUSIA	FL	12127	4.00	HALL	GA	13139	3.10
WAKULLA	FL	12129	3.70	HANCOCK	GA	13141	3.10
WALTON	FL	12131	3.45	HARALSON	GA	13143	3.10
WASHINGTON	FL	12133	3.70	HARRIS	GA	13145	3.30
APPLING	GA	13001	3.45	HART	GA	13147	3.10
ATKINSON	GA	13003	3.45	HEARD	GA	13149	3.10
BACON	GA	13005	3.45	HENRY	GA	13151	3.10
BAKER	GA	13007	3.45	HOUSTON	GA	13153	3.30
BALDWIN	GA	13009	3.10	IRWIN	GA	13155	3.45
BANKS	GA	13011	3.10	JACKSON	GA	13157	3.10
BARROW	GA	13013	3.10	JASPER	GA	13159	3.10
BARTOW	GA	13015	3.10	JEFF DAVIS	GA	13161	3.45
BEN HILL	GA	13017	3.45	JEFFERSON	GA	13163	3.30
BERRIEN	GA	13019	3.45	JENKINS	GA	13165	3.30
BIBB	GA	13021	3.30	JOHNSON	GA	13167	3.30
BLECKLEY	GA	13023	3.30	JONES	GA	13169	3.10
BRANTLEY	GA	13025	3.45	LAMAR	GA	13171	3.10
BROOKS	GA	13027	3.45	LANIER	GA	13173	3.45

County/parish/city	State	FIPS code	Class I differential adjusted for location	County/parish/city	State	FIPS code	Class I differential adjusted for location
LAURENS	GA	13175	3.30	WORTH	GA	13321	3.45
LEE	GA	13177	3.45	ADAIR	IA	19001	1.80
LIBERTY	GA	13179	3.45	ADAMS	IA	19003	1.80
LINCOLN	GA	13181	3.10	ALLAMAKEE	IA	19005	1.75
LONG	GA	13183	3.45	APPANOOSE	IA	19007	1.80
LOWNDES	GA	13185	3.45	AUDUBON	IA	19009	1.80
LUMPKIN	GA	13187	3.10	BENTON	IA	19011	1.80
MACON	GA	13193	3.30	BLACK HAWK	IA	19013	1.75
MADISON	GA	13195	3.10	BOONE	IA	19015	1.80
MARION	GA	13197	3.30	BREMER	IA	19017	1.75
MCDUFFIE	GA	13189	3.10	BUCHANAN	IA	19019	1.75
MCINTOSH	GA	13191	3.45	BUENA VISTA	IA	19021	1.75
MERIWETHER	GA	13199	3.10	BUTLER	IA	19023	1.75
MILLER	GA	13201	3.45	CALHOUN	IA	19025	1.75
MITCHELL	GA	13205	3.45	CARROLL	IA	19027	1.80
MONROE	GA	13207	3.10	CASS	IA	19029	1.80
MONTGOMERY	GA	13209	3.45	CEDAR	IA	19031	1.80
MORGAN	GA	13211	3.10	CERRO GORDO	IA	19033	1.75
MURRAY	GA	13213	2.80	CHEROKEE	IA	19035	1.75
MUSCOGEE	GA	13215	3.30	CHICKASAW	IA	19037	1.75
NEWTON	GA	13217	3.10	CLARKE	IA	19039	1.80
OCONEE	GA	13219	3.10	CLAY	IA	19041	1.75
OGLETHORPE	GA	13221	3.10	CLAYTON	IA	19043	1.75
PAULDING	GA	13223	3.10	CLINTON	IA	19045	1.80
PEACH	GA	13225	3.30	CRAWFORD	IA	19047	1.80
PICKENS	GA	13227	3.10	DALLAS	IA	19049	1.80
PIERCE	GA	13229	3.45	DAVIS	IA	19051	1.80
PIKE	GA	13231	3.10	DECATUR	IA	19053	1.80
POLK	GA	13233	3.10	DELAWARE	IA	19055	1.75
PULASKI	GA	13235	3.45	DES MOINES	IA	19057	1.80
PUTNAM	GA	13237	3.10	DICKINSON	IA	19059	1.75
QUITMAN	GA	13239	3.45	DUBUQUE	IA	19061	1.75
RABUN	GA	13241	3.10	EMMET	IA	19063	1.75
RANDOLPH	GA	13243	3.45	FAYETTE	IA	19065	1.75
RICHMOND	GA	13245	3.30	FLOYD	IA	19067	1.75
ROCKDALE	GA	13247	3.10	FRANKLIN	IA	19069	1.75
SCHLEY	GA	13249	3.30	FREMONT	IA	19071	1.85
SCREVEN	GA	13251	3.30	GREENE	IA	19073	1.80
SEMINOLE	GA	13253	3.45	GRUNDY	IA	19075	1.75
SPALDING	GA	13255	3.10	GUTHRIE	IA	19077	1.80
STEPHENS	GA	13257	3.10	HAMILTON	IA	19079	1.75
STEWART	GA	13259	3.45	HANCOCK	IA	19081	1.75
SUMTER	GA	13261	3.45	HARDIN	IA	19083	1.75
TALBOT	GA	13263	3.30	HARRISON	IA	19085	1.80
TALIAFERRO	GA	13265	3.10	HENRY	IA	19087	1.80
TATTNALL	GA	13267	3.45	HOWARD	IA	19089	1.75
TAYLOR	GA	13269	3.30	HUMBOLDT	IA	19091	1.75
TELFAIR	GA	13271	3.45	IDA	IA	19093	1.75
TERRELL	GA	13273	3.45	IOWA	IA	19095	1.80
THOMAS	GA	13275	3.45	JACKSON	IA	19097	1.80
TIFT	GA	13277	3.45	JASPER	IA	19099	1.80
TOOMBS	GA	13279	3.45	JEFFERSON	IA	19101	1.80
TOWNS	GA	13281	3.10	JOHNSON	IA	19103	1.80
TREUTLEN	GA	13283	3.30	JONES	IA	19105	1.80
TROUP	GA	13285	3.10	KEOKUK	IA	19107	1.80
TURNER	GA	13287	3.45	KOSSUTH	IA	19109	1.75
TWIGGS	GA	13289	3.30	LEE	IA	19111	1.80
UNION	GA	13291	3.10	LINN	IA	19113	1.80
UPSON	GA	13293	3.10	LOUISA	IA	19115	1.80
WALKER	GA	13295	2.80	LUCAS	IA	19117	1.80
WALTON	GA	13297	3.10	LYON	IA	19119	1.75
WARE	GA	13299	3.45	MADISON	IA	19121	1.80
WARREN	GA	13301	3.10	MAHASKA	IA	19123	1.80
WASHINGTON	GA	13303	3.30	MARION	IA	19125	1.80
WAYNE	GA	13305	3.45	MARSHALL	IA	19127	1.80
WEBSTER	GA	13307	3.45	MILLS	IA	19129	1.85
WHEELER	GA	13309	3.45	MITCHELL	IA	19131	1.75
WHITE	GA	13311	3.10	MONONA	IA	19133	1.80
WHITFIELD	GA	13313	2.80	MONROE	IA	19135	1.80
WILCOX	GA	13315	3.45	MONTGOMERY	IA	19137	1.80
WILKES	GA	13317	3.10	MUSCATINE	IA	19139	1.80
WILKINSON	GA	13319	3.30	O'BRIEN	IA	19141	1.75

County/parish/city	State	FIPS code	Class I differential adjusted for location	County/parish/city	State	FIPS code	Class I differential adjusted for location
OSCEOLA	IA	19143	1.75	ADAMS	IL	17001	1.80
PAGE	IA	19145	1.80	ALEXANDER	IL	17003	2.20
PALO ALTO	IA	19147	1.75	BOND	IL	17005	2.00
PLYMOUTH	IA	19149	1.75	BOONE	IL	17007	1.75
POCAHONTAS	IA	19151	1.75	BROWN	IL	17009	1.80
POLK	IA	19153	1.80	BUREAU	IL	17011	1.80
POTTAWATTAMIE	IA	19155	1.85	CALHOUN	IL	17013	2.00
POWESHIEK	IA	19157	1.80	CARROLL	IL	17015	1.80
RINGGOLD	IA	19159	1.80	CASS	IL	17017	1.80
SAC	IA	19161	1.75	CHAMPAIGN	IL	17019	1.80
SCOTT	IA	19163	1.80	CHRISTIAN	IL	17021	2.00
SHELBY	IA	19165	1.80	CLARK	IL	17023	2.00
SIOUX	IA	19167	1.75	CLAY	IL	17025	2.00
STORY	IA	19169	1.80	CLINTON	IL	17027	2.00
TAMA	IA	19171	1.80	COLES	IL	17029	2.00
TAYLOR	IA	19173	1.80	COOK	IL	17031	1.80
UNION	IA	19175	1.80	CRAWFORD	IL	17033	2.00
VAN BUREN	IA	19177	1.80	CUMBERLAND	IL	17035	2.00
WAPELLO	IA	19179	1.80	DE KALB	IL	17037	1.80
WARREN	IA	19181	1.80	DE WITT	IL	17039	1.80
WASHINGTON	IA	19183	1.80	DOUGLAS	IL	17041	2.00
WAYNE	IA	19185	1.80	DU PAGE	IL	17043	1.80
WEBSTER	IA	19187	1.75	EDGAR	IL	17045	2.00
WINNEBAGO	IA	19189	1.75	EDWARDS	IL	17047	2.20
WINNESHIEK	IA	19191	1.75	EFFINGHAM	IL	17049	2.00
WOODBURY	IA	19193	1.75	FAYETTE	IL	17051	2.00
WORTH	IA	19195	1.75	FORD	IL	17053	1.80
WRIGHT	IA	19197	1.75	FRANKLIN	IL	17055	2.20
ADA	ID	16001	1.60	FULTON	IL	17057	1.80
ADAMS	ID	16003	1.60	GALLATIN	IL	17059	2.20
BANNOCK	ID	16005	1.60	GREENE	IL	17061	2.00
BEAR LAKE	ID	16007	1.60	GRUNDY	IL	17063	1.80
BENEWAH	ID	16009	1.90	HAMILTON	IL	17065	2.20
BINGHAM	ID	16011	1.60	HANCOCK	IL	17067	1.80
BLAINE	ID	16013	1.60	HARDIN	IL	17069	2.20
BOISE	ID	16015	1.60	HENDERSON	IL	17071	1.80
BONNER	ID	16017	1.90	HENRY	IL	17073	1.80
BONNEVILLE	ID	16019	1.60	IROQUOIS	IL	17075	1.80
BOUNDARY	ID	16021	1.90	JACKSON	IL	17077	2.20
BUTTE	ID	16023	1.60	JASPER	IL	17079	2.00
CAMAS	ID	16025	1.60	JEFFERSON	IL	17081	2.00
CANYON	ID	16027	1.60	JERSEY	IL	17083	2.00
CARIBOU	ID	16029	1.60	JO DAVIESS	IL	17085	1.75
CASSIA	ID	16031	1.60	JOHNSON	IL	17087	2.20
CLARK	ID	16033	1.60	KANE	IL	17089	1.80
CLEARWATER	ID	16035	1.60	KANKAKEE	IL	17091	1.80
CUSTER	ID	16037	1.60	KENDALL	IL	17093	1.80
ELMORE	ID	16039	1.60	KNOX	IL	17095	1.80
FRANKLIN	ID	16041	1.60	LA SALLE	IL	17099	1.80
FREMONT	ID	16043	1.60	LAKE	IL	17097	1.80
GEM	ID	16045	1.60	LAWRENCE	IL	17101	2.00
GOODING	ID	16047	1.60	LEE	IL	17103	1.80
IDAHO	ID	16049	1.60	LIVINGSTON	IL	17105	1.80
JEFFERSON	ID	16051	1.60	LOGAN	IL	17107	1.80
JEROME	ID	16053	1.60	MACON	IL	17115	1.80
KOOTENAI	ID	16055	1.90	MACOUPIN	IL	17117	2.00
LATAH	ID	16057	1.90	MADISON	IL	17119	2.00
LEMHI	ID	16059	1.60	MARION	IL	17121	2.00
LEWIS	ID	16061	1.60	MARSHALL	IL	17123	1.80
LINCOLN	ID	16063	1.60	MASON	IL	17125	1.80
MADISON	ID	16065	1.60	MASSAC	IL	17127	2.20
MINIDOKA	ID	16067	1.60	MCDONOUGH	IL	17109	1.80
NEZ PERCE	ID	16069	1.60	MCHENRY	IL	17111	1.80
ONEIDA	ID	16071	1.60	MCLEAN	IL	17113	1.80
OWYHEE	ID	16073	1.60	MENARD	IL	17129	1.80
PAYETTE	ID	16075	1.60	MERCER	IL	17131	1.80
POWER	ID	16077	1.60	MONROE	IL	17133	2.00
SHOSHONE	ID	16079	1.90	MONTGOMERY	IL	17135	2.00
TETON	ID	16081	1.60	MORGAN	IL	17137	1.80
TWIN FALLS	ID	16083	1.60	MOULTRIE	IL	17139	2.00
VALLEY	ID	16085	1.60	OGLE	IL	17141	1.80
WASHINGTON	ID	16087	1.60	PEORIA	IL	17143	1.80

County/parish/city	State	FIPS code	Class I differential adjusted for location
PERRY	IL	17145	2.00
PIATT	IL	17147	1.80
PIKE	IL	17149	1.80
POPE	IL	17151	2.20
PULASKI	IL	17153	2.20
PUTNAM	IL	17155	1.80
RANDOLPH	IL	17157	2.00
RICHLAND	IL	17159	2.00
ROCK ISLAND	IL	17161	1.80
SALINE	IL	17165	2.20
SANGAMON	IL	17167	1.80
SCHUYLER	IL	17169	1.80
SCOTT	IL	17171	1.80
SHELBY	IL	17173	2.00
ST. CLAIR	IL	17163	2.00
STARK	IL	17175	1.80
STEPHENSON	IL	17177	1.75
TAZEWELL	IL	17179	1.80
UNION	IL	17181	2.20
VERMILION	IL	17183	1.80
WABASH	IL	17185	2.20
WARREN	IL	17187	1.80
WASHINGTON	IL	17189	2.00
WAYNE	IL	17191	2.20
WHITE	IL	17193	2.20
WHITESIDE	IL	17195	1.80
WILL	IL	17197	1.80
WILLIAMSON	IL	17199	2.20
WINNEBAGO	IL	17201	1.75
WOODFORD	IL	17203	1.80
ADAMS	IN	18001	1.80
ALLEN	IN	18003	1.80
BARTHOLOMEW	IN	18005	2.20
BENTON	IN	18007	1.80
BLACKFORD	IN	18009	1.80
BOONE	IN	18011	2.00
BROWN	IN	18013	2.20
CARROLL	IN	18015	1.80
CASS	IN	18017	1.80
CLARK	IN	18019	2.20
CLAY	IN	18021	2.00
CLINTON	IN	18023	1.80
CRAWFORD	IN	18025	2.20
DAVIESS	IN	18027	2.20
DEKALB	IN	18033	1.80
DEARBORN	IN	18029	2.20
DECATUR	IN	18031	2.20
DELAWARE	IN	18035	2.00
DUBOIS	IN	18037	2.20
ELKHART	IN	18039	1.80
FAYETTE	IN	18041	2.00
FLOYD	IN	18043	2.20
FOUNTAIN	IN	18045	1.80
FRANKLIN	IN	18047	2.00
FULTON	IN	18049	1.80
GIBSON	IN	18051	2.20
GRANT	IN	18053	1.80
GREENE	IN	18055	2.20
HAMILTON	IN	18057	2.00
HANCOCK	IN	18059	2.00
HARRISON	IN	18061	2.20
HENDRICKS	IN	18063	2.00
HENRY	IN	18065	2.00
HOWARD	IN	18067	1.80
HUNTINGTON	IN	18069	1.80
JACKSON	IN	18071	2.20
JASPER	IN	18073	1.80
JAY	IN	18075	1.80
JEFFERSON	IN	18077	2.20
JENNINGS	IN	18079	2.20
JOHNSON	IN	18081	2.00
KNOX	IN	18083	2.20

County/parish/city	State	FIPS code	Class I differential adjusted for location
KOSCIUSKO	IN	18085	1.80
LA PORTE	IN	18091	1.80
LAGRANGE	IN	18087	1.80
LAKE	IN	18089	1.80
LAWRENCE	IN	18093	2.20
MADISON	IN	18095	2.00
MARION	IN	18097	2.00
MARSHALL	IN	18099	1.80
MARTIN	IN	18101	2.20
MIAMI	IN	18103	1.80
MONROE	IN	18105	2.20
MONTGOMERY	IN	18107	2.00
MORGAN	IN	18109	2.00
NEWTON	IN	18111	1.80
NOBLE	IN	18113	1.80
OHIO	IN	18115	2.20
ORANGE	IN	18117	2.20
OWEN	IN	18119	2.00
PARKE	IN	18121	2.00
PERRY	IN	18123	2.20
PIKE	IN	18125	2.20
PORTER	IN	18127	1.80
POSEY	IN	18129	2.20
PULASKI	IN	18131	1.80
PUTNAM	IN	18133	2.00
RANDOLPH	IN	18135	2.00
RIPLEY	IN	18137	2.20
RUSH	IN	18139	2.00
SCOTT	IN	18143	2.20
SHELBY	IN	18145	2.00
SPENCER	IN	18147	2.20
ST. JOSEPH	IN	18141	1.80
STARKE	IN	18149	1.80
STEUBEN	IN	18151	1.80
SULLIVAN	IN	18153	2.20
SWITZERLAND	IN	18155	2.20
TIPPECANOE	IN	18157	1.80
TIPTON	IN	18159	1.80
UNION	IN	18161	2.00
VANDERBURGH	IN	18163	2.20
VERMILLION	IN	18165	2.00
VIGO	IN	18167	2.00
WABASH	IN	18169	1.80
WARREN	IN	18171	1.80
WARRICK	IN	18173	2.20
WASHINGTON	IN	18175	2.20
WAYNE	IN	18177	2.00
WELLS	IN	18179	1.80
WHITE	IN	18181	1.80
WHITLEY	IN	18183	1.80
ALLEN	KS	20001	2.20
ANDERSON	KS	20003	2.00
ATCHISON	KS	20005	2.00
BARBER	KS	20007	2.20
BARTON	KS	20009	2.20
BOURBON	KS	20011	2.20
BROWN	KS	20013	2.00
BUTLER	KS	20015	2.20
CHASE	KS	20017	2.20
CHAUTAUQUA	KS	20019	2.20
CHEROKEE	KS	20021	2.20
CHEYENNE	KS	20023	2.20
CLARK	KS	20025	2.20
CLAY	KS	20027	2.00
CLOUD	KS	20029	2.00
COFFEY	KS	20031	2.00
COMANCHE	KS	20033	2.20
COWLEY	KS	20035	2.20
CRAWFORD	KS	20037	2.20
DECATUR	KS	20039	2.00
DICKINSON	KS	20041	2.00
DONIPHAN	KS	20043	2.00

County/parish/city	State	FIPS code	Class I differential adjusted for location	County/parish/city	State	FIPS code	Class I differential adjusted for location
DOUGLAS	KS	20045	2.00	STEVENS	KS	20189	2.20
EDWARDS	KS	20047	2.20	SUMNER	KS	20191	2.20
ELK	KS	20049	2.20	THOMAS	KS	20193	2.00
ELLIS	KS	20051	2.00	TREGO	KS	20195	2.20
ELLSWORTH	KS	20053	2.00	WABAUNSEE	KS	20197	2.00
FINNEY	KS	20055	2.20	WALLACE	KS	20199	2.20
FORD	KS	20057	2.20	WASHINGTON	KS	20201	2.00
FRANKLIN	KS	20059	2.00	WICHITA	KS	20203	2.20
GEARY	KS	20061	2.00	WILSON	KS	20205	2.20
GOVE	KS	20063	2.20	WOODSON	KS	20207	2.20
GRAHAM	KS	20065	2.00	WYANDOTTE	KS	20209	2.00
GRANT	KS	20067	2.20	ADAIR	KY	21001	2.40
GRAY	KS	20069	2.20	ALLEN	KY	21003	2.40
GREELEY	KS	20071	2.20	ANDERSON	KY	21005	2.20
GREENWOOD	KS	20073	2.20	BALLARD	KY	21007	2.40
HAMILTON	KS	20075	2.20	BARREN	KY	21009	2.40
HARPER	KS	20077	2.20	BATH	KY	21011	2.20
HARVEY	KS	20079	2.20	BELL	KY	21013	2.40
HASKELL	KS	20081	2.20	BOONE	KY	21015	2.20
HODGEMAN	KS	20083	2.20	BOURBON	KY	21017	2.20
JACKSON	KS	20085	2.00	BOYD	KY	21019	2.20
JEFFERSON	KS	20087	2.00	BOYLE	KY	21021	2.20
JEWELL	KS	20089	2.00	BRACKEN	KY	21023	2.20
JOHNSON	KS	20091	2.00	BREATHITT	KY	21025	2.20
KEARNY	KS	20093	2.20	BRECKINRIDGE	KY	21027	2.20
KINGMAN	KS	20095	2.20	BULLITT	KY	21029	2.20
KIOWA	KS	20097	2.20	BUTLER	KY	21031	2.40
LABETTE	KS	20099	2.20	CALDWELL	KY	21033	2.40
LANE	KS	20101	2.20	CALLOWAY	KY	21035	2.40
LEAVENWORTH	KS	20103	2.00	CAMPBELL	KY	21037	2.20
LINCOLN	KS	20105	2.00	CARLISLE	KY	21039	2.40
LINN	KS	20107	2.20	CARROLL	KY	21041	2.20
LOGAN	KS	20109	2.20	CARTER	KY	21043	2.20
LYON	KS	20111	2.00	CASEY	KY	21045	2.40
MARION	KS	20115	2.20	CHRISTIAN	KY	21047	2.40
MARSHALL	KS	20117	2.00	CLARK	KY	21049	2.20
MCPHERSON	KS	20113	2.20	CLAY	KY	21051	2.40
MEADE	KS	20119	2.20	CLINTON	KY	21053	2.40
MIAMI	KS	20121	2.00	CRITTENDEN	KY	21055	2.40
MITCHELL	KS	20123	2.00	CUMBERLAND	KY	21057	2.40
MONTGOMERY	KS	20125	2.20	DAVIESS	KY	21059	2.20
MORRIS	KS	20127	2.00	EDMONSON	KY	21061	2.40
MORTON	KS	20129	2.20	ELLIOTT	KY	21063	2.20
NEMAHA	KS	20131	2.00	ESTILL	KY	21065	2.20
NEOSHO	KS	20133	2.20	FAYETTE	KY	21067	2.20
NESS	KS	20135	2.20	FLEMING	KY	21069	2.20
NORTON	KS	20137	2.00	FLOYD	KY	21071	2.20
OSAGE	KS	20139	2.00	FRANKLIN	KY	21073	2.20
OSBORNE	KS	20141	2.00	FULTON	KY	21075	2.40
OTTAWA	KS	20143	2.00	GALLATIN	KY	21077	2.20
PAWNEE	KS	20145	2.20	GARRARD	KY	21079	2.20
PHILLIPS	KS	20147	2.00	GRANT	KY	21081	2.20
POTTAWATOMIE	KS	20149	2.00	GRAVES	KY	21083	2.40
PRATT	KS	20151	2.20	GRAYSON	KY	21085	2.40
RAWLINS	KS	20153	2.00	GREEN	KY	21087	2.40
RENO	KS	20155	2.20	GREENUP	KY	21089	2.20
REPUBLIC	KS	20157	2.00	HANCOCK	KY	21091	2.20
RICE	KS	20159	2.20	HARDIN	KY	21093	2.20
RILEY	KS	20161	2.00	HARLAN	KY	21095	2.40
ROOKS	KS	20163	2.00	HARRISON	KY	21097	2.20
RUSH	KS	20165	2.20	HART	KY	21099	2.40
RUSSELL	KS	20167	2.00	HENDERSON	KY	21101	2.20
SALINE	KS	20169	2.00	HENRY	KY	21103	2.20
SCOTT	KS	20171	2.20	HICKMAN	KY	21105	2.40
SEDGWICK	KS	20173	2.20	HOPKINS	KY	21107	2.40
SEWARD	KS	20175	2.20	JACKSON	KY	21109	2.20
SHAWNEE	KS	20177	2.00	JEFFERSON	KY	21111	2.20
SHERIDAN	KS	20179	2.00	JESSAMINE	KY	21113	2.20
SHERMAN	KS	20181	2.20	JOHNSON	KY	21115	2.20
SMITH	KS	20183	2.00	KENTON	KY	21117	2.20
STAFFORD	KS	20185	2.20	KNOTT	KY	21119	2.40
STANTON	KS	20187	2.20	KNOX	KY	21121	2.40

County/parish/city	State	FIPS code	Class I differential adjusted for location	County/parish/city	State	FIPS code	Class I differential adjusted for location
LARUE	KY	21123	2.20	CLAIBORNE	LA	22027	3.10
LAUREL	KY	21125	2.40	CONCORDIA	LA	22029	3.40
LAWRENCE	KY	21127	2.20	DE SOTO	LA	22031	3.30
LEE	KY	21129	2.20	EAST BATON ROUGE	LA	22033	3.60
LESLIE	KY	21131	2.40	EAST CARROLL	LA	22035	3.10
LETCHER	KY	21133	2.40	EAST FELICIANA	LA	22037	3.50
LEWIS	KY	21135	2.20	EVANGELINE	LA	22039	3.50
LINCOLN	KY	21137	2.20	FRANKLIN	LA	22041	3.30
LIVINGSTON	KY	21139	2.40	GRANT	LA	22043	3.40
LOGAN	KY	21141	2.40	IBERIA	LA	22045	3.60
LYON	KY	21143	2.40	IBERVILLE	LA	22047	3.60
MADISON	KY	21151	2.20	JACKSON	LA	22049	3.30
MAGOFFIN	KY	21153	2.20	JEFFERSON	LA	22051	3.60
MARION	KY	21155	2.20	JEFFERSON DAVIS	LA	22053	3.50
MARSHALL	KY	21157	2.40	LA SALLE	LA	22059	3.40
MARTIN	KY	21159	2.20	LAFAYETTE	LA	22055	3.60
MASON	KY	21161	2.20	LAFOURCHE	LA	22057	3.60
MCCRACKEN	KY	21145	2.40	LINCOLN	LA	22061	3.10
MCCREARY	KY	21147	2.40	LIVINGSTON	LA	22063	3.60
MCLEAN	KY	21149	2.20	MADISON	LA	22065	3.30
MEADE	KY	21163	2.20	MOREHOUSE	LA	22067	3.10
MENIFEE	KY	21165	2.20	NATCHITOCHES	LA	22069	3.30
MERCER	KY	21167	2.20	ORLEANS	LA	22071	3.60
METCALFE	KY	21169	2.40	OUACHITA	LA	22073	3.10
MONROE	KY	21171	2.40	PLAQUEMINES	LA	22075	3.60
MONTGOMERY	KY	21173	2.20	POINTE COUPEE	LA	22077	3.50
MORGAN	KY	21175	2.20	RAPIDES	LA	22079	3.40
MUHLENBERG	KY	21177	2.40	RED RIVER	LA	22081	3.30
NELSON	KY	21179	2.20	RICHLAND	LA	22083	3.10
NICHOLAS	KY	21181	2.20	SABINE	LA	22085	3.30
OHIO	KY	21183	2.40	ST. BERNARD	LA	22087	3.60
OLDHAM	KY	21185	2.20	ST. CHARLES	LA	22089	3.60
OWEN	KY	21187	2.20	ST. HELENA	LA	22091	3.50
OWSLEY	KY	21189	2.20	ST. JAMES	LA	22093	3.60
PENDLETON	KY	21191	2.20	ST. JOHN THE BAPTIST	LA	22095	3.60
PERRY	KY	21193	2.40	ST. LANDRY	LA	22097	3.50
PIKE	KY	21195	2.40	ST. MARTIN	LA	22099	3.60
POWELL	KY	21197	2.20	ST. MARY	LA	22101	3.60
PULASKI	KY	21199	2.40	ST. TAMMANY	LA	22103	3.50
ROBERTSON	KY	21201	2.20	TANGIPAHOA	LA	22105	3.60
ROCKCASTLE	KY	21203	2.20	TENSAS	LA	22107	3.30
ROWAN	KY	21205	2.20	TERREBONNE	LA	22109	3.60
RUSSELL	KY	21207	2.40	UNION	LA	22111	3.10
SCOTT	KY	21209	2.20	VERMILION	LA	22113	3.60
SHELBY	KY	21211	2.20	VERNON	LA	22115	3.40
SIMPSON	KY	21213	2.40	WASHINGTON	LA	22117	3.50
SPENCER	KY	21215	2.20	WEBSTER	LA	22119	3.10
TAYLOR	KY	21217	2.40	WEST BATON ROUGE	LA	22121	3.60
TODD	KY	21219	2.40	WEST CARROLL	LA	22123	3.10
TRIGG	KY	21221	2.40	WEST FELICIANA	LA	22125	3.50
TRIMBLE	KY	21223	2.20	WINN	LA	22127	3.30
UNION	KY	21225	2.20	BARNSTABLE	MA ...	25001	3.25
WARREN	KY	21227	2.40	BERKSHIRE	MA ...	25003	2.80
WASHINGTON	KY	21229	2.20	BRISTOL	MA ...	25005	3.25
WAYNE	KY	21231	2.40	DUKES	MA ...	25007	3.25
WEBSTER	KY	21233	2.40	ESSEX	MA ...	25009	3.25
WHITLEY	KY	21235	2.40	FRANKLIN	MA ...	25011	3.00
WOLFE	KY	21237	2.20	HAMPDEN	MA ...	25013	3.00
WOODFORD	KY	21239	2.20	HAMPSHIRE	MA ...	25015	3.00
ACADIA	LA	22001	3.50	MIDDLESEX	MA ...	25017	3.25
ALLEN	LA	22003	3.50	NANTUCKET	MA ...	25019	3.25
ASCENSION	LA	22005	3.60	NORFOLK	MA ...	25021	3.25
ASSUMPTION	LA	22007	3.60	PLYMOUTH	MA ...	25023	3.25
AVOYELLES	LA	22009	3.40	SUFFOLK	MA ...	25025	3.25
BEAUREGARD	LA	22011	3.50	WORCESTER	MA ...	25027	3.10
BIENVILLE	LA	22013	3.30	ALLEGANY	MD ...	24001	2.60
BOSSIER	LA	22015	3.10	ANNE ARUNDEL	MD ...	24003	3.00
CADDO	LA	22017	3.10	BALTIMORE	MD ...	24005	3.00
CALCASIEU	LA	22019	3.50	BALTIMORE CITY	MD ...	24510	3.00
CALDWELL	LA	22021	3.30	CALVERT	MD ...	24009	3.00
CAMERON	LA	22023	3.60	CAROLINE	MD ...	24011	3.00
CATAHOULA	LA	22025	3.40	CARROLL	MD ...	24013	2.90

County/parish/city	State	FIPS code	Class I differential adjusted for location	County/parish/city	State	FIPS code	Class I differential adjusted for location
CECIL	MD ...	24015	3.05	KALKASKA	MI	26079	1.80
CHARLES	MD ...	24017	3.00	KENT	MI	26081	1.80
DORCHESTER	MD ...	24019	3.00	KEWEENAW	MI	26083	1.70
FREDERICK	MD ...	24021	2.90	LAKE	MI	26085	1.80
GARRETT	MD ...	24023	2.60	LAPEER	MI	26087	1.80
HARFORD	MD ...	24025	3.00	LEELANAU	MI	26089	1.80
HOWARD	MD ...	24027	3.00	LENAWEE	MI	26091	1.80
KENT	MD ...	24029	3.00	LIVINGSTON	MI	26093	1.80
MONTGOMERY	MD ...	24031	3.00	LUCE	MI	26095	1.80
PRINCE GEORGE'S	MD ...	24033	3.00	MACKINAC	MI	26097	1.80
QUEEN ANNE'S	MD ...	24035	3.00	MACOMB	MI	26099	1.80
SOMERSET	MD ...	24039	3.00	MANISTEE	MI	26101	1.80
ST. MARY'S	MD ...	24037	3.00	MARQUETTE	MI	26103	1.80
TALBOT	MD ...	24041	3.00	MASON	MI	26105	1.80
WASHINGTON	MD ...	24043	2.80	MECOSTA	MI	26107	1.80
WICOMICO	MD ...	24045	3.00	MENOMINEE	MI	26109	1.70
WORCESTER	MD ...	24047	3.00	MIDLAND	MI	26111	1.80
ANDROSCOGGIN	ME ...	23001	2.80	MISSAUKEE	MI	26113	1.80
AROOSTOOK	ME ...	23003	2.60	MONROE	MI	26115	1.80
CUMBERLAND	ME ...	23005	3.00	MONTCALM	MI	26117	1.80
FRANKLIN	ME ...	23007	2.60	MONTMORENCY	MI	26119	1.80
HANCOCK	ME ...	23009	2.80	MUSKEGON	MI	26121	1.80
KENNEBEC	ME ...	23011	2.80	NEWAYGO	MI	26123	1.80
KNOX	ME ...	23013	2.80	OAKLAND	MI	26125	1.80
LINCOLN	ME ...	23015	2.80	OCEANA	MI	26127	1.80
OXFORD	ME ...	23017	2.80	OGEMAW	MI	26129	1.80
PENOBSCOT	ME ...	23019	2.80	ONTONAGON	MI	26131	1.70
PISCATAQUIS	ME ...	23021	2.60	OSCEOLA	MI	26133	1.80
SAGADAHOC	ME ...	23023	2.80	OSCODA	MI	26135	1.80
SOMERSET	ME ...	23025	2.60	OTSEGO	MI	26137	1.80
WALDO	ME ...	23027	2.80	OTTAWA	MI	26139	1.80
WASHINGTON	ME ...	23029	2.80	PRESQUE ISLE	MI	26141	1.80
YORK	ME ...	23031	3.00	ROSCOMMON	MI	26143	1.80
ALCONA	MI	26001	1.80	SAGINAW	MI	26145	1.80
ALGER	MI	26003	1.80	SANILAC	MI	26151	1.80
ALLEGAN	MI	26005	1.80	SCHOOLCRAFT	MI	26153	1.80
ALPENA	MI	26007	1.80	SHIAWASSEE	MI	26155	1.80
ANTRIM	MI	26009	1.80	ST. CLAIR	MI	26147	1.80
ARENAC	MI	26011	1.80	ST. JOSEPH	MI	26149	1.80
BARAGA	MI	26013	1.70	TUSCOLA	MI	26157	1.80
BARRY	MI	26015	1.80	VAN BUREN	MI	26159	1.80
BAY	MI	26017	1.80	WASHTENAW	MI	26161	1.80
BENZIE	MI	26019	1.80	WAYNE	MI	26163	1.80
BERRIEN	MI	26021	1.80	WEXFORD	MI	26165	1.80
BRANCH	MI	26023	1.80	AITKIN	MN ...	27001	1.65
CALHOUN	MI	26025	1.80	ANOKA	MN ...	27003	1.70
CASS	MI	26027	1.80	BECKER	MN ...	27005	1.65
CHARLEVOIX	MI	26029	1.80	BELTRAMI	MN ...	27007	1.65
CHEBOYGAN	MI	26031	1.80	BENTON	MN ...	27009	1.70
CHIPPEWA	MI	26033	1.80	BIG STONE	MN ...	27011	1.70
CLARE	MI	26035	1.80	BLUE EARTH	MN ...	27013	1.70
CLINTON	MI	26037	1.80	BROWN	MN ...	27015	1.70
CRAWFORD	MI	26039	1.80	CARLTON	MN ...	27017	1.65
DELTA	MI	26041	1.70	CARVER	MN ...	27019	1.70
DICKINSON	MI	26043	1.70	CASS	MN ...	27021	1.65
EATON	MI	26045	1.80	CHIPPEWA	MN ...	27023	1.70
EMMET	MI	26047	1.80	CHISAGO	MN ...	27025	1.70
GENESEE	MI	26049	1.80	CLAY	MN ...	27027	1.65
GLADWIN	MI	26051	1.80	CLEARWATER	MN ...	27029	1.65
GOGEBIC	MI	26053	1.70	COOK	MN ...	27031	1.65
GRAND TRAVERSE	MI	26055	1.80	COTTONWOOD	MN ...	27033	1.70
GRATIOT	MI	26057	1.80	CROW WING	MN ...	27035	1.65
HILLSDALE	MI	26059	1.80	DAKOTA	MN ...	27037	1.70
HOUGHTON	MI	26061	1.70	DODGE	MN ...	27039	1.70
HURON	MI	26063	1.80	DOUGLAS	MN ...	27041	1.70
INGHAM	MI	26065	1.80	FARIBAULT	MN ...	27043	1.70
IONIA	MI	26067	1.80	FILLMORE	MN ...	27045	1.70
IOSCO	MI	26069	1.80	FREEBORN	MN ...	27047	1.70
IRON	MI	26071	1.70	GOODHUE	MN ...	27049	1.70
ISABELLA	MI	26073	1.80	GRANT	MN ...	27051	1.70
JACKSON	MI	26075	1.80	HENNEPIN	MN ...	27053	1.70
KALAMAZOO	MI	26077	1.80	HOUSTON	MN ...	27055	1.70

County/parish/city	State	FIPS code	Class I differential adjusted for location	County/parish/city	State	FIPS code	Class I differential adjusted for location
HUBBARD	MN ...	27057	1.65	CALLAWAY	MO ...	29027	2.00
ISANTI	MN ...	27059	1.70	CAMDEN	MO ...	29029	2.00
ITASCA	MN ...	27061	1.65	CAPE GIRARDEAU	MO ...	29031	2.20
JACKSON	MN ...	27063	1.70	CARROLL	MO ...	29033	1.80
KANABEC	MN ...	27065	1.70	CARTER	MO ...	29035	2.20
KANDIYOHI	MN ...	27067	1.70	CASS	MO ...	29037	2.00
KITTSON	MN ...	27069	1.60	CEDAR	MO ...	29039	2.20
KOOCHICHING	MN ...	27071	1.65	CHARITON	MO ...	29041	1.80
LAC QUI PARLE	MN ...	27073	1.70	CHRISTIAN	MO ...	29043	2.20
LAKE	MN ...	27075	1.65	CLARK	MO ...	29045	1.80
LAKE OF THE WOODS	MN ...	27077	1.60	CLAY	MO ...	29047	1.80
LE SUEUR	MN ...	27079	1.70	CLINTON	MO ...	29049	1.80
LINCOLN	MN ...	27081	1.70	COLE	MO ...	29051	2.00
LYON	MN ...	27083	1.70	COOPER	MO ...	29053	2.00
MAHNOMEN	MN ...	27087	1.65	CRAWFORD	MO ...	29055	2.00
MARSHALL	MN ...	27089	1.65	DADE	MO ...	29057	2.20
MARTIN	MN ...	27091	1.70	DALLAS	MO ...	29059	2.20
MCLEOD	MN ...	27085	1.70	DAVIESS	MO ...	29061	1.80
MEEKER	MN ...	27093	1.70	DE KALB	MO ...	29063	1.80
MILLE LACS	MN ...	27095	1.70	DENT	MO ...	29065	2.00
MORRISON	MN ...	27097	1.70	DOUGLAS	MO ...	29067	2.20
MOWER	MN ...	27099	1.70	DUNKLIN	MO ...	29069	2.20
MURRAY	MN ...	27101	1.70	FRANKLIN	MO ...	29071	2.00
NICOLLET	MN ...	27103	1.70	GASCONADE	MO ...	29073	2.00
NOBLES	MN ...	27105	1.70	GENTRY	MO ...	29075	1.80
NORMAN	MN ...	27107	1.65	GREENE	MO ...	29077	2.20
OLMSTED	MN ...	27109	1.70	GRUNDY	MO ...	29079	1.80
OTTER TAIL	MN ...	27111	1.65	HARRISON	MO ...	29081	1.80
PENNINGTON	MN ...	27113	1.65	HENRY	MO ...	29083	2.00
PINE	MN ...	27115	1.70	HICKORY	MO ...	29085	2.00
PIPESTONE	MN ...	27117	1.70	HOLT	MO ...	29087	1.80
POLK	MN ...	27119	1.65	HOWARD	MO ...	29089	2.00
POPE	MN ...	27121	1.70	HOWELL	MO ...	29091	2.20
RAMSEY	MN ...	27123	1.70	IRON	MO ...	29093	2.00
RED LAKE	MN ...	27125	1.65	JACKSON	MO ...	29095	2.00
REDWOOD	MN ...	27127	1.70	JASPER	MO ...	29097	2.20
RENVILLE	MN ...	27129	1.70	JEFFERSON	MO ...	29099	2.00
RICE	MN ...	27131	1.70	JOHNSON	MO ...	29101	2.00
ROCK	MN ...	27133	1.70	KNOX	MO ...	29103	1.80
ROSEAU	MN ...	27135	1.60	LACLEDE	MO ...	29105	2.20
SCOTT	MN ...	27139	1.70	LAFAYETTE	MO ...	29107	2.00
SHERBURNE	MN ...	27141	1.70	LAWRENCE	MO ...	29109	2.20
SIBLEY	MN ...	27143	1.70	LEWIS	MO ...	29111	1.80
ST. LOUIS	MN ...	27137	1.65	LINCOLN	MO ...	29113	2.00
STEARNS	MN ...	27145	1.70	LINN	MO ...	29115	1.80
STEELE	MN ...	27147	1.70	LIVINGSTON	MO ...	29117	1.80
STEVENS	MN ...	27149	1.70	MACON	MO ...	29121	1.80
SWIFT	MN ...	27151	1.70	MADISON	MO ...	29123	2.20
TODD	MN ...	27153	1.70	MARIES	MO ...	29125	2.00
TRAVERSE	MN ...	27155	1.70	MARION	MO ...	29127	1.80
WABASHA	MN ...	27157	1.70	MCDONALD	MO ...	29119	2.20
WADENA	MN ...	27159	1.65	MERCER	MO ...	29129	1.80
WASECA	MN ...	27161	1.70	MILLER	MO ...	29131	2.00
WASHINGTON	MN ...	27163	1.70	MISSISSIPPI	MO ...	29133	2.20
WATONWAN	MN ...	27165	1.70	MONITEAU	MO ...	29135	2.00
WILKIN	MN ...	27167	1.65	MONROE	MO ...	29137	1.80
WINONA	MN ...	27169	1.70	MONTGOMERY	MO ...	29139	2.00
WRIGHT	MN ...	27171	1.70	MORGAN	MO ...	29141	2.00
YELLOW MEDICINE	MN ...	27173	1.70	NEW MADRID	MO ...	29143	2.20
ADAIR	MO ...	29001	1.80	NEWTON	MO ...	29145	2.20
ANDREW	MO ...	29003	1.80	NODAWAY	MO ...	29147	1.80
ATCHISON	MO ...	29005	1.80	OREGON	MO ...	29149	2.20
AUDRAIN	MO ...	29007	2.00	OSAGE	MO ...	29151	2.00
BARRY	MO ...	29009	2.20	OZARK	MO ...	29153	2.20
BARTON	MO ...	29011	2.20	PEMISCOT	MO ...	29155	2.20
BATES	MO ...	29013	2.00	PERRY	MO ...	29157	2.20
BENTON	MO ...	29015	2.00	PETTIS	MO ...	29159	2.00
BOLLINGER	MO ...	29017	2.20	PHELPS	MO ...	29161	2.00
BOONE	MO ...	29019	2.00	PIKE	MO ...	29163	2.00
BUCHANAN	MO ...	29021	1.80	PLATTE	MO ...	29165	1.80
BUTLER	MO ...	29023	2.20	POLK	MO ...	29167	2.00
CALDWELL	MO ...	29025	1.80	PULASKI	MO ...	29169	2.20

County/parish/city	State	FIPS code	Class I differential adjusted for location	County/parish/city	State	FIPS code	Class I differential adjusted for location
PUTNAM	MO ...	29171	1.80	LINCOLN	MS ...	28085	3.40
RALLS	MO ...	29173	2.00	LOWNDES	MS ...	28087	3.10
RANDOLPH	MO ...	29175	1.80	MADISON	MS ...	28089	3.10
RAY	MO ...	29177	1.80	MARION	MS ...	28091	3.40
REYNOLDS	MO ...	29179	2.20	MARSHALL	MS ...	28093	2.90
RIPLEY	MO ...	29181	2.20	MONROE	MS ...	28095	3.10
SALINE	MO ...	29195	2.00	MONTGOMERY	MS ...	28097	3.10
SCHUYLER	MO ...	29197	1.80	NESHOBA	MS ...	28099	3.10
SCOTLAND	MO ...	29199	1.80	NEWTON	MS ...	28101	3.30
SCOTT	MO ...	29201	2.20	NOXUBEE	MS ...	28103	3.10
SHANNON	MO ...	29203	2.20	OKTIBBEHA	MS ...	28105	3.10
SHELBY	MO ...	29205	1.80	PANOLA	MS ...	28107	2.90
ST. CHARLES	MO ...	29183	2.00	PEARL RIVER	MS ...	28109	3.40
ST. CLAIR	MO ...	29185	2.00	PERRY	MS ...	28111	3.40
ST. FRANCOIS	MO ...	29187	2.00	PIKE	MS ...	28113	3.40
ST. LOUIS	MO ...	29189	2.00	PONTOTOC	MS ...	28115	2.90
ST. LOUIS CITY	MO ...	29510	2.00	PRENTISS	MS ...	28117	2.90
STE. GENEVIEVE	MO ...	29186	2.00	QUITMAN	MS ...	28119	2.90
STODDARD	MO ...	29207	2.20	RANKIN	MS ...	28121	3.30
STONE	MO ...	29209	2.20	SCOTT	MS ...	28123	3.30
SULLIVAN	MO ...	29211	1.80	SHARKEY	MS ...	28125	3.10
TANEY	MO ...	29213	2.20	SIMPSON	MS ...	28127	3.30
TEXAS	MO ...	29215	2.20	SMITH	MS ...	28129	3.30
VERNON	MO ...	29217	2.20	STONE	MS ...	28131	3.40
WARREN	MO ...	29219	2.00	SUNFLOWER	MS ...	28133	3.10
WASHINGTON	MO ...	29221	2.00	TALLAHATCHIE	MS ...	28135	3.10
WAYNE	MO ...	29223	2.20	TATE	MS ...	28137	2.90
WEBSTER	MO ...	29225	2.20	TIPPAH	MS ...	28139	2.90
WORTH	MO ...	29227	1.80	TISHOMINGO	MS ...	28141	2.90
WRIGHT	MO ...	29229	2.20	TUNICA	MS ...	28143	2.90
ADAMS	MS ...	28001	3.40	UNION	MS ...	28145	2.90
ALCORN	MS ...	28003	2.90	WALTHALL	MS ...	28147	3.40
AMITE	MS ...	28005	3.40	WARREN	MS ...	28149	3.30
ATTALA	MS ...	28007	3.10	WASHINGTON	MS ...	28151	3.10
BENTON	MS ...	28009	2.90	WAYNE	MS ...	28153	3.40
BOLIVAR	MS ...	28011	3.10	WEBSTER	MS ...	28155	3.10
CALHOUN	MS ...	28013	3.10	WILKINSON	MS ...	28157	3.40
CARROLL	MS ...	28015	3.10	WINSTON	MS ...	28159	3.10
CHICKASAW	MS ...	28017	3.10	YALOBUSHA	MS ...	28161	3.10
CHOCTAW	MS ...	28019	3.10	YAZOO	MS ...	28163	3.10
CLAIBORNE	MS ...	28021	3.30	BEAVERHEAD	MT	30001	1.60
CLARKE	MS ...	28023	3.30	BIG HORN	MT	30003	1.60
CLAY	MS ...	28025	3.10	BLAINE	MT	30005	1.60
COAHOMA	MS ...	28027	2.90	BROADWATER	MT	30007	1.60
COPIAH	MS ...	28029	3.30	CARBON	MT	30009	1.60
COVINGTON	MS ...	28031	3.40	CARTER	MT	30011	1.65
DE SOTO	MS ...	28033	2.90	CASCADE	MT	30013	1.60
FORREST	MS ...	28035	3.40	CHOUTEAU	MT	30015	1.60
FRANKLIN	MS ...	28037	3.40	CUSTER	MT	30017	1.60
GEORGE	MS ...	28039	3.40	DANIELS	MT	30019	1.60
GREENE	MS ...	28041	3.40	DAWSON	MT	30021	1.60
GRENADA	MS ...	28043	3.10	DEER LODGE	MT	30023	1.60
HANCOCK	MS ...	28045	3.50	FALLON	MT	30025	1.65
HARRISON	MS ...	28047	3.50	FERGUS	MT	30027	1.60
HINDS	MS ...	28049	3.30	FLATHEAD	MT	30029	1.60
HOLMES	MS ...	28051	3.10	GALLATIN	MT	30031	1.60
HUMPHREYS	MS ...	28053	3.10	GARFIELD	MT	30033	1.60
ISSAQUENA	MS ...	28055	3.10	GLACIER	MT	30035	1.60
ITAWAMBA	MS ...	28057	2.90	GOLDEN VALLEY	MT	30037	1.60
JACKSON	MS ...	28059	3.50	GRANITE	MT	30039	1.60
JASPER	MS ...	28061	3.30	HILL	MT	30041	1.60
JEFFERSON	MS ...	28063	3.40	JEFFERSON	MT	30043	1.60
JEFFERSON DAVIS	MS ...	28065	3.40	JUDITH BASIN	MT	30045	1.60
JONES	MS ...	28067	3.40	LAKE	MT	30047	1.60
KEMPER	MS ...	28069	3.10	LEWIS AND CLARK	MT	30049	1.60
LAFAYETTE	MS ...	28071	2.90	LIBERTY	MT	30051	1.60
LAMAR	MS ...	28073	3.40	LINCOLN	MT	30053	1.80
LAUDERDALE	MS ...	28075	3.30	MADISON	MT	30057	1.60
LAWRENCE	MS ...	28077	3.40	MCCONE	MT	30055	1.60
LEAKE	MS ...	28079	3.10	MEAGHER	MT	30059	1.60
LEE	MS ...	28081	2.90	MINERAL	MT	30061	1.80
LEFLORE	MS ...	28083	3.10	MISSOULA	MT	30063	1.60

County/parish/city	State	FIPS code	Class I differential adjusted for location	County/parish/city	State	FIPS code	Class I differential adjusted for location
MUSSELSHELL	MT	30065	1.60	HYDE	NC	37095	3.20
PARK	MT	30067	1.60	IREDELL	NC	37097	3.10
PETROLEUM	MT	30069	1.60	JACKSON	NC	37099	2.95
PHILLIPS	MT	30071	1.60	JOHNSTON	NC	37101	3.20
PONDERA	MT	30073	1.60	JONES	NC	37103	3.20
POWDER RIVER	MT	30075	1.60	LEE	NC	37105	3.10
POWELL	MT	30077	1.60	LENOIR	NC	37107	3.20
PRAIRIE	MT	30079	1.60	LINCOLN	NC	37109	3.10
RAVALLI	MT	30081	1.60	MACON	NC	37113	2.95
RICHLAND	MT	30083	1.60	MADISON	NC	37115	2.95
ROOSEVELT	MT	30085	1.60	MARTIN	NC	37117	3.20
ROSEBUD	MT	30087	1.60	MCDOWELL	NC	37111	2.95
SANDERS	MT	30089	1.80	MECKLENBURG	NC	37119	3.10
SHERIDAN	MT	30091	1.60	MITCHELL	NC	37121	2.95
SILVER BOW	MT	30093	1.60	MONTGOMERY	NC	37123	3.10
STILLWATER	MT	30095	1.60	MOORE	NC	37125	3.10
SWEET GRASS	MT	30097	1.60	NASH	NC	37127	3.10
TETON	MT	30099	1.60	NEW HANOVER	NC	37129	3.30
TOOLE	MT	30101	1.60	NORTHAMPTON	NC	37131	3.10
TREASURE	MT	30103	1.60	ONSLOW	NC	37133	3.30
VALLEY	MT	30105	1.60	ORANGE	NC	37135	3.10
WHEATLAND	MT	30107	1.60	PAMLICO	NC	37137	3.20
WIBAUX	MT	30109	1.60	PASQUOTANK	NC	37139	3.20
YELLOWSTONE	MT	30111	1.60	PENDER	NC	37141	3.30
YELLOWSTONE NAT. PARK	MT	30113	1.60	PERQUIMANS	NC	37143	3.20
ALAMANCE	NC	37001	3.10	PERSON	NC	37145	3.10
ALEXANDER	NC	37003	2.95	PITT	NC	37147	3.20
ALLEGHANY	NC	37005	2.95	POLK	NC	37149	3.10
ANSON	NC	37007	3.10	RANDOLPH	NC	37151	3.10
ASHE	NC	37009	2.95	RICHMOND	NC	37153	3.10
AVERY	NC	37011	2.95	ROBESON	NC	37155	3.30
BEAUFORT	NC	37013	3.20	ROCKINGHAM	NC	37157	2.95
BERTIE	NC	37015	3.20	ROWAN	NC	37159	3.10
BLADEN	NC	37017	3.30	RUTHERFORD	NC	37161	3.10
BRUNSWICK	NC	37019	3.30	SAMPSON	NC	37163	3.30
BUNCOMBE	NC	37021	2.95	SCOTLAND	NC	37165	3.30
BURKE	NC	37023	2.95	STANLY	NC	37167	3.10
CABARRUS	NC	37025	3.10	STOKES	NC	37169	2.95
CALDWELL	NC	37027	2.95	SURRY	NC	37171	2.95
CAMDEN	NC	37029	3.20	SWAIN	NC	37173	2.95
CARTERET	NC	37031	3.20	TRANSYLVANIA	NC	37175	2.95
CASWELL	NC	37033	3.10	TYRRELL	NC	37177	3.20
CATAWBA	NC	37035	3.10	UNION	NC	37179	3.10
CHATHAM	NC	37037	3.10	VANCE	NC	37181	3.10
CHEROKEE	NC	37039	2.95	WAKE	NC	37183	3.10
CHOWAN	NC	37041	3.20	WARREN	NC	37185	3.10
CLAY	NC	37043	2.95	WASHINGTON	NC	37187	3.20
CLEVELAND	NC	37045	3.10	WATAUGA	NC	37189	2.95
COLUMBUS	NC	37047	3.30	WAYNE	NC	37191	3.20
CRAVEN	NC	37049	3.20	WILKES	NC	37193	2.95
CUMBERLAND	NC	37051	3.30	WILSON	NC	37195	3.20
CURRITUCK	NC	37053	3.20	YADKIN	NC	37197	3.10
DARE	NC	37055	3.20	YANCEY	NC	37199	2.95
DAVIDSON	NC	37057	3.10	ADAMS	ND	38001	1.65
DAVIE	NC	37059	3.10	BARNES	ND	38003	1.65
DUPLIN	NC	37061	3.30	BENSON	ND	38005	1.60
DURHAM	NC	37063	3.10	BILLINGS	ND	38007	1.60
EDGECOMBE	NC	37065	3.20	BOTTINEAU	ND	38009	1.60
FORSYTH	NC	37067	3.10	BOWMAN	ND	38011	1.65
FRANKLIN	NC	37069	3.10	BURKE	ND	38013	1.60
GASTON	NC	37071	3.10	BURLEIGH	ND	38015	1.65
GATES	NC	37073	3.20	CASS	ND	38017	1.65
GRAHAM	NC	37075	2.95	CAVALIER	ND	38019	1.60
GRANVILLE	NC	37077	3.10	DICKEY	ND	38021	1.65
GREENE	NC	37079	3.20	DIVIDE	ND	38023	1.60
GUILFORD	NC	37081	3.10	DUNN	ND	38025	1.60
HALIFAX	NC	37083	3.10	EDDY	ND	38027	1.65
HARNETT	NC	37085	3.30	EMMONS	ND	38029	1.65
HAYWOOD	NC	37087	2.95	FOSTER	ND	38031	1.65
HENDERSON	NC	37089	2.95	GOLDEN VALLEY	ND	38033	1.60
HERTFORD	NC	37091	3.20	GRAND FORKS	ND	38035	1.65
HOKE	NC	37093	3.30	GRANT	ND	38037	1.65

County/parish/city	State	FIPS code	Class I differential adjusted for location	County/parish/city	State	FIPS code	Class I differential adjusted for location
GRIGGS	ND	38039	1.65	GREELEY	NE	31077	1.80
HETTINGER	ND	38041	1.65	HALL	NE	31079	1.80
KIDDER	ND	38043	1.65	HAMILTON	NE	31081	1.80
LA MOURE	ND	38045	1.65	HARLAN	NE	31083	1.80
LOGAN	ND	38047	1.65	HAYES	NE	31085	1.80
MCHENRY	ND	38049	1.60	HITCHCOCK	NE	31087	1.80
MCINTOSH	ND	38051	1.65	HOLT	NE	31089	1.75
MCKENZIE	ND	38053	1.60	HOOKER	NE	31091	1.75
MCLEAN	ND	38055	1.60	HOWARD	NE	31093	1.80
MERCER	ND	38057	1.60	JEFFERSON	NE	31095	1.80
MORTON	ND	38059	1.65	JOHNSON	NE	31097	1.85
MOUNTRAIL	ND	38061	1.60	KEARNEY	NE	31099	1.80
NELSON	ND	38063	1.65	KEITH	NE	31101	1.80
OLIVER	ND	38065	1.60	KEYA PAHA	NE	31103	1.75
PEMBINA	ND	38067	1.60	KIMBALL	NE	31105	1.80
PIERCE	ND	38069	1.60	KNOX	NE	31107	1.75
RAMSEY	ND	38071	1.60	LANCASTER	NE	31109	1.85
RANSOM	ND	38073	1.65	LINCOLN	NE	31111	1.80
RENVILLE	ND	38075	1.60	LOGAN	NE	31113	1.80
RICHLAND	ND	38077	1.65	LOUP	NE	31115	1.75
ROLETTE	ND	38079	1.60	MADISON	NE	31119	1.80
SARGENT	ND	38081	1.65	MCPHERSON	NE	31117	1.80
SHERIDAN	ND	38083	1.60	MERRICK	NE	31121	1.80
SIOUX	ND	38085	1.65	MORRILL	NE	31123	1.80
SLOPE	ND	38087	1.65	NANCE	NE	31125	1.80
STARK	ND	38089	1.60	NEMAHA	NE	31127	1.85
STEELE	ND	38091	1.65	NUCKOLLS	NE	31129	1.80
STUTSMAN	ND	38093	1.65	OTOE	NE	31131	1.85
TOWNER	ND	38095	1.60	PAWNEE	NE	31133	1.85
TRAILL	ND	38097	1.65	PERKINS	NE	31135	1.80
WALSH	ND	38099	1.60	PHELPS	NE	31137	1.80
WARD	ND	38101	1.60	PIERCE	NE	31139	1.75
WELLS	ND	38103	1.65	PLATTE	NE	31141	1.80
WILLIAMS	ND	38105	1.60	POLK	NE	31143	1.80
ADAMS	NE	31001	1.80	RED WILLOW	NE	31145	1.80
ANTELOPE	NE	31003	1.75	RICHARDSON	NE	31147	1.85
ARTHUR	NE	31005	1.80	ROCK	NE	31149	1.75
BANNER	NE	31007	1.80	SALINE	NE	31151	1.80
BLAINE	NE	31009	1.75	SARPY	NE	31153	1.85
BOONE	NE	31011	1.80	SAUNDERS	NE	31155	1.85
BOX BUTTE	NE	31013	1.80	SCOTTS BLUFF	NE	31157	1.80
BOYD	NE	31015	1.75	SEWARD	NE	31159	1.80
BROWN	NE	31017	1.75	SHERIDAN	NE	31161	1.80
BUFFALO	NE	31019	1.80	SHERMAN	NE	31163	1.80
BURT	NE	31021	1.80	SIOUX	NE	31165	1.80
BUTLER	NE	31023	1.80	STANTON	NE	31167	1.80
CASS	NE	31025	1.85	THAYER	NE	31169	1.80
CEDAR	NE	31027	1.75	THOMAS	NE	31171	1.75
CHASE	NE	31029	1.80	THURSTON	NE	31173	1.75
CHERRY	NE	31031	1.75	VALLEY	NE	31175	1.80
CHEYENNE	NE	31033	1.80	WASHINGTON	NE	31177	1.85
CLAY	NE	31035	1.80	WAYNE	NE	31179	1.75
COLFAX	NE	31037	1.80	WEBSTER	NE	31181	1.80
CUMING	NE	31039	1.80	WHEELER	NE	31183	1.75
CUSTER	NE	31041	1.80	YORK	NE	31185	1.80
DAKOTA	NE	31043	1.75	BELKNAP	NH	33001	2.80
DAWES	NE	31045	1.80	CARROLL	NH	33003	2.80
DAWSON	NE	31047	1.80	CHESHIRE	NH	33005	2.80
DEUEL	NE	31049	1.80	COOS	NH	33007	2.60
DIXON	NE	31051	1.75	GRAFTON	NH	33009	2.60
DODGE	NE	31053	1.80	HILLSBOROUGH	NH	33011	3.00
DOUGLAS	NE	31055	1.85	MERRIMACK	NH	33013	3.00
DUNDY	NE	31057	1.80	ROCKINGHAM	NH	33015	3.00
FILLMORE	NE	31059	1.80	STRAFFORD	NH	33017	3.00
FRANKLIN	NE	31061	1.80	SULLIVAN	NH	33019	2.80
FRONTIER	NE	31063	1.80	ATLANTIC	NJ	34001	3.05
FURNAS	NE	31065	1.80	BERGEN	NJ	34003	3.15
GAGE	NE	31067	1.85	BURLINGTON	NJ	34005	3.05
GARDEN	NE	31069	1.80	CAMDEN	NJ	34007	3.05
GARFIELD	NE	31071	1.75	CAPE MAY	NJ	34009	3.05
GOSPER	NE	31073	1.80	CUMBERLAND	NJ	34011	3.05
GRANT	NE	31075	1.75	ESSEX	NJ	34013	3.15

County/parish/city	State	FIPS code	Class I differential adjusted for location	County/parish/city	State	FIPS code	Class I differential adjusted for location
GLOUCESTER	NJ	34015	3.05	CHENANGO	NY	36017	2.50
HUDSON	NJ	34017	3.15	CLINTON	NY	36019	2.30
HUNTERDON	NJ	34019	3.10	COLUMBIA	NY	36021	2.70
MERCER	NJ	34021	3.10	CORTLAND	NY	36023	2.50
MIDDLESEX	NJ	34023	3.10	DELAWARE	NY	36025	2.70
MONMOUTH	NJ	34025	3.10	DUTCHESS	NY	36027	2.80
MORRIS	NJ	34027	3.10	ERIE	NY	36029	2.20
OCEAN	NJ	34029	3.10	ESSEX	NY	36031	2.30
PASSAIC	NJ	34031	3.15	FRANKLIN	NY	36033	2.30
SALEM	NJ	34033	3.05	FULTON	NY	36035	2.50
SOMERSET	NJ	34035	3.10	GENESEE	NY	36037	2.20
SUSSEX	NJ	34037	3.10	GREENE	NY	36039	2.70
UNION	NJ	34039	3.15	HAMILTON	NY	36041	2.50
WARREN	NJ	34041	3.10	HERKIMER	NY	36043	2.50
BERNALILLO	NM ...	35001	2.35	JEFFERSON	NY	36045	2.30
CATRON	NM ...	35003	2.10	KINGS	NY	36047	3.15
CHAVES	NM ...	35005	2.10	LEWIS	NY	36049	2.30
CIBOLA	NM ...	35006	1.90	LIVINGSTON	NY	36051	2.30
COLFAX	NM ...	35007	2.35	MADISON	NY	36053	2.50
CURRY	NM ...	35009	2.10	MONROE	NY	36055	2.30
DE BACA	NM ...	35011	2.10	MONTGOMERY	NY	36057	2.70
DONA ANA	NM ...	35013	2.10	NASSAU	NY	36059	3.15
EDDY	NM ...	35015	2.10	NEW YORK	NY	36061	3.15
GRANT	NM ...	35017	2.10	NIAGARA	NY	36063	2.20
GUADALUPE	NM ...	35019	2.35	ONEIDA	NY	36065	2.50
HARDING	NM ...	35021	2.35	ONONDAGA	NY	36067	2.50
HIDALGO	NM ...	35023	2.10	ONTARIO	NY	36069	2.30
LEA	NM ...	35025	2.10	ORANGE	NY	36071	3.00
LINCOLN	NM ...	35027	2.10	ORLEANS	NY	36073	2.20
LOS ALAMOS	NM ...	35028	2.35	OSWEGO	NY	36075	2.30
LUNA	NM ...	35029	2.10	OTSEGO	NY	36077	2.50
MCKINLEY	NM ...	35031	1.90	PUTNAM	NY	36079	3.00
MORA	NM ...	35033	2.35	QUEENS	NY	36081	3.15
OTERO	NM ...	35035	2.10	RENSSELAER	NY	36083	2.70
QUAY	NM ...	35037	2.35	RICHMOND	NY	36085	3.15
RIO ARRIBA	NM ...	35039	1.90	ROCKLAND	NY	36087	3.15
ROOSEVELT	NM ...	35041	2.10	SARATOGA	NY	36091	2.70
SAN JUAN	NM ...	35045	1.90	SCHENECTADY	NY	36093	2.70
SAN MIGUEL	NM ...	35047	2.35	SCHOHARIE	NY	36095	2.70
SANDOVAL	NM ...	35043	2.35	SCHUYLER	NY	36097	2.30
SANTA FE	NM ...	35049	2.35	SENECA	NY	36099	2.30
SIERRA	NM ...	35051	2.10	ST. LAWRENCE	NY	36089	2.30
SOCORRO	NM ...	35053	2.10	STEUBEN	NY	36101	2.30
TAOS	NM ...	35055	1.90	SUFFOLK	NY	36103	3.15
TORRANCE	NM ...	35057	2.35	SULLIVAN	NY	36105	2.80
UNION	NM ...	35059	2.35	TIOGA	NY	36107	2.50
VALENCIA	NM ...	35061	2.35	TOMPKINS	NY	36109	2.50
CARSON CITY	NV	32510	1.70	ULSTER	NY	36111	2.80
CHURCHILL	NV	32001	1.70	WARREN	NY	36113	2.50
CLARK	NV	32003	2.00	WASHINGTON	NY	36115	2.60
DOUGLAS	NV	32005	1.70	WAYNE	NY	36117	2.30
ELKO	NV	32007	1.90	WESTCHESTER	NY	36119	3.15
ESMERALDA	NV	32009	1.60	WYOMING	NY	36121	2.20
EUREKA	NV	32011	1.70	YATES	NY	36123	2.30
HUMBOLDT	NV	32013	1.70	ADAMS	OH ...	39001	2.20
LANDER	NV	32015	1.70	ALLEN	OH ...	39003	2.00
LINCOLN	NV	32017	1.60	ASHLAND	OH ...	39005	2.00
LYON	NV	32019	1.70	ASHTABULA	OH ...	39007	2.00
MINERAL	NV	32021	1.60	ATHENS	OH ...	39009	2.00
NYE	NV	32023	1.60	AUGLAIZE	OH ...	39011	2.00
PERSHING	NV	32027	1.70	BELMONT	OH ...	39013	2.00
STOREY	NV	32029	1.70	BROWN	OH ...	39015	2.20
WASHOE	NV	32031	1.70	BUTLER	OH ...	39017	2.00
WHITE PINE	NV	32033	1.90	CARROLL	OH ...	39019	2.00
ALBANY	NY	36001	2.70	CHAMPAIGN	OH ...	39021	2.00
ALLEGANY	NY	36003	2.30	CLARK	OH ...	39023	2.00
BRONX	NY	36005	3.15	CLERMONT	OH ...	39025	2.20
BROOME	NY	36007	2.70	CLINTON	OH ...	39027	2.00
CATTARAUGUS	NY	36009	2.10	COLUMBIANA	OH ...	39029	2.00
CAYUGA	NY	36011	2.30	COSHOCTON	OH ...	39031	2.00
CHAUTAUQUA	NY	36013	2.10	CRAWFORD	OH ...	39033	2.00
CHEMUNG	NY	36015	2.50	CUYAHOGA	OH ...	39035	2.00

County/parish/city	State	FIPS code	Class I differential adjusted for location	County/parish/city	State	FIPS code	Class I differential adjusted for location
DARKE	OH ...	39037	2.00	ATOKA	OK	40005	2.80
DEFIANCE	OH ...	39039	1.80	BEAVER	OK	40007	2.40
DELAWARE	OH ...	39041	2.00	BECKHAM	OK	40009	2.40
ERIE	OH ...	39043	2.00	BLAINE	OK	40011	2.40
FAIRFIELD	OH ...	39045	2.00	BRYAN	OK	40013	2.80
FAYETTE	OH · ...	39047	2.00	CADDO	OK	40015	2.60
FRANKLIN	OH ...	39049	2.00	CANADIAN	OK	40017	2.60
FULTON	OH ...	39051	1.80	CARTER	OK	40019	2.80
GALLIA	OH ...	39053	2.20	CHEROKEE	OK	40021	2.60
GEAUGA	OH ...	39055	2.00	CHOCTAW	OK	40023	2.80
GREENE	OH ...	39057	2.00	CIMARRON	OK	40025	2.40
GUERNSEY	OH ...	39059	2.00	CLEVELAND	OK	40027	2.60
HAMILTON	OH ...	39061	2.20	COAL	OK	40029	2.80
HANCOCK	OH ...	39063	2.00	COMANCHE	OK	40031	2.60
HARDIN	OH ...	39065	2.00	COTTON	OK	40033	2.80
HARRISON	OH ...	39067	2.00	CRAIG	OK	40035	2.40
HENRY	OH ...	39069	1.80	CREEK	OK	40037	2.60
HIGHLAND	OH ...	39071	2.20	CUSTER	OK	40039	2.40
HOCKING	OH ...	39073	2.00	DELAWARE	OK	40041	2.40
HOLMES	OH ...	39075	2.00	DEWEY	OK	40043	2.40
HURON	OH ...	39077	2.00	ELLIS	OK	40045	2.40
JACKSON	OH ...	39079	2.20	GARFIELD	OK	40047	2.40
JEFFERSON	OH ...	39081	2.00	GARVIN	OK	40049	2.60
KNOX	OH ...	39083	2.00	GRADY	OK	40051	2.60
LAKE	OH ...	39085	2.00	GRANT	OK	40053	2.40
LAWRENCE	OH ...	39087	2.20	GREER	OK	40055	2.60
LICKING	OH ...	39089	2.00	HARMON	OK	40057	2.60
LOGAN	OH ...	39091	2.00	HARPER	OK	40059	2.40
LORAIN	OH ...	39093	2.00	HASKELL	OK	40061	2.80
LUCAS	OH ...	39095	1.80	HUGHES	OK	40063	2.60
MADISON	OH ...	39097	2.00	JACKSON	OK	40065	2.60
MAHONING	OH ...	39099	2.00	JEFFERSON	OK	40067	2.80
MARION	OH ...	39101	2.00	JOHNSTON	OK	40069	2.80
MEDINA	OH ...	39103	2.00	KAY	OK	40071	2.40
MEIGS	OH ...	39105	2.00	KINGFISHER	OK	40073	2.40
MERCER	OH ...	39107	2.00	KIOWA	OK	40075	2.60
MIAMI	OH ...	39109	2.00	LATIMER	OK	40077	2.80
MONROE	OH ...	39111	2.00	LE FLORE	OK	40079	2.80
MONTGOMERY	OH ...	39113	2.00	LINCOLN	OK	40081	2.60
MORGAN	OH ...	39115	2.00	LOGAN	OK	40083	2.40
MORROW	OH ...	39117	2.00	LOVE	OK	40085	2.80
MUSKINGUM	OH ...	39119	2.00	MAJOR	OK	40093	2.40
NOBLE	OH ...	39121	2.00	MARSHALL	OK	40095	2.80
OTTAWA	OH ...	39123	2.00	MAYES	OK	40097	2.40
PAULDING	OH ...	39125	1.80	MCCLAIN	OK	40087	2.60
PERRY	OH ...	39127	2.00	MCCURTAIN	OK	40089	2.80
PICKAWAY	OH ...	39129	2.00	MCINTOSH	OK	40091	2.60
PIKE	OH ...	39131	2.20	MURRAY	OK	40099	2.80
PORTAGE	OH ...	39133	2.00	MUSKOGEE	OK	40101	2.60
PREBLE	OH ...	39135	2.00	NOBLE	OK	40103	2.40
PUTNAM	OH ...	39137	1.80	NOWATA	OK	40105	2.40
RICHLAND	OH ...	39139	2.00	OKFUSKEE	OK	40107	2.60
ROSS	OH ...	39141	2.00	OKLAHOMA	OK	40109	2.60
SANDUSKY	OH ...	39143	2.00	OKMULGEE	OK	40111	2.60
SCIOTO	OH ...	39145	2.20	OSAGE	OK	40113	2.40
SENECA	OH ...	39147	2.00	OTTAWA	OK	40115	2.40
SHELBY	OH ...	39149	2.00	PAWNEE	OK	40117	2.40
STARK	OH ...	39151	2.00	PAYNE	OK	40119	2.40
SUMMIT	OH ...	39153	2.00	PITTSBURG	OK	40121	2.80
TRUMBULL	OH ...	39155	2.00	PONTOTOC	OK	40123	2.80
TUSCARAWAS	OH ...	39157	2.00	POTTAWATOMIE	OK	40125	2.60
UNION	OH ...	39159	2.00	PUSHMATAHA	OK	40127	2.80
VAN WERT	OH ...	39161	1.80	ROGER MILLS	OK	40129	2.40
VINTON	OH ...	39163	2.00	ROGERS	OK	40131	2.40
WARREN	OH ...	39165	2.00	SEMINOLE	OK	40133	2.60
WASHINGTON	OH ...	39167	2.00	SEQUOYAH	OK	40135	2.80
WAYNE	OH ...	39169	2.00	STEPHENS	OK	40137	2.80
WILLIAMS	OH ...	39171	1.80	TEXAS	OK	40139	2.40
WOOD	OH ...	39173	2.00	TILLMAN	OK	40141	2.60
WYANDOT	OH ...	39175	2.00	TULSA	OK	40143	2.60
ADAIR	OK	40001	2.60	WAGONER	OK	40145	2.60
ALFALFA	OK	40003	2.40	WASHINGTON	OK	40147	2.40

County/parish/city	State	FIPS code	Class I differential adjusted for location	County/parish/city	State	FIPS code	Class I differential adjusted for location
WASHITA	OK	40149	2.40	JUNIATA	PA	42067	2.70
WOODS	OK	40151	2.40	LACKAWANNA	PA	42069	2.70
WOODWARD	OK	40153	2.40	LANCASTER	PA	42071	2.90
BAKER	OR ...	41001	1.60	LAWRENCE	PA	42073	2.10
BENTON	OR ...	41003	1.90	LEBANON	PA	42075	2.80
CLACKAMAS	OR ...	41005	1.90	LEHIGH	PA	42077	2.80
CLATSOP	OR ...	41007	1.90	LUZERNE	PA	42079	2.70
COLUMBIA	OR ...	41009	1.90	LYCOMING	PA	42081	2.50
COOS	OR ...	41011	1.90	MCKEAN	PA	42083	2.30
CROOK	OR ...	41013	1.75	MERCER	PA	42085	2.10
CURRY	OR ...	41015	1.90	MIFFLIN	PA	42087	2.70
DESCHUTES	OR ...	41017	1.75	MONROE	PA	42089	2.80
DOUGLAS	OR ...	41019	1.90	MONTGOMERY	PA	42091	3.05
GILLIAM	OR ...	41021	1.75	MONTOUR	PA	42093	2.70
GRANT	OR ...	41023	1.60	NORTHAMPTON	PA	42095	2.80
HARNEY	OR ...	41025	1.60	NORTHUMBERLAND	PA	42097	2.70
HOOD RIVER	OR ...	41027	1.90	PERRY	PA	42099	2.70
JACKSON	OR ...	41029	1.90	PHILADELPHIA	PA	42101	3.05
JEFFERSON	OR ...	41031	1.75	PIKE	PA	42103	2.80
JOSEPHINE	OR ...	41033	1.90	POTTER	PA	42105	2.50
KLAMATH	OR ...	41035	1.75	SCHUYLKILL	PA	42107	2.80
LAKE	OR ...	41037	1.75	SNYDER	PA	42109	2.70
LANE	OR ...	41039	1.90	SOMERSET	PA	42111	2.30
LINCOLN	OR ...	41041	1.90	SULLIVAN	PA	42113	2.50
LINN	OR ...	41043	1.90	SUSQUEHANNA	PA	42115	2.50
MALHEUR	OR ...	41045	1.60	TIOGA	PA	42117	2.50
MARION	OR ...	41047	1.90	UNION	PA	42119	2.70
MORROW	OR ...	41049	1.75	VENANGO	PA	42121	2.10
MULTNOMAH	OR ...	41051	1.90	WARREN	PA	42123	2.10
POLK	OR ...	41053	1.90	WASHINGTON	PA	42125	2.10
SHERMAN	OR ...	41055	1.75	WAYNE	PA	42127	2.70
TILLAMOOK	OR ...	41057	1.90	WESTMORELAND	PA	42129	2.30
UMATILLA	OR ...	41059	1.75	WYOMING	PA	42131	2.50
UNION	OR ...	41061	1.60	YORK	PA	42133	2.90
WALLOWA	OR ...	41063	1.60	BRISTOL	RI	44001	3.25
WASCO	OR ...	41065	1.75	KENT	RI	44003	3.25
WASHINGTON	OR ...	41067	1.90	NEWPORT	RI	44005	3.25
WHEELER	OR ...	41069	1.75	PROVIDENCE	RI	44007	3.25
YAMHILL	OR ...	41071	1.90	WASHINGTON	RI	44009	3.25
ADAMS	PA	42001	2.80	ABBEVILLE	SC	45001	3.10
ALLEGHENY	PA	42003	2.10	AIKEN	SC	45003	3.30
ARMSTRONG	PA	42005	2.30	ALLENDALE	SC	45005	3.30
BEAVER	PA	42007	2.10	ANDERSON	SC	45007	3.10
BEDFORD	PA	42009	2.30	BAMBERG	SC	45009	3.30
BERKS	PA	42011	2.80	BARNWELL	SC	45011	3.30
BLAIR	PA	42013	2.30	BEAUFORT	SC	45013	3.30
BRADFORD	PA	42015	2.50	BERKELEY	SC	45015	3.30
BUCKS	PA	42017	3.05	CALHOUN	SC	45017	3.30
BUTLER	PA	42019	2.10	CHARLESTON	SC	45019	3.30
CAMBRIA	PA	42021	2.30	CHEROKEE	SC	45021	3.10
CAMERON	PA	42023	2.30	CHESTER	SC	45023	3.10
CARBON	PA	42025	2.80	CHESTERFIELD	SC	45025	3.30
CENTRE	PA	42027	2.50	CLARENDON	SC	45027	3.30
CHESTER	PA	42029	3.05	COLLETON	SC	45029	3.30
CLARION	PA	42031	2.30	DARLINGTON	SC	45031	3.30
CLEARFIELD	PA	42033	2.30	DILLON	SC	45033	3.30
CLINTON	PA	42035	2.50	DORCHESTER	SC	45035	3.30
COLUMBIA	PA	42037	2.70	EDGEFIELD	SC	45037	3.30
CRAWFORD	PA	42039	2.10	FAIRFIELD	SC	45039	3.30
CUMBERLAND	PA	42041	2.80	FLORENCE	SC	45041	3.30
DAUPHIN	PA	42043	2.80	GEORGETOWN	SC	45043	3.30
DELAWARE	PA	42045	3.05	GREENVILLE	SC	45045	3.10
ELK	PA	42047	2.30	GREENWOOD	SC	45047	3.10
ERIE	PA	42049	2.10	HAMPTON	SC	45049	3.30
FAYETTE	PA	42051	2.30	HORRY	SC	45051	3.30
FOREST	PA	42053	2.30	JASPER	SC	45053	3.30
FRANKLIN	PA	42055	2.80	KERSHAW	SC	45055	3.30
FULTON	PA	42057	2.70	LANCASTER	SC	45057	3.10
GREENE	PA	42059	2.10	LAURENS	SC	45059	3.10
HUNTINGDON	PA	42061	2.30	LEE	SC	45061	3.30
INDIANA	PA	42063	2.30	LEXINGTON	SC	45063	3.30
JEFFERSON	PA	42065	2.30	MARION	SC	45067	3.30

County/parish/city	State	FIPS code	Class I differential adjusted for location	County/parish/city	State	FIPS code	Class I differential adjusted for location
MARLBORO	SC	45069	3.30	TODD	SD	46121	1.70
MCCORMICK	SC	45065	3.10	TRIPP	SD	46123	1.70
NEWBERRY	SC	45071	3.30	TURNER	SD	46125	1.75
OCONEE	SC	45073	3.10	UNION	SD	46127	1.75
ORANGEBURG	SC	45075	3.30	WALWORTH	SD	46129	1.70
PICKENS	SC	45077	3.10	YANKTON	SD	46135	1.75
RICHLAND	SC	45079	3.30	ZIEBACH	SD	46137	1.65
SALUDA	SC	45081	3.30	ANDERSON	TN	47001	2.80
SPARTANBURG	SC	45083	3.10	BEDFORD	TN	47003	2.60
SUMTER	SC	45085	3.30	BENTON	TN	47005	2.60
UNION	SC	45087	3.10	BLEDSOE	TN	47007	2.60
WILLIAMSBURG	SC	45089	3.30	BLOUNT	TN	47009	2.80
YORK	SC	45091	3.10	BRADLEY	TN	47011	2.80
AURORA	SD	46003	1.70	CAMPBELL	TN	47013	2.80
BEADLE	SD	46005	1.70	CANNON	TN	47015	2.60
BENNETT	SD	46007	1.70	CARROLL	TN	47017	2.60
BON HOMME	SD	46009	1.75	CARTER	TN	47019	2.80
BROOKINGS	SD	46011	1.70	CHEATHAM	TN	47021	2.60
BROWN	SD	46013	1.70	CHESTER	TN	47023	2.80
BRULE	SD	46015	1.70	CLAIBORNE	TN	47025	2.80
BUFFALO	SD	46017	1.70	CLAY	TN	47027	2.60
BUTTE	SD	46019	1.65	COCKE	TN	47029	2.80
CAMPBELL	SD	46021	1.65	COFFEE	TN	47031	2.60
CHARLES MIX	SD	46023	1.75	CROCKETT	TN	47033	2.60
CLARK	SD	46025	1.70	CUMBERLAND	TN	47035	2.80
CLAY	SD	46027	1.75	DAVIDSON	TN	47037	2.60
CODINGTON	SD	46029	1.70	DE KALB	TN	47041	2.60
CORSON	SD	46031	1.65	DECATUR	TN	47039	2.60
CUSTER	SD	46033	1.80	DICKSON	TN	47043	2.60
DAVISON	SD	46035	1.70	DYER	TN	47045	2.60
DAY	SD	46037	1.70	FAYETTE	TN	47047	2.80
DEUEL	SD	46039	1.70	FENTRESS	TN	47049	2.60
DEWEY	SD	46041	1.65	FRANKLIN	TN	47051	2.80
DOUGLAS	SD	46043	1.75	GIBSON	TN	47053	2.60
EDMUNDS	SD	46045	1.70	GILES	TN	47055	2.80
FALL RIVER	SD	46047	1.80	GRAINGER	TN	47057	2.80
FAULK	SD	46049	1.70	GREENE	TN	47059	2.80
GRANT	SD	46051	1.70	GRUNDY	TN	47061	2.60
GREGORY	SD	46053	1.75	HAMBLEN	TN	47063	2.80
HAAKON	SD	46055	1.70	HAMILTON	TN	47065	2.80
HAMLIN	SD	46057	1.70	HANCOCK	TN	47067	2.80
HAND	SD	46059	1.70	HARDEMAN	TN	47069	2.80
HANSON	SD	46061	1.70	HARDIN	TN	47071	2.80
HARDING	SD	46063	1.65	HAWKINS	TN	47073	2.80
HUGHES	SD	46065	1.70	HAYWOOD	TN	47075	2.60
HUTCHINSON	SD	46067	1.75	HENDERSON	TN	47077	2.60
HYDE	SD	46069	1.70	HENRY	TN	47079	2.60
JACKSON	SD	46071	1.70	HICKMAN	TN	47081	2.60
JERAULD	SD	46073	1.70	HOUSTON	TN	47083	2.60
JONES	SD	46075	1.70	HUMPHREYS	TN	47085	2.60
KINGSBURY	SD	46077	1.70	JACKSON	TN	47087	2.60
LAKE	SD	46079	1.70	JEFFERSON	TN	47089	2.80
LAWRENCE	SD	46081	1.80	JOHNSON	TN	47091	2.80
LINCOLN	SD	46083	1.75	KNOX	TN	47093	2.80
LYMAN	SD	46085	1.70	LAKE	TN	47095	2.60
MARSHALL	SD	46091	1.70	LAUDERDALE	TN	47097	2.60
MCCOOK	SD	46087	1.70	LAWRENCE	TN	47099	2.80
MCPHERSON	SD	46089	1.70	LEWIS	TN	47101	2.60
MEADE	SD	46093	1.65	LINCOLN	TN	47103	2.80
MELLETTE	SD	46095	1.70	LOUDON	TN	47105	2.80
MINER	SD	46097	1.70	MACON	TN	47111	2.60
MINNEHAHA	SD	46099	1.70	MADISON	TN	47113	2.60
MOODY	SD	46101	1.70	MARION	TN	47115	2.80
PENNINGTON	SD	46103	1.80	MARSHALL	TN	47117	2.60
PERKINS	SD	46105	1.65	MAURY	TN	47119	2.60
POTTER	SD	46107	1.70	MCMINN	TN	47107	2.80
ROBERTS	SD	46109	1.70	MCNAIRY	TN	47109	2.80
SANBORN	SD	46111	1.70	MEIGS	TN	47121	2.80
SHANNON	SD	46113	1.80	MONROE	TN	47123	2.80
SPINK	SD	46115	1.70	MONTGOMERY	TN	47125	2.60
STANLEY	SD	46117	1.70	MOORE	TN	47127	2.80
SULLY	SD	46119	1.70	MORGAN	TN	47129	2.80

County/parish/city	State	FIPS code	Class I differential adjusted for location
OBION	TN	47131	2.60
OVERTON	TN	47133	2.60
PERRY	TN	47135	2.60
PICKETT	TN	47137	2.60
POLK	TN	47139	2.80
PUTNAM	TN	47141	2.60
RHEA	TN	47143	2.80
ROANE	TN	47145	2.80
ROBERTSON	TN	47147	2.60
RUTHERFORD	TN	47149	2.60
SCOTT	TN	47151	2.80
SEQUATCHIE	TN	47153	2.80
SEVIER	TN	47155	2.80
SHELBY	TN	47157	2.80
SMITH	TN	47159	2.60
STEWART	TN	47161	2.60
SULLIVAN	TN	47163	2.80
SUMNER	TN	47165	2.60
TIPTON	TN	47167	2.80
TROUSDALE	TN	47169	2.60
UNICOI	TN	47171	2.80
UNION	TN	47173	2.80
VAN BUREN	TN	47175	2.60
WARREN	TN	47177	2.60
WASHINGTON	TN	47179	2.80
WAYNE	TN	47181	2.80
WEAKLEY	TN	47183	2.60
WHITE	TN	47185	2.60
WILLIAMSON	TN	47187	2.60
WILSON	TN	47189	2.60
ANDERSON	TX	48001	3.15
ANDREWS	TX	48003	2.40
ANGELINA	TX	48005	3.15
ARANSAS	TX	48007	3.65
ARCHER	TX	48009	2.80
ARMSTRONG	TX	48011	2.40
ATASCOSA	TX	48013	3.45
AUSTIN	TX	48015	3.60
BAILEY	TX	48017	2.40
BANDERA	TX	48019	3.30
BASTROP	TX	48021	3.30
BAYLOR	TX	48023	2.60
BEE	TX	48025	3.65
BELL	TX	48027	3.15
BEXAR	TX	48029	3.45
BLANCO	TX	48031	3.30
BORDEN	TX	48033	2.40
BOSQUE	TX	48035	3.15
BOWIE	TX	48037	3.00
BRAZORIA	TX	48039	3.60
BRAZOS	TX	48041	3.30
BREWSTER	TX	48043	2.40
BRISCOE	TX	48045	2.40
BROOKS	TX	48047	3.65
BROWN	TX	48049	2.80
BURLESON	TX	48051	3.30
BURNET	TX	48053	3.30
CALDWELL	TX	48055	3.45
CALHOUN	TX	48057	3.65
CALLAHAN	TX	48059	2.80
CAMERON	TX	48061	3.65
CAMP	TX	48063	3.00
CARSON	TX	48065	2.40
CASS	TX	48067	3.00
CASTRO	TX	48069	2.40
CHAMBERS	TX	48071	3.60
CHEROKEE	TX	48073	3.15
CHILDRESS	TX	48075	2.40
CLAY	TX	48077	2.80
COCHRAN	TX	48079	2.40
COKE	TX	48081	2.60
COLEMAN	TX	48083	2.80

County/parish/city	State	FIPS code	Class I differential adjusted for location
COLLIN	TX	48085	3.00
COLLINGSWORTH	TX	48087	2.40
COLORADO	TX	48089	3.60
COMAL	TX	48091	3.45
COMANCHE	TX	48093	2.80
CONCHO	TX	48095	2.80
COOKE	TX	48097	3.00
CORYELL	TX	48099	3.15
COTTLE	TX	48101	2.40
CRANE	TX	48103	2.40
CROCKETT	TX	48105	2.60
CROSBY	TX	48107	2.40
CULBERSON	TX	48109	2.40
DALLAM	TX	48111	2.40
DALLAS	TX	48113	3.00
DAWSON	TX	48115	2.40
DE WITT	TX	48123	3.60
DEAF SMITH	TX	48117	2.40
DELTA	TX	48119	3.00
DENTON	TX	48121	3.00
DICKENS	TX	48125	2.40
DIMMIT	TX	48127	3.45
DONLEY	TX	48129	2.40
DUVAL	TX	48131	3.65
EASTLAND	TX	48133	2.80
ECTOR	TX	48135	2.40
EDWARDS	TX	48137	2.80
EL PASO	TX	48141	2.25
ELLIS	TX	48139	3.00
ERATH	TX	48143	3.00
FALLS	TX	48145	3.15
FANNIN	TX	48147	3.00
FAYETTE	TX	48149	3.60
FISHER	TX	48151	2.60
FLOYD	TX	48153	2.40
FOARD	TX	48155	2.60
FORT BEND	TX	48157	3.60
FRANKLIN	TX	48159	3.00
FREESTONE	TX	48161	3.15
FRIO	TX	48163	3.45
GAINES	TX	48165	2.40
GALVESTON	TX	48167	3.60
GARZA	TX	48169	2.40
GILLESPIE	TX	48171	3.30
GLASSCOCK	TX	48173	2.60
GOLIAD	TX	48175	3.65
GONZALES	TX	48177	3.45
GRAY	TX	48179	2.40
GRAYSON	TX	48181	3.00
GREGG	TX	48183	3.00
GRIMES	TX	48185	3.30
GUADALUPE	TX	48187	3.45
HALE	TX	48189	2.40
HALL	TX	48191	2.40
HAMILTON	TX	48193	3.15
HANSFORD	TX	48195	2.40
HARDEMAN	TX	48197	2.60
HARDIN	TX	48199	3.60
HARRIS	TX	48201	3.60
HARRISON	TX	48203	3.00
HARTLEY	TX	48205	2.40
HASKELL	TX	48207	2.60
HAYS	TX	48209	3.45
HEMPHILL	TX	48211	2.40
HENDERSON	TX	48213	3.00
HIDALGO	TX	48215	3.65
HILL	TX	48217	3.15
HOCKLEY	TX	48219	2.40
HOOD	TX	48221	3.00
HOPKINS	TX	48223	3.00
HOUSTON	TX	48225	3.15
HOWARD	TX	48227	2.40

County/parish/city	State	FIPS code	Class I differential adjusted for location	County/parish/city	State	FIPS code	Class I differential adjusted for location
HUDSPETH	TX	48229	2.25	POLK	TX	48373	3.30
HUNT	TX	48231	3.00	POTTER	TX	48375	2.40
HUTCHINSON	TX	48233	2.40	PRESIDIO	TX	48377	2.40
IRION	TX	48235	2.60	RAINS	TX	48379	3.00
JACK	TX	48237	2.80	RANDALL	TX	48381	2.40
JACKSON	TX	48239	3.60	REAGAN	TX	48383	2.60
JASPER	TX	48241	3.30	REAL	TX	48385	3.30
JEFF DAVIS	TX	48243	2.40	RED RIVER	TX	48387	3.00
JEFFERSON	TX	48245	3.60	REEVES	TX	48389	2.40
JIM HOGG	TX	48247	3.65	REFUGIO	TX	48391	3.65
JIM WELLS	TX	48249	3.65	ROBERTS	TX	48393	2.40
JOHNSON	TX	48251	3.00	ROBERTSON	TX	48395	3.30
JONES	TX	48253	2.60	ROCKWALL	TX	48397	3.00
KARNES	TX	48255	3.85	RUNNELS	TX	48399	2.80
KAUFMAN	TX	48257	3.00	RUSK	TX	48401	3.00
KENDALL	TX	48259	3.30	SABINE	TX	48403	3.15
KENEDY	TX	48261	3.65	SAN AUGUSTINE	TX	48405	3.15
KENT	TX	48263	2.60	SAN JACINTO	TX	48407	3.30
KERR	TX	48265	3.30	SAN PATRICIO	TX	48409	3.65
KIMBLE	TX	48267	2.80	SAN SABA	TX	48411	2.80
KING	TX	48269	2.60	SCHLEICHER	TX	48413	2.80
KINNEY	TX	48271	3.30	SCURRY	TX	48415	2.60
KLEBERG	TX	48273	3.65	SHACKELFORD	TX	48417	2.80
KNOX	TX	48275	2.60	SHELBY	TX	48419	3.15
LA SALLE	TX	48283	3.45	SHERMAN	TX	48421	2.40
LAMAR	TX	48277	3.00	SMITH	TX	48423	3.00
LAMB	TX	48279	2.40	SOMERVELL	TX	48425	3.00
LAMPASAS	TX	48281	3.15	STARR	TX	48427	3.65
LAVACA	TX	48285	3.60	STEPHENS	TX	48429	2.80
LEE	TX	48287	3.30	STERLING	TX	48431	2.60
LEON	TX	48289	3.15	STONEWALL	TX	48433	2.60
LIBERTY	TX	48291	3.60	SUTTON	TX	48435	2.80
LIMESTONE	TX	48293	3.15	SWISHER	TX	48437	2.40
LIPSCOMB	TX	48295	2.40	TARRANT	TX	48439	3.00
LIVE OAK	TX	48297	3.65	TAYLOR	TX	48441	2.60
LLANO	TX	48299	3.30	TERRELL	TX	48443	2.60
LOVING	TX	48301	2.40	TERRY	TX	48445	2.40
LUBBOCK	TX	48303	2.40	THROCKMORTON	TX	48447	2.80
LYNN	TX	48305	2.40	TITUS	TX	48449	3.00
MADISON	TX	48313	3.30	TOM GREEN	TX	48451	2.80
MARION	TX	48315	3.00	TRAVIS	TX	48453	3.30
MARTIN	TX	48317	2.40	TRINITY	TX	48455	3.30
MASON	TX	48319	2.80	TYLER	TX	48457	3.30
MATAGORDA	TX	48321	3.60	UPSHUR	TX	48459	3.00
MAVERICK	TX	48323	3.30	UPTON	TX	48461	2.40
MCCULLOCH	TX	48307	2.80	UVALDE	TX	48463	3.30
MCLENNAN	TX	48309	3.15	VAL VERDE	TX	48465	2.80
MCMULLEN	TX	48311	3.45	VAN ZANDT	TX	48467	3.00
MEDINA	TX	48325	3.30	VICTORIA	TX	48469	3.65
MENARD	TX	48327	2.80	WALKER	TX	48471	3.30
MIDLAND	TX	48329	2.40	WALLER	TX	48473	3.60
MILAM	TX	48331	3.30	WARD	TX	48475	2.40
MILLS	TX	48333	2.80	WASHINGTON	TX	48477	3.30
MITCHELL	TX	48335	2.60	WEBB	TX	48479	3.45
MONTAGUE	TX	48337	2.80	WHARTON	TX	48481	3.60
MONTGOMERY	TX	48339	3.60	WHEELER	TX	48483	2.40
MOORE	TX	48341	2.40	WICHITA	TX	48485	2.80
MORRIS	TX	48343	3.00	WILBARGER	TX	48487	2.60
MOTLEY	TX	48345	2.40	WILLACY	TX	48489	3.65
NACOGDOCHES	TX	48347	3.15	WILLIAMSON	TX	48491	3.30
NAVARRO	TX	48349	3.15	WILSON	TX	48493	3.45
NEWTON	TX	48351	3.30	WINKLER	TX	48495	2.40
NOLAN	TX	48353	2.60	WISE	TX	48497	3.00
NUECES	TX	48355	3.65	WOOD	TX	48499	3.00
OCHILTREE	TX	48357	2.40	YOAKUM	TX	48501	2.40
OLDHAM	TX	48359	2.40	YOUNG	TX	48503	2.80
ORANGE	TX	48361	3.60	ZAPATA	TX	48505	3.65
PALO PINTO	TX	48363	2.80	ZAVALA	TX	48507	3.30
PANOLA	TX	48365	3.00	BEAVER	UT	49001	1.60
PARKER	TX	48367	3.00	BOX ELDER	UT	49003	1.90
PARMER	TX	48369	2.40	CACHE	UT	49005	1.90
PECOS	TX	48371	2.40	CARBON	UT	49007	1.90

40

County/parish/city	State	FIPS code	Class I differential adjusted for location
DAGGETT	UT	49009	1.90
DAVIS	UT	49011	1.90
DUCHESNE	UT	49013	1.90
EMERY	UT	49015	1.90
GARFIELD	UT	49017	1.60
GRAND	UT	49019	1.90
IRON	UT	49021	1.60
JUAB	UT	49023	1.90
KANE	UT	49025	1.60
MILLARD	UT	49027	1.90
MORGAN	UT	49029	1.90
PIUTE	UT	49031	1.60
RICH	UT	49033	1.90
SALT LAKE	UT	49035	1.90
SAN JUAN	UT	49037	1.60
SANPETE	UT	49039	1.90
SEVIER	UT	49041	1.90
SUMMIT	UT	49043	1.90
TOOELE	UT	49045	1.90
UINTAH	UT	49047	1.90
UTAH	UT	49049	1.90
WASATCH	UT	49051	1.90
WASHINGTON	UT	49053	1.60
WAYNE	UT	49055	1.60
WEBER	UT	49057	1.90
ACCOMACK	VA	51001	3.00
ALBEMARLE	VA	51003	2.80
ALEXANDRIA CITY	VA	51510	3.00
ALLEGHANY	VA	51005	2.80
AMELIA	VA	51007	3.10
AMHERST	VA	51009	2.80
APPOMATTOX	VA	51011	2.80
ARLINGTON	VA	51013	3.00
AUGUSTA	VA	51015	2.80
BATH	VA	51017	2.80
BEDFORD	VA	51019	2.80
BEDFORD CITY	VA	51515	2.80
BLAND	VA	51021	2.80
BOTETOURT	VA	51023	2.80
BRISTOL CITY	VA	51520	2.80
BRUNSWICK	VA	51025	3.10
BUCHANAN	VA	51027	2.80
BUCKINGHAM	VA	51029	2.80
BUENA VISTA CITY	VA	51530	2.80
CAMPBELL	VA	51031	2.80
CAROLINE	VA	51033	3.10
CARROLL	VA	51035	2.80
CHARLES CITY	VA	51036	3.10
CHARLOTTE	VA	51037	3.10
CHARLOTTESVILLE CITY	VA	51540	2.80
CHESAPEAKE CITY	VA	51550	3.20
CHESTERFIELD	VA	51041	3.10
CLARKE	VA	51043	2.80
CLIFTON FORGE CITY	VA	51560	2.80
COLONIAL HEIGHTS CITY	VA	51570	3.10
COVINGTON CITY	VA	51580	2.80
CRAIG	VA	51045	2.80
CULPEPER	VA	51047	2.80
CUMBERLAND	VA	51049	2.80
DANVILLE CITY	VA	51590	2.80
DICKENSON	VA	51051	2.80
DINWIDDIE	VA	51053	3.10
EMPORIA CITY	VA	51595	3.10
ESSEX	VA	51057	3.10
FAIRFAX	VA	51059	3.00
FAIRFAX CITY	VA	51600	3.00
FALLS CHURCH CITY	VA	51610	3.00
FAUQUIER	VA	51061	3.00
FLOYD	VA	51063	2.80
FLUVANNA	VA	51065	2.80
FRANKLIN	VA	51067	2.80
FRANKLIN CITY	VA	51620	3.10

County/parish/city	State	FIPS code	Class I differential adjusted for location
FREDERICK	VA	51069	2.80
FREDERICKSBURG CITY	VA	51630	2.80
GALAX CITY	VA	51640	2.80
GILES	VA	51071	2.80
GLOUCESTER	VA	51073	3.20
GOOCHLAND	VA	51075	3.10
GRAYSON	VA	51077	2.80
GREENE	VA	51079	2.80
GREENSVILLE	VA	51081	3.10
HALIFAX	VA	51083	3.10
HAMPTON CITY	VA	51650	3.20
HANOVER	VA	51085	3.10
HARRISONBURG CITY	VA	51660	2.80
HENRICO	VA	51087	3.10
HENRY	VA	51089	2.80
HIGHLAND	VA	51091	2.80
HOPEWELL CITY	VA	51670	3.10
ISLE OF WIGHT	VA	51093	3.20
JAMES CITY	VA	51095	3.10
KING AND QUEEN	VA	51097	3.10
KING GEORGE	VA	51099	3.10
KING WILLIAM	VA	51101	3.10
LANCASTER	VA	51103	3.10
LEE	VA	51105	2.80
LEXINGTON CITY	VA	51678	2.80
LOUDOUN	VA	51107	3.00
LOUISA	VA	51109	2.80
LUNENBURG	VA	51111	3.10
LYNCHBURG CITY	VA	51680	2.80
MADISON	VA	51113	2.80
MANASSAS CITY	VA	51683	3.00
MANASSAS PARK CITY	VA	51685	3.00
MARTINSVILLE CITY	VA	51690	2.80
MATHEWS	VA	51115	3.20
MECKLENBURG	VA	51117	3.10
MIDDLESEX	VA	51119	3.10
MONTGOMERY	VA	51121	2.80
NELSON	VA	51125	2.80
NEW KENT	VA	51127	3.10
NEWPORT NEWS CITY	VA	51700	3.20
NORFOLK CITY	VA	51710	3.20
NORTHAMPTON	VA	51131	3.00
NORTHUMBERLAND	VA	51133	3.10
NORTON CITY	VA	51720	2.80
NOTTOWAY	VA	51135	3.10
ORANGE	VA	51137	2.80
PAGE	VA	51139	2.80
PATRICK	VA	51141	2.80
PETERSBURG CITY	VA	51730	3.10
PITTSYLVANIA	VA	51143	2.80
POQUOSON CITY	VA	51735	3.20
PORTSMOUTH CITY	VA	51740	3.20
POWHATAN	VA	51145	3.10
PRINCE EDWARD	VA	51147	3.10
PRINCE GEORGE	VA	51149	3.10
PRINCE WILLIAM	VA	51153	3.00
PULASKI	VA	51155	2.80
RADFORD CITY	VA	51750	2.80
RAPPAHANNOCK	VA	51157	2.80
RICHMOND	VA	51159	3.10
RICHMOND CITY	VA	51760	3.10
ROANOKE	VA	51161	2.80
ROANOKE CITY	VA	51770	2.80
ROCKBRIDGE	VA	51163	2.80
ROCKINGHAM	VA	51165	2.80
RUSSELL	VA	51167	2.80
SALEM CITY	VA	51775	2.80
SCOTT	VA	51169	2.80
SHENANDOAH	VA	51171	2.80
SMYTH	VA	51173	2.80
SOUTHAMPTON	VA	51175	3.10
SPOTSYLVANIA	VA	51177	2.80

County/parish/city	State	FIPS code	Class I differential adjusted for location	County/parish/city	State	FIPS code	Class I differential adjusted for location
STAFFORD	VA	51179	3.00	BAYFIELD	WI	55007	1.70
STAUNTON CITY	VA	51790	2.80	BROWN	WI	55009	1.75
SUFFOLK CITY	VA	51800	3.20	BUFFALO	WI	55011	1.70
SURRY	VA	51181	3.10	BURNETT	WI	55013	1.70
SUSSEX	VA	51183	3.10	CALUMET	WI	55015	1.75
TAZEWELL	VA	51185	2.80	CHIPPEWA	WI	55017	1.70
VIRGINIA BEACH CITY	VA	51810	3.20	CLARK	WI	55019	1.70
WARREN	VA	51187	2.80	COLUMBIA	WI	55021	1.75
WASHINGTON	VA	51191	2.80	CRAWFORD	WI	55023	1.75
WAYNESBORO CITY	VA	51820	2.80	DANE	WI	55025	1.75
WESTMORELAND	VA	51193	3.10	DODGE	WI	55027	1.75
WILLIAMSBURG CITY	VA	51830	3.10	DOOR	WI	55029	1.75
WINCHESTER CITY	VA	51840	2.80	DOUGLAS	WI	55031	1.70
WISE	VA	51195	2.80	DUNN	WI	55033	1.70
WYTHE	VA	51197	2.80	EAU CLAIRE	WI	55035	1.70
YORK	VA	51199	3.20	FLORENCE	WI	55037	1.70
ADDISON	VT	50001	2.60	FOND DU LAC	WI	55039	1.75
BENNINGTON	VT	50003	2.80	FOREST	WI	55041	1.70
CALEDONIA	VT	50005	2.60	GRANT	WI	55043	1.75
CHITTENDEN	VT	50007	2.50	GREEN	WI	55045	1.75
ESSEX	VT	50009	2.40	GREEN LAKE	WI	55047	1.70
FRANKLIN	VT	50011	2.40	IOWA	WI	55049	1.75
GRAND ISLE	VT	50013	2.40	IRON	WI	55051	1.70
LAMOILLE	VT	50015	2.50	JACKSON	WI	55053	1.70
ORANGE	VT	50017	2.60	JEFFERSON	WI	55055	1.75
ORLEANS	VT	50019	2.40	JUNEAU	WI	55057	1.70
RUTLAND	VT	50021	2.60	KENOSHA	WI	55059	1.75
WASHINGTON	VT	50023	2.60	KEWAUNEE	WI	55061	1.75
WINDHAM	VT	50025	2.80	LA CROSSE	WI	55063	1.70
WINDSOR	VT	50027	2.80	LAFAYETTE	WI	55065	1.75
ADAMS	WA ...	53001	1.75	LANGLADE	WI	55067	1.70
ASOTIN	WA ...	53003	1.75	LINCOLN	WI	55069	1.70
BENTON	WA ...	53005	1.75	MANITOWOC	WI	55071	1.75
CHELAN	WA ...	53007	1.75	MARATHON	WI	55073	1.70
CLALLAM	WA ...	53009	1.90	MARINETTE	WI	55075	1.70
CLARK	WA ...	53011	1.90	MARQUETTE	WI	55077	1.70
COLUMBIA	WA ...	53013	1.75	MENOMINEE	WI	55078	1.70
COWLITZ	WA ...	53015	1.90	MILWAUKEE	WI	55079	1.75
DOUGLAS	WA ...	53017	1.75	MONROE	WI	55081	1.70
FERRY	WA ...	53019	1.90	OCONTO	WI	55083	1.70
FRANKLIN	WA ...	53021	1.75	ONEIDA	WI	55085	1.70
GARFIELD	WA ...	53023	1.75	OUTAGAMIE	WI	55087	1.75
GRANT	WA ...	53025	1.75	OZAUKEE	WI	55089	1.75
GRAYS HARBOR	WA ...	53027	1.90	PEPIN	WI	55091	1.70
ISLAND	WA ...	53029	1.90	PIERCE	WI	55093	1.70
JEFFERSON	WA ...	53031	1.90	POLK	WI	55095	1.70
KING	WA ...	53033	1.90	PORTAGE	WI	55097	1.70
KITSAP	WA ...	53035	1.90	PRICE	WI	55099	1.70
KITTITAS	WA ...	53037	1.75	RACINE	WI	55101	1.75
KLICKITAT	WA ...	53039	1.75	RICHLAND	WI	55103	1.75
LEWIS	WA ...	53041	1.90	ROCK	WI	55105	1.75
LINCOLN	WA ...	53043	1.90	RUSK	WI	55107	1.70
MASON	WA ...	53045	1.90	SAUK	WI	55111	1.75
OKANOGAN	WA ...	53047	1.75	SAWYER	WI	55113	1.70
PACIFIC	WA ...	53049	1.90	SHAWANO	WI	55115	1.70
PEND OREILLE	WA ...	53051	1.90	SHEBOYGAN	WI	55117	1.75
PIERCE	WA ...	5303	1.90	ST. CROIX	WI	55109	1.70
SAN JUAN	WA ...	53055	1.90	TAYLOR	WI	55119	1.70
SKAGIT	WA ...	53057	1.90	TREMPEALEAU	WI	55121	1.70
SKAMANIA	WA ...	53059	1.90	VERNON	WI	55123	1.75
SNOHOMISH	WA ...	53061	1.90	VILAS	WI	55125	1.70
SPOKANE	WA ...	53063	1.90	WALWORTH	WI	55127	1.75
STEVENS	WA ...	53065	1.90	WASHBURN	WI	55129	1.70
THURSTON	WA ...	53067	1.90	WASHINGTON	WI	55131	1.75
WAHKIAKUM	WA ...	53069	1.90	WAUKESHA	WI	55133	1.75
WALLA WALLA	WA ...	53071	1.75	WAUPACA	WI	55135	1.75
WHATCOM	WA ...	53073	1.90	WAUSHARA	WI	55137	1.70
WHITMAN	WA ...	53075	1.90	WINNEBAGO	WI	55139	1.75
YAKIMA	WA ...	53077	1.75	WOOD	WI	55141	1.70
ADAMS	WI	55001	1.70	BARBOUR	WV ...	54001	2.30
ASHLAND	WI	55003	1.70	BERKELEY	WV ...	54003	2.60
BARRON	WI	55005	1.70	BOONE	WV ...	54005	2.20

County/parish/city	State	FIPS code	Class I differential adjusted for location
BRAXTON	WV ...	54007	2.20
BROOKE	WV ...	54009	2.10
CABELL	WV ...	54011	2.20
CALHOUN	WV ...	54013	2.20
CLAY	WV ...	54015	2.20
DODDRIDGE	WV ...	54017	2.10
FAYETTE	WV ...	54019	2.20
GILMER	WV ...	54021	2.20
GRANT	WV ...	54023	2.60
GREENBRIER	WV ...	54025	2.20
HAMPSHIRE	WV ...	54027	2.60
HANCOCK	WV ...	54029	2.10
HARDY	WV ...	54031	2.60
HARRISON	WV ...	54033	2.10
JACKSON	WV ...	54035	2.20
JEFFERSON	WV ...	54037	2.60
KANAWHA	WV ...	54039	2.20
LEWIS	WV ...	54041	2.10
LINCOLN	WV ...	54043	2.20
LOGAN	WV ...	54045	2.20
MARION	WV ...	54049	2.10
MARSHALL	WV ...	54051	2.10
MASON	WV ...	54053	2.20
MCDOWELL	WV ...	54047	2.80
MERCER	WV ...	54055	2.80
MINERAL	WV ...	54057	2.60
MINGO	WV ...	54059	2.20
MONONGALIA	WV ...	54061	2.10
MONROE	WV ...	54063	2.20
MORGAN	WV ...	54065	2.60
NICHOLAS	WV ...	54067	2.20
OHIO	WV ...	54069	2.10
PENDLETON	WV ...	54071	2.60
PLEASANTS	WV ...	54073	2.20
POCAHONTAS	WV ...	54075	2.20
PRESTON	WV ...	54077	2.30
PUTNAM	WV ...	54079	2.20
RALEIGH	WV ...	54081	2.20
RANDOLPH	WV ...	54083	2.30
RITCHIE	WV ...	54085	2.20
ROANE	WV ...	54087	2.20
SUMMERS	WV ...	54089	2.20
TAYLOR	WV ...	54091	2.30
TUCKER	WV ...	54093	2.30
TYLER	WV ...	54095	2.10
UPSHUR	WV ...	54097	2.30
WAYNE	WV ...	54099	2.20
WEBSTER	WV ...	54101	2.20
WETZEL	WV ...	54103	2.10
WIRT	WV ...	54105	2.20
WOOD	WV ...	54107	2.20
WYOMING	WV ...	54109	2.20
ALBANY	WY ...	56001	1.90
BIG HORN	WY ...	56003	1.60
CAMPBELL	WY ...	56005	1.65
CARBON	WY ...	56007	1.90
CONVERSE	WY ...	56009	1.70
CROOK	WY ...	56011	1.65
FREMONT	WY ...	56013	1.60
GOSHEN	WY ...	56015	1.90
HOT SPRINGS	WY ...	56017	1.60
JOHNSON	WY ...	56019	1.65
LARAMIE	WY ...	56021	2.45
LINCOLN	WY ...	56023	1.60
NATRONA	WY ...	56025	1.70
NIOBRARA	WY ...	56027	1.70
PARK	WY ...	56029	1.60
PLATTE	WY ...	56031	1.90
SHERIDAN	WY ...	56033	1.60
SUBLETTE	WY ...	56035	1.60
SWEETWATER	WY ...	56037	1.90
TETON	WY ...	56039	1.60
UINTA	WY ...	56041	1.90
WASHAKIE	WY ...	56043	1.60
WESTON	WY ...	56045	1.70

[64 FR 70869, Dec. 17, 1999; 64 FR 73386, Dec. 30, 1999, as amended at 68 FR 48771, Aug. 15, 2003]

§ 1000.53 Announcement of class prices, component prices, and advanced pricing factors.

(a) On or before the 5th day of the month, the market administrator for each Federal milk marketing order shall announce the following prices (as applicable to that order) for the preceding month:

(1) The Class II price;

(2) The Class II butterfat price;

(3) The Class III price;

(4) The Class III skim milk price;

(5) The Class IV price;

(6) The Class IV skim milk price;

(7) The butterfat price;

(8) The nonfat solids price;

(9) The protein price;

(10) The other solids price; and

(11) The somatic cell adjustment rate.

(b) On or before the 23rd day of the month, the market administrator for each Federal milk marketing order shall announce the following prices and pricing factors for the following month:

(1) The Class I price;

(2) The Class I skim milk price;

(3) The Class I butterfat price;

(4) The Class II skim milk price;

(5) The Class II nonfat solids price; and

(6) The advanced pricing factors described in § 1000.50(q).

§ 1000.54 Equivalent price.

If for any reason a price or pricing constituent required for computing the prices described in § 1000.50 is not available, the market administrator shall use a price or pricing constituent determined by the Deputy Administrator, Dairy Programs, Agricultural Marketing Service, to be equivalent to the price or pricing constituent that is required.

Subpart H—Payments for Milk

§ 1000.70   Producer-settlement fund.

The market administrator shall establish and maintain a separate fund known as the producer-settlement fund into which the market administrator shall deposit all payments made by handlers pursuant to §§_____.71, _____.76, and _____.77 of each Federal milk order and out of which the market administrator shall make all payments pursuant to §§_____.72 and _____.77 of each Federal milk order. Payments due any handler shall be offset by any payments due from that handler.

§ 1000.76   Payments by a handler operating a partially regulated distributing plant.

On or before the 25th day after the end of the month (except as provided in § 1000.90), the operator of a partially regulated distributing plant, other than a plant that is subject to marketwide pooling of producer returns under a State government's milk classification and pricing program, shall pay to the market administrator for the producer-settlement fund the amount computed pursuant to paragraph (a) of this section or, if the handler submits the information specified in §§_____.30(b) and _____.31(b) of the order, the handler may elect to pay the amount computed pursuant to paragraph (b) of this section. A partially regulated distributing plant that is subject to marketwide pooling of producer returns under a State government's milk classification and pricing program shall pay the amount computed pursuant to paragraph (c) of this section.

(a) The payment under this paragraph shall be an amount resulting from the following computations:

(1) From the plant's route disposition in the marketing area:

(i) Subtract receipts of fluid milk products classified as Class I milk from pool plants, plants fully regulated under other Federal orders, and handlers described in § 1000.9(c) and § 1135.11 of this chapter, except those receipts subtracted under a similar provision of another Federal milk order;

(ii) Subtract receipts of fluid milk products from another nonpool plant that is not a plant fully regulated under another Federal order to the extent that an equivalent amount of fluid milk products disposed of to the nonpool plant by handlers fully regulated under any Federal order is classified and priced as Class I milk and is not used as an offset for any payment obligation under any order; and

(iii) Subtract the pounds of reconstituted milk made from nonfluid milk products which are disposed of as route disposition in the marketing area;

(2) For orders with multiple component pricing, compute a Class I differential price by subtracting Class III price from the current month's Class I price. Multiply the pounds remaining after the computation in paragraph (a)(1)(iii) of this section by the amount by which the Class I differential price exceeds the producer price differential, both prices to be applicable at the location of the partially regulated distributing plant except that neither the adjusted Class I differential price nor the adjusted producer price differential shall be less than zero;

(3) For orders with skim milk and butterfat pricing, multiply the remaining pounds by the amount by which the Class I price exceeds the uniform price, both prices to be applicable at the location of the partially regulated distributing plant except that neither the adjusted Class I price nor the adjusted uniform price differential shall be less than the lowest announced class price; and

(4) Unless the payment option described in paragraph (d) is selected, add the amount obtained from multiplying the pounds of labeled reconstituted milk included in paragraph (a)(1)(iii) of this section by any positive difference between the Class I price applicable at the location of the partially regulated distributing plant (less $1.00 if the reconstituted milk is labeled as such) and the Class IV price.

(b) The payment under this paragraph shall be the amount resulting from the following computations:

(1) Determine the value that would have been computed pursuant to §_____.60 of the order for the partially regulated distributing plant if

the plant had been a pool plant, subject to the following modifications:

(i) Fluid milk products and bulk fluid cream products received at the plant from a pool plant, a plant fully regulated under another Federal order, and handlers described in §1000.9(c) and §1135.11 of this chapter shall be allocated at the partially regulated distributing plant to the same class in which such products were classified at the fully regulated plant;

(ii) Fluid milk products and bulk fluid cream products transferred from the partially regulated distributing plant to a pool plant or a plant fully regulated under another Federal order shall be classified at the partially regulated distributing plant in the class to which allocated at the fully regulated plant. Such transfers shall be allocated to the extent possible to those receipts at the partially regulated distributing plant from the pool plant and plants fully regulated under other Federal orders that are classified in the corresponding class pursuant to paragraph (b)(1)(i) of this section. Any such transfers remaining after the above allocation which are in Class I and for which a value is computed pursuant to §_____.60 of the order for the partially regulated distributing plant shall be priced at the statistical uniform price or uniform price, whichever is applicable, of the respective order regulating the handling of milk at the receiving plant, with such statistical uniform price or uniform price adjusted to the location of the nonpool plant (but not to be less than the lowest announced class price of the respective order); and

(iii) If the operator of the partially regulated distributing plant so requests, the handler's value of milk determined pursuant to §_____.60 of the order shall include a value of milk determined for each nonpool plant that is not a plant fully regulated under another Federal order which serves as a supply plant for the partially regulated distributing plant by making shipments to the partially regulated distributing plant during the month equivalent to the requirements of §_____. 7(c) of the order subject to the following conditions:

(A) The operator of the partially regulated distributing plant submits with its reports filed pursuant to §§_____.30(b) and _____.31(b) of the order similar reports for each such nonpool supply plant;

(B) The operator of the nonpool plant maintains books and records showing the utilization of all skim milk and butterfat received at the plant which are made available if requested by the market administrator for verification purposes; and

(C) The value of milk determined pursuant to §_____.60 for the unregulated supply plant shall be determined in the same manner prescribed for computing the obligation of the partially regulated distributing plant; and

(2) From the partially regulated distributing plant's value of milk computed pursuant to paragraph (b)(1) of this section, subtract:

(i) The gross payments that were made for milk that would have been producer milk had the plant been fully regulated;

(ii) If paragraph (b)(1)(iii) of this section applies, the gross payments by the operator of the nonpool supply plant for milk received at the plant during the month that would have been producer milk if the plant had been fully regulated; and

(iii) The payments by the operator of the partially regulated distributing plant to the producer-settlement fund of another Federal order under which the plant is also a partially regulated distributing plant and, if paragraph (b)(1)(iii) of this section applies, payments made by the operator of the nonpool supply plant to the producer-settlement fund of any order.

(c) The operator of a partially regulated distributing plant that is subject to marketwide pooling of returns under a milk classification and pricing program that is imposed under the authority of a State government shall pay on or before the 25th day after the end of the month (except as provided in §1000.90) to the market administrator for the producer-settlement fund an amount computed as follows:

After completing the computations described in paragraphs (a)(1)(i) and (ii) of this section, determine the value of the remaining pounds of fluid milk products disposed of as route disposition in the marketing area by multiplying the hundredweight of such pounds by the amount, if greater than zero, that remains after subtracting the State program's class prices applicable to such products at the plant's location from the Federal order Class I price applicable at the location of the plant.

(d) Any handler may elect partially regulated distributing plant status for any plant with respect to receipts of nonfluid milk ingredients that are reconstituted for fluid use. Payments may be made to the producer-settlement fund of the order regulating the producer milk used to produce the nonfluid milk ingredients at the positive difference between the Class I price applicable under the other order at the location of the plant where the nonfluid milk ingredients were processed and the Class IV price. This payment option shall apply only if a majority of the total milk received at the plant that processed the nonfluid milk ingredients is regulated under one or more Federal orders and payment may only be made to the producer-settlement fund of the order pricing a plurality of the milk used to produce the nonfluid milk ingredients. This payment option shall not apply if the source of the nonfluid ingredients used in reconstituted fluid milk products cannot be determined by the market administrator.

§ 1000.77 Adjustment of accounts.

Whenever audit by the market administrator of any handler's reports, books, records, or accounts, or other verification discloses errors resulting in money due the market administrator from a handler, or due a handler from the market administrator, or due a producer or cooperative association from a handler, the market administrator shall promptly notify such handler of any amount so due and payment thereof shall be made on or before the next date for making payments as set forth in the provisions under which the error(s) occurred.

§ 1000.78 Charges on overdue accounts.

Any unpaid obligation due the market administrator, producers, or cooperative associations from a handler pursuant to the provisions of the order shall be increased 1.0 percent each month beginning with the day following the date such obligation was due under the order. Any remaining amount due shall be increased at the same rate on the corresponding day of each succeeding month until paid. The amounts payable pursuant to this section shall be computed monthly on each unpaid obligation and shall include any unpaid charges previously computed pursuant to this section. The late charges shall accrue to the administrative assessment fund. For the purpose of this section, any obligation that was determined at a date later than prescribed by the order because of a handler's failure to submit a report to the market administrator when due shall be considered to have been payable by the date it would have been due if the report had been filed when due.

Subpart I—Administrative Assessment and Marketing Service Deduction

§ 1000.85 Assessment for order administration.

On or before the payment receipt date specified under § _____.71 of each Federal milk order each handler shall pay to the market administrator its pro rata share of the expense of administration of the order at a rate specified by the market administrator that is no more than 5 cents per hundredweight with respect to:

(a) Receipts of producer milk (including the handler's own production) other than such receipts by a handler described in § 1000.9(c) that were delivered to pool plants of other handlers;

(b) Receipts from a handler described in § 1000.9(c);

(c) Receipts of concentrated fluid milk products from unregulated supply plants and receipts of nonfluid milk products assigned to Class I use pursuant to § 1000.43(d) and other source milk allocated to Class I pursuant to

§ 1000.44(a) (3) and (8) and the corresponding steps of § 1000.44(b), except other source milk that is excluded from the computations pursuant to § _____ .60 (d) and (e) of parts 1005, 1006, and 1007 of this chapter or § _____ .60 (h) and (i) of parts 1001, 1030, 1032, 1033, 1124, 1126, 1131, and 1135 of this chapter; and

(d) Route disposition in the marketing area from a partially regulated distributing plant that exceeds the skim milk and butterfat subtracted pursuant to § 1000.76(a)(1) (i) and (ii).

§ 1000.86 Deduction for marketing services.

(a) Except as provided in paragraph (b) of this section, each handler in making payments to producers for milk (other than milk of such handler's own production) pursuant to § _____ .73 of each Federal milk order shall deduct an amount specified by the market administrator that is no more than 7 cents per hundredweight and shall pay the amount deducted to the market administrator not later than the payment receipt date specified under § _____ .71 of each Federal milk order. The money shall be used by the market administrator to verify or establish weights, samples and tests of producer milk and provide market information for producers who are not receiving such services from a cooperative association. The services shall be performed in whole or in part by the market administrator or an agent engaged by and responsible to the market administrator.

(b) In the case of producers for whom the market administrator has determined that a cooperative association is actually performing the services set forth in paragraph (a) of this section, each handler shall make deductions from the payments to be made to producers as may be authorized by the membership agreement or marketing contract between the cooperative association and the producers. On or before the 15th day after the end of the month (except as provided in § 1000.90), such deductions shall be paid to the cooperative association rendering the services accompanied by a statement showing the amount of any deductions and the amount of milk for which the deduc-

tion was computed for each producer. These deductions shall be made in lieu of the deduction specified in paragraph (a) of this section.

Subpart J—Miscellaneous Provisions

§ 1000.90 Dates.

If a date required for a payment contained in a Federal milk order falls on a Saturday, Sunday, or national holiday, such payment will be due on the next day that the market administrator's office is open for public business.

§§ 1000.91–1000.92 [Reserved]

§ 1000.93 OMB control number assigned pursuant to the Paperwork Reduction Act.

The information collection requirements contained in this part have been approved by the Office of Management and Budget (OMB) under the provisions of Title 44 U.S.C. chapter 35 and have been assigned OMB control number 0581–0032.

PART 1001—MILK IN THE NORTHEAST MARKETING AREA

Subpart—Order Regulating Handling

GENERAL PROVISIONS

Sec.
1001.1 General provisions.

DEFINITIONS

1001.2 Northeast marketing area.
1001.3 Route disposition.
1001.4 Plant.
1001.5 Distributing plant.
1001.6 Supply plant.
1001.7 Pool plant.
1001.8 Nonpool plant.
1001.9 Handler.
1001.10 Producer-handler.
1001.11 [Reserved]
1001.12 Producer.
1001.13 Producer milk.
1001.14 Other source milk.
1001.15 Fluid milk product.
1001.16 Fluid cream product.
1001.17 [Reserved]
1001.18 Cooperative association.
1001.19 Commercial food processing establishment.

HANDLER REPORTS

1001.30 Reports of receipts and utilization.

AUTHORITY: 7 U.S.C. 601–674, and 7253.

SOURCE: 64 FR 47954, Sept. 1, 1999, unless otherwise noted.

Subpart—Order Regulating Handling

GENERAL PROVISIONS

§ 1001.1 General provisions.

The terms, definitions, and provisions in part 1000 of this chapter apply to this part 1001. In this part 1001, all references to sections in part 1000 refer to part 1000 of this chapter.

DEFINITIONS

§ 1001.2 Northeast marketing area.

The marketing area means all the territory within the bounds of the following states and political subdivisions, including all piers, docks and wharves connected therewith and all craft moored thereat, and all territory occupied by government (municipal, State or Federal) reservations, installations, institutions, or other similar establishments if any part thereof is within any of the listed states or political subdivisions:

CONNECTICUT, DELAWARE, MASSACHUSETTS, NEW HAMPSHIRE, NEW JERSEY, RHODE ISLAND, VERMONT AND DISTRICT OF COLUMBIA

All of the States of Connecticut, Delaware, Massachusetts, New Hampshire, New Jersey, Rhode Island, Vermont and the District of Columbia.

MARYLAND COUNTIES

All of the State of Maryland except the counties of Allegany and Garrett.

NEW YORK COUNTIES, CITIES, AND TOWNSHIPS

All counties within the State of New York except Allegany, Cattaraugus, Chatauqua, Erie, Genessee, Livingston, Monroe, Niagara, Ontario, Orleans, Seneca, Wayne, and Wyoming; the townships of Conquest, Montezuma, Sterling and Victory in Cayuga County; the city of Hornell, and the townships of Avoca, Bath, Bradford, Canisteo, Cohocton, Dansville, Fremont, Pulteney, Hartsville, Hornellsville, Howard, Prattsburg, Urbana, Wayland, Wayne and Wheeler in Steuben County; and the townships of Italy, Middlesex, and Potter in Yates County.

PENNSYLVANIA COUNTIES

Adams, Bucks, Chester, Cumberland, Dauphin, Delaware, Franklin, Fulton, Juniata, Lancaster, Lebanon, Montgomery, Perry, Philadelphia, and York.

VIRGINIA COUNTIES AND CITIES

Arlington, Fairfax, Loudoun, and Prince William, and the cities of Alexandria, Fairfax, Falls Church, Manassas, and Manassas Park.

§ 1001.3 Route disposition.

See § 1000.3.

§1001.4 Plant.

(a) Except as provided in paragraph (b) of this section, plant means the land, buildings, facilities, and equipment constituting a single operating unit or establishment at which milk or milk products are received, processed, or packaged, including a facility described in paragraph (b)(2) of this section if the facility receives the milk of more than one dairy farmer.

(b) Plant shall not include:

(1) A separate building without stationary storage tanks that is used only as a reload point for transferring bulk milk from one tank truck to another or a separate building used only as a distribution point for storing packaged fluid milk products in transit for route disposition;

(2) An on-farm facility operated as part of a single dairy farm entity for the separation of cream and skim milk or the removal of water from milk; or

(3) Bulk reload points where milk is transferred from one tank truck to another while en route from dairy farmers' farms to a plant. If stationary storage tanks are used for transferring milk at the premises, the operator of the facility shall make an advance written request to the market administrator that the facility shall be treated as a reload point. The cooling of milk, collection of samples, and washing and sanitizing of tank trucks at the premises shall not disqualify it as a bulk reload point.

§1001.5 Distributing plant.

See §1000.5.

§1001.6 Supply plant.

See §1000.6.

§1001.7 Pool plant.

Pool plant means a plant, unit of plants, or system of plants as specified in paragraphs (a) through (f) of this section, but excluding a plant described in paragraph (h) of this section. The pooling standards described in paragraphs (c) and (f) of this section are subject to modification pursuant to paragraph (g) of this section.

(a) A distributing plant, other than a plant qualified as a pool plant pursuant to paragraph (b) of this section or §_____.7(b) of any other Federal milk order, from which during the month 25 percent or more of the total quantity of fluid milk products physically received at the plant (excluding concentrated milk received from another plant by agreement for other than Class I use) are disposed of as route disposition or are transferred in the form of packaged fluid milk products to other distributing plants. At least 25 percent of such route disposition and transfers must be to outlets in the marketing area.

(b) Any distributing plant located in the marketing area which during the month processed at least 25 percent of the total quantity of fluid milk products physically received at the plant (excluding concentrated milk received from another plant by agreement for other than Class I use) into ultra-pasteurized or aseptically-processed fluid milk products.

(c) A supply plant from which fluid milk products are transferred or diverted to plants described in paragraph (a) or (b) of this section subject to the additional conditions described in this paragraph. In the case of a supply plant operated by a cooperative association handler described in §1000.9(c), fluid milk products that the cooperative delivers to pool plants directly from producers' farms shall be treated as if transferred from the cooperative association's plant for the purpose of meeting the shipping requirements of this paragraph.

(1) In each of the months of January through August and December, such shipments and transfers to distributing plants must not equal less than 10 percent of the total quantity of milk (except the milk of a producer described in §1001.12(b)) that is received at the plant or diverted from it pursuant to §1001.13 during the month;

(2) In each of the months of September through November, such shipments and transfers to distributing plants must equal not less than 20 percent of the total quantity of milk (except the milk of a producer described in §1001.12(b)) that is received at the plant or diverted from it pursuant to §1001.13 during the month;

(3) If milk is delivered directly from producers' farms that are located outside of the states included in the marketing area or outside Maine or West Virginia, such producers must be grouped by state into reporting units and each reporting unit must independently meet the shipping requirements of this paragraph; and

(4) Concentrated milk transferred from the supply plant to a distributing plant for an agreed-upon use other than Class I shall be excluded from the supply plant's shipments in computing the percentages in paragraphs (c)(1) and (2) of this section.

(d) Any distributing plant, located within the marketing area as described on May 1, 2006, in § 1001.2;

(1) From which there is route disposition and/or transfers of packaged fluid milk products in any non-Federally regulated marketing area(s) located within one or more States that require handlers to pay minimum prices for raw milk provided that 25 percent or more of the total quantity of fluid milk products physically received at such plant (excluding concentrated milk received from another plant by agreement for other than Class I use) is disposed of as route disposition and/or is transferred in the form of packaged fluid milk products to other plants. At least 25 percent of such route disposition and/or transfers, in aggregate, are in any non-Federally regulated marketing area(s) located within one or more States that require handlers to pay minimum prices for raw milk. Subject to the following exclusions:

(i) The plant is described in § 1001.7(a), (b), or (e);

(ii) The plant is subject to the pricing provisions of a State-operated milk pricing plan which provides for the payment of minimum class prices for raw milk;

(iii) The plant is described in § 1000.8(a) or (e); or

(iv) A producer-handler described in § 1001.10 with less than three million pounds during the month of route dispositions and/or transfers of packaged fluid milk products to other plants.

(2) [Reserved]

(e) Two or more plants that are located in the marketing area and operated by the same handler may qualify as a unit by meeting the total and in-area route distribution requirements specified in paragraph (a) of this section subject to the following additional requirements:

(1) At least one of the plants in the unit qualifies as a pool distributing plant pursuant to paragraph (a) of this section;

(2) Other plants in the unit must process at least 60 percent of monthly receipts of producer milk only as Class I or Class II products and must be located in the Northeast marketing area, as defined in § 1001.2, in a pricing zone providing the same or a lower Class I price than the price applicable at the distributing plant(s) included in the unit; and

(3) A written request to form a unit, or to add or remove plants from a unit, or to cancel a unit, must be filed with the market administrator prior to the first day of the month for which unit formation is to be effective.

(f) Two or more supply plants operated by the same handler, or by one or more cooperative associations, may qualify for pooling as a system of plants by meeting the applicable percentage requirements of paragraph (c) of this section in the same manner as a single plant subject to the following additional requirements:

(1) A supply plant system will be effective for the period of August 1 through July 31 of the following year. Written notification must be given to the market administrator listing the plants to be included in the system prior to the first day of July preceding the effective date of the system. The plants included in the system shall be listed in the sequence in which they shall qualify for pool plant status based on the minimum deliveries required. If the deliveries made are insufficient to qualify the entire system for pooling, the last listed plant shall be excluded from the system, followed by the plant next-to-last on the list, and continuing in this sequence until remaining listed plants have met the minimum shipping requirements; and

(2) Each plant that qualifies as a pool plant within a system shall continue each month as a plant in the system through the following July unless the plant subsequently fails to qualify for

pooling, the handler submits a written notification to the market administrator prior to the first day of the month that the plant be deleted from the system, or that the system be discontinued. Any plant that has been so deleted from the system, or that has failed to qualify as a pool plant in any month, will not be part of the system for the remaining months through July. For any system that qualifies in August, no plant may be added in any subsequent month through the following July unless the plant replaces another plant in the system that has ceased operations and the market administrator is notified of such replacement prior to the first day of the month for which it is to be effective.

(g) The applicable shipping percentages of paragraphs (c) and (f) of this section may be increased or decreased by the market administrator if the market administrator finds that such adjustment is necessary to encourage needed shipments or to prevent uneconomic shipments. Before making such a finding, the market administrator shall investigate the need for adjustment either on the market administrator's own initiative or at the request of interested parties if the request is made in writing at least 15 days prior to the month for which the requested revision is desired effective. If the investigation shows that an adjustment of the shipping percentages might be appropriate, the market administrator shall issue a notice stating that an adjustment is being considered and invite data, views and arguments. Any decision to revise an applicable shipping percentage must be issued in writing at least one day before the effective date.

(h) The term pool plant shall not apply to the following plants:

(1) A producer-handler plant;

(2) An exempt plant as defined in §1000.8(e);

(3) A plant qualified pursuant to paragraph (a) of this section that is located within the marketing area if the plant also meets the pooling requirements of another Federal order and more than 50 percent of its route distribution has been in such other Federal order marketing area for 3 consecutive months;

(4) A plant qualified pursuant to paragraph (a) of this section which is not located within any Federal order marketing area that meets the pooling requirements of another Federal order and has had greater route disposition in such other Federal order's marketing area for 3 consecutive months;

(5) A plant qualified pursuant to paragraph (a) of this section that is located in another Federal order marketing area if the plant meets the pooling requirements of such other Federal order and does not have a majority of its route distribution in this marketing area for 3 consecutive months or if the plant is required to be regulated under such other Federal order without regard to its route disposition in any other Federal order marketing area; and

(6) A plant qualified pursuant to paragraph (c) of this section which also meets the pooling requirements of another Federal order and from which greater qualifying shipments are made to plants regulated under the other Federal order than are made to plants regulated under the order in this part, or the plant has automatic pooling status under the other Federal order.

[64 FR 47954, Sept. 1, 1999, as amended at 70 FR 18962, Apr. 12, 2005; 71 FR 25497, May 1, 2006; 71 FR 28249, May 16, 2006]

§1001.8 Nonpool plant.

See §1000.8.

§1001.9 Handler.

See §1000.9.

§1001.10 Producer-handler.

Producer-handler means a person who:

(a) Operates a dairy farm and a distributing plant from which there is route disposition in the marketing area, and from which total route disposition and packaged sales of fluid milk products to other plants during the month does not exceed 3 million pounds;

(b) Receives milk solely from own farm production or receives milk that is fully subject to the pricing and pooling provisions of this or any other Federal order;

(c) Receives at its plant or acquires for route disposition no more than 150,000 pounds of fluid milk products from handlers fully regulated under any Federal order. This limitation shall not apply if the producer-handler's own farm production is less than 150,000 pounds during the month;

(d) Disposes of no other source milk as Class I milk except by increasing the nonfat milk solids content of the fluid milk products; and

(e) Provides proof satisfactory to the market administrator that the care and management of the dairy animals and other resources necessary to produce all Class I milk handled (excluding receipts from handlers fully regulated under any Federal order) and the processing and packaging operations are the producer-handler's own enterprise and at its own risk.

(f) Any producer-handler with Class I route dispositions and/or transfers of packaged fluid milk products in the marketing area described in § 1131.2 of this chapter shall be subject to payments into the Order 1131 producer settlement fund on such dispositions pursuant to § 1000.76(a) and payments into the Order 1131 administrative fund provided such dispositions are less than three million pounds in the current month and such producer-handler had total Class I route dispositions and/or transfers of packaged fluid milk products from own farm production of three million pounds or more the previous month. If the producer-handler has Class I route dispositions and/or transfers of packaged fluid milk products into the marketing area described in § 1131.2 of this chapter of three million pounds or more during the current month, such producer-handler shall be subject to the provisions described in § 1131.7 of this chapter or § 1000.76(a).

[64 FR 47954, Sept. 1, 1999, as amended at 71 FR 25497, May 1, 2006; 75 FR 21160, Apr. 23, 2010]

§ 1001.11 [Reserved]

§ 1001.12 Producer.

(a) Except as provided in paragraph (b) of this section, *producer* means any person who produces milk approved by a duly constituted regulatory agency for fluid consumption as Grade A milk and whose milk (or components of milk) is:

(1) Received at a pool plant directly from the producer or diverted by the plant operator in accordance with § 1001.13; or

(2) Received by a handler described in § 1000.9(c).

(b) Producer shall not include a dairy farmer described in paragraphs (b)(1) through (6) of this section. A dairy farmer described in paragraphs (b)(5) or (6) of this section shall be known as a *dairy farmer for other markets.*

(1) A producer-handler as defined in any Federal order;

(2) A dairy farmer whose milk is received at an exempt plant, excluding producer milk diverted to the exempt plant pursuant to § 1001.13(d);

(3) A dairy farmer whose milk is received by diversion at a pool plant from a handler regulated under another Federal order if the other Federal order designates the dairy farmer as a producer under that order and that milk is allocated by request to a utilization other than Class I;

(4) A dairy farmer whose milk is reported as diverted to a plant fully regulated under another Federal order with respect to that portion of the milk so diverted that is assigned to Class I under the provisions of such other order;

(5) For any month of December through June, any dairy farmer whose milk is received at a pool plant or by a cooperative association handler described in § 1000.9(c) if the pool plant operator or the cooperative association caused milk from the same farm to be delivered to any plant as other than producer milk, as defined under the order in this part or any other Federal milk order, during the same month, either of the 2 preceding months, or during any of the preceding months of July through November; and

(6) For any month of July through November, any dairy farmer whose milk is received at a pool plant or by a cooperative association handler described in § 1000.9(c) if the pool plant operator or the cooperative association caused milk from the same farm to be delivered to any plant as other than producer milk, as defined under the order in this part or any other Federal milk order, during the same month.

§ 1001.13 Producer milk.

Producer milk means the skim milk (or the skim equivalent of components of skim milk) and butterfat contained in milk of a producer that is:

(a) Received by the operator of a pool plant directly from a producer or from a handler described in §1000.9(c). Any milk which is picked up from the producer's farm in a tank truck under the control of the operator of a pool plant or a handler described in §1000.9(c) but which is not received at a plant until the following month shall be considered as having been received by the handler during the month in which it is picked up at the farm. All milk received pursuant to this paragraph shall be priced at the location of the plant where it is first physically received;

(b) Received by the operator of a pool plant or a handler described in §1000.9(c) in excess of the quantity delivered to pool plants subject to the following conditions:

(1) The producers whose farms are outside of the states included in the marketing area and outside the states of Maine or West Virginia shall be organized into state units and each such unit shall be reported separately; and

(2) For pooling purposes, each reporting unit must satisfy the shipping standards specified for a supply plant pursuant to §1001.7(c);

(c) Diverted by a proprietary pool plant operator to another pool plant. Milk so diverted shall be priced at the location of the plant to which diverted; or

(d) Diverted by the operator of a pool plant or by a handler described in §1000.9(c) to a nonpool plant, subject to the following conditions:

(1) Milk of a dairy farmer shall not be eligible for diversion unless one day's milk production of such dairy farmer was physically received as producer milk and the dairy farmer has continuously retained producer status since that time. If a dairy farmer loses producer status under the order in this part (except as a result of a temporary loss of Grade A approval), the dairy farmer's milk shall not be eligible for diversion unless milk of the dairy farmer has been physically received as producer milk at a pool plant during the month;

(2) Of the total quantity of producer milk received during the month (including diversion but excluding the quantity of producer milk received from a handler described in §1000.9(c) or which is diverted to another pool plant), the handler diverted to nonpool plants not more than 80 percent during each of the months of September through November and 90 percent during each of the months of January through August and December. In the event that a handler causes the milk of a producer to be over diverted, a dairy farmer will not lose producer status;

(3) Diverted milk shall be priced at the location of the plant to which diverted.

(4) Any milk diverted in excess of the limits set forth in paragraph (d)(2) of this section shall not be producer milk. The diverting handler shall designate the dairy farmer deliveries that shall not be producer milk. If the handler fails to designate the dairy farmer deliveries which are ineligible, producer milk status shall be forfeited with respect to all milk diverted to nonpool plants by such handler; and

(5) The delivery day requirement and the diversion percentages in paragraphs (d)(1) and (d)(2) of this section may be increased or decreased by the Market Administrator if the Market Administrator finds that such revision is necessary to assure orderly marketing and efficient handling of milk in the marketing area. Before making such a finding, the Market Administrator shall investigate the need for the revision either on the Market Administrator's own initiative or at the request of interested persons if the request is made in writing at least 15 days prior to the month for which the requested revision is desired to be effective. If the investigation shows that a revision might be appropriate, the Market Administrator shall issue a notice stating that the revision is being considered and inviting written data, views, and arguments. Any decision to revise an applicable percentage or delivery day requirement must be issued in writing at least one day before the effective date.

(e) Producer milk shall not include milk of a producer that is subject to inclusion and participation in a marketwide equalization pool under a milk classification and pricing program imposed under the authority of another government entity.

[64 FR 47954, Sept. 1, 1999, as amended at 70 FR 18962, Apr. 12, 2005]

§ 1001.14   Other source milk.

See § 1000.14.

§ 1001.15   Fluid milk product.

See § 1000.15.

§ 1001.16   Fluid cream product.

See § 1000.16.

§ 1001.17   [Reserved]

§ 1001.18   Cooperative association.

See § 1000.18.

§ 1001.19   Commercial food processing establishment.

See § 1000.19.

HANDLER REPORTS

§ 1001.30   Reports of receipts and utilization.

Each handler shall report monthly so that the Market Administrator's office receives the report on or before the 10th day after the end of the month, in the detail and on prescribed forms, as follows:

(a) Each pool plant operator shall report for each of its operations the following information:

(1) Product pounds, pounds of butterfat, pounds of protein, and pounds of nonfat solids other than protein (other solids) contained in or represented by:

(i) Receipts of producer milk, including producer milk diverted by the reporting handler, from sources other than handlers described in § 1000.9(c); and

(ii) Receipts of milk from handlers described in § 1000.9(c);

(2) Product pounds and pounds of butterfat contained in:

(i) Receipts of fluid milk products and bulk fluid cream products from other pool plants;

(ii) Receipts of other source milk; and

(iii) Inventories at the beginning and end of the month of fluid milk products and bulk fluid cream products;

(3) The utilization or disposition of all milk and milk products required to be reported pursuant to this paragraph; and

(4) Such other information with respect to the receipts and utilization of skim milk, butterfat, milk protein, and other nonfat solids as the market administrator may prescribe.

(b) Each handler operating a partially regulated distributing plant shall report with respect to such plant in the same manner as prescribed for reports required by paragraph (a) of this section. Receipts of milk that would have been producer milk if the plant had been fully regulated shall be reported in lieu of producer milk. The report shall show also the quantity of any reconstituted skim milk in route disposition in the marketing area.

(c) Each handler described in § 1000.9(c) shall report:

(1) The product pounds, pounds of butterfat, pounds of protein, and the pounds of solids-not-fat other than protein (other solids) contained in receipts of milk from producers; and

(2) The utilization or disposition of such receipts.

(d) Each handler not specified in paragraph (a) or (b) of this section shall report with respect to its receipts and utilization of milk and milk products in such manner as the market administrator may prescribe.

[64 FR 47954, Sept. 1, 1999, as amended at 70 FR 18963, Apr. 12, 2005]

§ 1001.31   Payroll reports.

(a) On or before the 22nd day after the end of each month, each handler that operates a pool plant pursuant to § 1001.7 and each handler described in § 1000.9(c) shall report to the market administrator its producer payroll for the month, in detail prescribed by the market administrator, showing for each producer the information specified in § 1001.73(e).

(b) Each handler operating a partially regulated distributing plant who elects to make payment pursuant to § 1000.76(b) shall report for each dairy farmer who would have been a producer if the plant had been fully regulated in the same manner as prescribed for reports required by paragraph (a) of this section.

§ 1001.32 Other reports.

In addition to the reports required pursuant to §§ 1001.30 and 1001.31, each handler shall report any information the market administrator deems necessary to verify or establish each handler's obligation under the order.

CLASSIFICATION OF MILK

§ 1001.40 Classes of utilization.

See § 1000.40.

§ 1001.41 [Reserved]

§ 1001.42 Classification of transfers and diversions.

See § 1000.42.

§ 1001.43 General classification rules.

See § 1000.43.

§ 1001.44 Classification of producer milk.

See § 1000.44.

§ 1001.45 Market administrator's reports and announcements concerning classification.

See § 1000.45.

CLASS PRICES

§ 1001.50 Class prices, component prices, and advanced pricing factors.

See § 1000.50.

§ 1001.51 Class I differential and price.

The Class I differential shall be the differential established for Suffolk County, Massachusetts, which is reported in § 1000.52. The Class I price shall be the price computed pursuant to § 1000.50(a) for Suffolk County, Massachusetts.

§ 1001.52 Adjusted Class I differentials.

See § 1000.52.

§ 1001.53 Announcement of class prices, component prices, and advanced pricing factors.

See § 1000.53.

§ 1001.54 Equivalent price.

See § 1000.54.

PRODUCER PRICE DIFFERENTIAL

§ 1001.60 Handler's value of milk.

For the purpose of computing a handler's obligation for producer milk, the market administrator shall determine for each month the value of milk of each handler with respect to each of the handler's pool plants and of each handler described in § 1000.9(c) with respect to milk that was not received at a pool plant by adding the amounts computed in paragraphs (a) through (h) of this section and subtracting from that total amount the value computed in paragraph (i) of this section. Unless otherwise specified, the skim milk, butterfat, and the combined pounds of skim milk and butterfat referred to in this section shall result from the steps set forth in § 1000.44(a), (b), and (c), respectively, and the nonfat components of producer milk in each class shall be based upon the proportion of such components in producer skim milk. Receipts of nonfluid milk products that are distributed as labeled reconstituted milk for which payments are made to the producer-settlement fund of another Federal order under § 1000.76(a)(4) or (d) shall be excluded from pricing under this section.

(a) Class I value. (1) Multiply the pounds of skim milk in Class I by the Class I skim milk price; and

(2) Add an amount obtained by multiplying the pounds of butterfat in Class I by the Class I butterfat price.

(b) Class II value. (1) Multiply the pounds of nonfat solids in Class II skim milk by the Class II nonfat solids price; and

(2) Add an amount obtained by multiplying the pounds of butterfat in Class II times the Class II butterfat price.

(c) Class III value. (1) Multiply the pounds of protein in Class III skim milk by the protein price;

(2) Add an amount obtained by multiplying the pounds of other solids in Class III skim milk by the other solids price; and

(3) Add an amount obtained by multiplying the pounds of butterfat in Class III by the butterfat price.

(d) Class IV value. (1) Multiply the pounds of nonfat solids in Class IV skim milk by the nonfat solids price; and

(2) Add an amount obtained by multiplying the pounds of butterfat in Class IV by the butterfat price.

(e) Multiply the pounds of skim milk and butterfat overage assigned to each class pursuant to § 1000.44(a)(11) and the corresponding step of § 1000.44(b) by the skim milk prices and butterfat prices applicable to each class.

(f) Multiply the difference between the current month's Class I, II, or III price, as the case may be, and the Class IV price for the preceding month by the hundredweight of skim milk and butterfat subtracted from Class I, II, or III, respectively, pursuant to § 1000.44(a)(7) and the corresponding step of § 1000.44(b);

(g) Multiply the difference between the Class I price applicable at the location of the pool plant and the Class IV price by the hundredweight of skim milk and butterfat assigned to Class I pursuant to § 1000.43(d) and the hundredweight of skim milk and butterfat subtracted from Class I pursuant to § 1000.44(a)(3)(i) through (vi) and the corresponding step of § 1000.44(b), excluding receipts of bulk fluid cream products from a plant regulated under other Federal orders and bulk concentrated fluid milk products from pool plants, plants regulated under other Federal orders, and unregulated supply plants.

(h) Multiply the difference between the Class I price applicable at the location of the nearest unregulated supply plants from which an equivalent volume was received and the Class III price by the pounds of skim milk and butterfat in receipts of concentrated fluid milk products assigned to Class I pursuant to § 1000.43(d) and § 1000.44(a)(3)(i) and the corresponding step of § 1000.44(b) and the pounds of skim milk and butterfat subtracted from Class I pursuant to § 1000.44(a)(8) and the corresponding step of § 1000.44(b), excluding such skim milk and butterfat in receipts of fluid milk products from an unregulated supply plant to the extent that an equivalent amount of skim milk or butterfat disposed of to such plant by handlers fully regulated under any Federal milk order is classified and priced as Class I milk and is not used as an offset for any other payment obligation under any order.

(i) For reconstituted milk made from receipts of nonfluid milk products, multiply $1.00 (but not more than the difference between the Class I price applicable at the location of the pool plant and the Class IV price) by the hundredweight of skim milk and butterfat contained in receipts of nonfluid milk products that are allocated to Class I use pursuant to § 1000.43(d).

[64 FR 47954, Sept. 1, 1999, as amended at 65 FR 82834, Dec. 28, 2000; 68 FR 7065, Feb. 12, 2003]

§ 1001.61 Computation of producer price differential.

For each month, the market administrator shall compute a producer price differential per hundredweight. The report of any handler who has not made payments required pursuant to § 1001.71 for the preceding month shall not be included in the computation of the producer price differential, and such handler's report shall not be included in the computation for succeeding months until the handler has made full payment of outstanding monthly obligations. Subject to the conditions in this paragraph, the market administrator shall compute the producer price differential in the following manner:

(a) Combine into one total the values computed pursuant to § 1001.60 for all handlers required to file reports prescribed in § 1001.30;

(b) Subtract the total of the values obtained by multiplying each handler's total pounds of protein, other solids, and butterfat contained in the milk for which an obligation was computed pursuant to § 1001.60 by the protein price, other solids price, and the butterfat price, respectively;

(c) Add an amount equal to the minus location adjustments and subtract an amount equal to the plus location adjustments computed pursuant to § 1001.75;

(d) Add an amount equal to not less than one-half of the unobligated balance in the producer-settlement fund;

(e) Divide the resulting amount by the sum of the following for all handlers included in these computations:

(1) The total hundredweight of producer milk; and

(2) The total hundredweight for which a value is computed pursuant to §1001.60(h); and

(f) Subtract not less than 4 cents nor more than 5 cents from the price computed pursuant to paragraph (e) of this section. The result, rounded to the nearest cent, shall be known as the producer price differential for the month.

[68 FR 7065, Feb. 12, 2003]

§1001.62 Announcement of producer prices.

On or before the 14th day after the end of the month, the Market Administrator shall announce the following prices and information:

(a) The producer price differential;
(b) The protein price;
(c) The nonfat solids price;
(d) The other solids price;
(e) The butterfat price;
(f) The average butterfat, protein, nonfat solids, and other solids content of producer milk; and
(g) The statistical uniform price for milk containing 3.5 percent butterfat computed by combining the Class III price and the producer price differential.

(h) If the 14th falls on a Saturday, Sunday, or national holiday, the Market Administrator may have up to two additional business days to announce the producer price differential and the statistical uniform price.

[64 FR 47954, Sept. 1, 1999, as amended at 65 FR 82834, Dec. 28, 2000; 68 FR 7065, Feb. 12, 2003; 70 FR 18963, Apr. 12, 2005]

PAYMENTS FOR MILK

§1001.70 Producer-settlement fund.

See §1000.70.

§1001.71 Payments to the producer-settlement fund.

Each handler shall make payment to the producer-settlement fund in a manner that provides receipt of the funds by the Market Administrator no later than two days after the announcement of the producer price differential and the statistical uniform price pursuant to §1001.62 (except as provided for in §1000.90). Payment shall be the amount, if any, by which the amount specified in paragraph (a) of this sec-

tion exceeds the amount specified in paragraph (b) of this section:

(a) The total value of milk to the handler for the month as determined pursuant to §1001.60.

(b) The sum of:

(1) An amount obtained by multiplying the total hundredweight of producer milk as determined pursuant to §1000.44(c) by the producer price differential as adjusted pursuant to §1001.75;

(2) An amount obtained by multiplying the total pounds of protein, other solids, and butterfat contained in producer milk by the protein, other solids, and butterfat prices respectively; and

(3) An amount obtained by multiplying the pounds of skim milk and butterfat for which a value was computed pursuant to §1001.60(h) by the producer price differential as adjusted pursuant to §1001.75 for the location of the plant from which received.

[64 FR 47954, Sept. 1, 1999, as amended at 65 FR 82834, Dec. 28, 2000; 68 FR 7066, Feb. 12, 2003; 70 FR 18963, Apr. 12, 2005]

§1001.72 Payments from the producer—settlement fund.

No later than the day after the due date required for payment to the Market Administrator pursuant to §1001.71 (except as provided in §1001.90), the Market Administrator shall pay to each handler the amount, if any, by which the amount computed pursuant to §1001.71(b) exceeds the amount computed pursuant to §1001.71(a). If, at such time, the balance in the producer-settlement fund is insufficient to make all payments pursuant to this section, the Market Administrator shall reduce uniformly such payments and shall complete the payments as soon as the funds are available.

[70 FR 18963, Apr. 12, 2005]

§1001.73 Payments to producers and to cooperative associations.

(a) Each handler that is not paying a cooperative association for producer milk shall pay each producer as follows:

(1) *Partial payment.* For each producer who has not discontinued shipments as of the 23rd day of the month, payment shall be made so that it is received by

the producer on or before the 26th day of the month (except as provided in § 1000.90) for milk received during the first 15 days of the month at not less than the lowest announced class price for the preceding month, less proper deductions authorized in writing by the producer.

(2) *Final payment.* For milk received during the month, payment shall be made during the following month so it is received by each producer no later than the day after the required date of payment by the Market Administrator, pursuant to § 1001.72, in an amount computed as follows:

(i) Multiply the hundredweight of producer milk received by the producer price differential for the month as adjusted pursuant to § 1001.75;

(ii) Multiply the pounds of butterfat received by the butterfat price for the month;

(iii) Multiply the pounds of protein received by the protein price for the month;

(iv) Multiply the pounds of other solids received by the other solids price for the month; and

(v) Add the amounts computed in paragraphs (a)(2)(i) through (iv) of this section, and from that sum:

(A) Subtract the partial payment made pursuant to paragraph (a)(1) of this section;

(B) Subtract the deduction for marketing services pursuant to § 1000.86;

(C) Add or subtract for errors made in previous payments to the producer; and

(D) Subtract proper deductions authorized in writing by the producer.

(b) One day before partial and final payments are due pursuant to paragraph (a) of this section, each handler shall pay a cooperative association for milk received as follows:

(1) *Partial payment to a cooperative association for bulk milk received directly from producers' farms.* For bulk milk (including the milk of producers who are not members of such association and who the market administrator determines have authorized the cooperative association to collect payment for their milk) received during the first 15 days of the month from a cooperative association in any capacity, except as the operator of a pool plant, the payment shall be equal to the hundredweight of milk received multiplied by the lowest announced class price for the preceding month.

(2) *Partial payment to a cooperative association for milk transferred from its pool plant.* For bulk milk/skimmed milk products received during the first 15 days of the month from a cooperative association in its capacity as the operator of a pool plant, the partial payment shall be at the pool plant operator's estimated use value of the milk using the most recent class prices available at the receiving plant's location.

(3) *Final payment to a cooperative association for milk transferred from its pool plant.* Following the classification of bulk fluid milk products and bulk fluid cream products received during the month from a cooperative association in its capacity as the operator of a pool plant, the final payment for such receipts shall be determined as follows:

(i) Multiply the hundredweight of Class I skim milk by the Class I skim milk price for the month at the receiving plant;

(ii) Multiply the pounds of Class I butterfat by the Class I butterfat price for the month at the receiving plant;

(iii) Multiply the pounds of nonfat solids in Class II skim milk by the Class II nonfat solids price;

(iv) Multiply the pounds of butterfat in Class II times the Class II butterfat price;

(v) Multiply the pounds of nonfat solids in Class IV milk by the nonfat solids price for the month;

(vi) Multiply the pounds of butterfat in Class III and Class IV milk by the butterfat price for the month;

(vii) Multiply the pounds of protein in Class III milk by the protein price for the month;

(viii) Multiply the pounds of other solids in Class III milk by the other solids price for the month; and

(ix) Add together the amounts computed in paragraphs (b)(3)(i) through (viii) of this section and from that sum deduct any payment made pursuant to paragraph (b)(2) of this section.

(4) *Final payment to a cooperative association for bulk milk received directly from producers' farms.* For bulk milk received from a cooperative association

during the month, including the milk of producers who are not members of such association and who the market administrator determines have authorized the cooperative association to collect payment for their milk, the final payment for such milk shall be an amount equal to the sum of the individual payments otherwise payable for such milk pursuant to paragraph (a)(2) of this section.

(c) If a handler has not received full payment from the market administrator pursuant to § 1001.72 by the payment date specified in paragraph (a) or (b) of this section, the handler may reduce payments pursuant to paragraphs (a) and (b) of this section, but by not more than the amount of the underpayment. The payments shall be completed on the next scheduled payment date after receipt of the balance due from the market administrator.

(d) If a handler claims that a required payment to a producer cannot be made because the producer is deceased or cannot be located, or because the cooperative association or its lawful successor or assignee is no longer in existence, the payment shall be made to the producer-settlement fund, and in the event that the handler subsequently locates and pays the producer or a lawful claimant, or in the event that the handler no longer exists and a lawful claim is later established, the market administrator shall make the required payment from the producer-settlement fund to the handler or to the lawful claimant as the case may be.

(e) In making payments to producers pursuant to this section, each handler shall furnish each producer (except for a producer whose milk was received from a cooperative association handler described in § 1000.9(a) or 9(c)), a supporting statement in such form that it may be retained by the recipient which shall show:

(1) The name, address, Grade A identifier assigned by a duly constituted regulatory agency, and the payroll number of the producer;

(2) The month and dates that milk was received from the producer, including the daily and total pounds of milk received;

(3) The total pounds of butterfat, protein, and other solids contained in the producer's milk;

(4) The minimum rate or rates at which payment to the producer is required pursuant to the order in this part;

(5) The rate used in making payment if the rate is other than the applicable minimum rate;

(6) The amount, or rate per hundredweight, or rate per pound of component, and the nature of each deduction claimed by the handler; and

(7) The net amount of payment to the producer or cooperative association.

[64 FR 47954, Sept. 1, 1999, as amended at 65 FR 32010, May 22, 2000; 65 FR 82835, Dec. 28, 2000; 68 FR 7066, Feb. 12, 2003; 70 FR 18963, Apr. 12, 2005]

§ 1001.74 [Reserved]

§ 1001.75 Plant location adjustments for producer milk and nonpool milk.

For purposes of making payments for producer milk and nonpool milk, a plant location adjustment shall be determined by subtracting the Class I price specified in § 1001.51 from the Class I price at the plant's location. The difference, plus or minus as the case may be, shall be used to adjust the payments required pursuant to §§ 1001.73 and 1000.76.

§ 1001.76 Payments by a handler operating a partially regulated distributing plant.

See § 1000.76.

§ 1001.77 Adjustment of accounts.

See § 1000.77.

§ 1001.78 Charges on overdue accounts.

See § 1000.78.

ADMINISTRATIVE ASSESSMENT AND MARKETING SERVICE DEDUCTION

§ 1001.85 Assessment for order administration.

See § 1000.85.

§ 1001.86 Deduction for marketing services.

See § 1000.86.

PARTS 1002-1004 [RESERVED]

PART 1005—MILK IN THE APPALACHIAN MARKETING AREA

Subpart—Order Regulating Handling

AUTHORITY: 7 U.S.C. 601–674, and 7253.

SOURCE: 64 FR 47960, Sept. 1, 1999, unless otherwise noted.

Subpart—Order Regulating Handling

GENERAL PROVISIONS

§ 1005.1   General provisions.

The terms, definitions, and provisions in part 1000 of this chapter apply to this part 1005. In this part 1005, all references to sections in part 1000 refer to part 1000 of this chapter.

DEFINITIONS

§ 1005.2   Appalachian marketing area.

The marketing area means all the territory within the bounds of the following states and political subdivisions, including all piers, docks and wharves connected therewith and all craft moored thereat, and all territory occupied by government (municipal, State or Federal) reservations, installations, institutions, or other similar establishments if any part thereof is within any of the listed states or political subdivisions:

GEORGIA COUNTIES

Catoosa, Chattooga, Dade, Fannin, Murray, Walker, and Whitfield.

INDIANA COUNTIES

Clark, Crawford, Daviess, Dubois, Floyd, Gibson, Greene, Harrison, Knox, Martin, Orange, Perry, Pike, Posey, Scott, Spencer, Sullivan, Vanderburgh, Warrick, and Washington.

KENTUCKY COUNTIES

Adair, Anderson, Bath, Bell, Bourbon, Boyle, Breathitt, Breckinridge, Bullitt, Butler, Carroll, Carter, Casey, Clark, Clay, Clinton, Cumberland, Daviess, Edmonson, Elliott, Estill, Fayette, Fleming, Franklin, Gallatin, Garrard, Grayson, Green, Hancock, Hardin, Harlan, Hart, Henderson, Henry, Hopkins, Jackson, Jefferson, Jessamine, Knott, Knox, Larue, Laurel, Lee, Leslie, Letcher, Lincoln, Madison, Marion, McCreary, McLean, Meade, Menifee, Mercer, Montgomery, Morgan, Muhlenberg, Nelson, Nicholas, Ohio, Oldham, Owen, Owsley, Perry, Powell, Pulaski, Rockcastle, Rowan, Russell, Scott, Shelby, Spencer, Taylor, Trimble, Union, Washington, Wayne, Webster, Whitley, Wolfe, and Woodford.

NORTH CAROLINA AND SOUTH CAROLINA

All of the States of North Carolina and South Carolina.

TENNESSEE COUNTIES

Anderson, Blount, Bradley, Campbell, Carter, Claiborne, Cocke, Cumberland, Grainger, Greene, Hamblen, Hamilton, Hancock, Hawkins, Jefferson, Johnson, Knox, Loudon, Marion, McMinn, Meigs, Monroe, Morgan, Polk, Rhea, Roane, Scott, Sequatchie, Sevier, Sullivan, Unicoi, Union, and Washington.

VIRGINIA COUNTIES AND CITIES

Alleghany, Amherst, Augusta, Bath, Bedford, Bland, Botetourt, Buchanan, Campbell, Carroll, Craig, Dickenson, Floyd, Franklin, Giles, Grayson, Henry, Highland, Lee, Montgomery, Patrick, Pittsylvania, Pulaski, Roanoke, Rockbridge, Rockingham, Russell, Scott, Smyth, Tazewell, Washington, Wise, and Wythe; and the cities of Bedford, Bristol, Buena Vista, Clifton Forge, Covington, Danville, Galax, Harrisonburg, Lexington, Lynchburg, Martinsville, Norton, Radford, Roanoke, Salem, Staunton, and Waynesboro.

WEST VIRGINIA COUNTIES

McDowell and Mercer.

[64 FR 47960, Sept. 1, 1999, as amended at 70 FR 59223, Oct. 12, 2005]

§1005.3 Route disposition.

See §1000.3.

§1005.4 Plant.

See §1000.4.

§1005.5 Distributing plant.

See §1000.5.

§1005.6 Supply plant.

See §1000.6.

§1005.7 Pool plant.

Pool plant means a plant specified in paragraphs (a) through (d) of this section, a unit of plants as specified in paragraph (e) of this section, or a plant specified in paragraph (g) of this section but excluding a plant specified in paragraph (h) of this section. The pooling standards described in paragraphs (c) and (d) of this section are subject to modification pursuant to paragraph (f) of this section:

(a) A distributing plant, other than a plant qualified as a pool plant pursuant to paragraph (b) of this section or §_____.7(b) of any other Federal milk order, from which during the month 50 percent or more of the fluid milk products physically received at such plant (excluding concentrated milk received from another plant by agreement for other than Class I use) are disposed of as route disposition or are transferred in the form of packaged fluid milk products to other distributing plants. At least 25 percent of such route disposition and transfers must be to outlets in the marketing area.

(b) Any distributing plant located in the marketing area which during the month processed at least 50 percent of the total quantity of fluid milk products physically received at the plant (excluding concentrated milk received from another plant by agreement for other than Class I use) into ultra-pasteurized or aseptically-processed fluid milk products.

(c) A supply plant from which 50 percent or more of the total quantity of milk that is physically received during the month from dairy farmers and handlers described in § 1000.9(c), including milk that is diverted from the plant, is transferred to pool distributing plants. Concentrated milk transferred from the supply plant to a distributing plant for an agreed-upon use other than Class I shall be excluded from the supply plant's shipments in computing the plant's shipping percentage.

(d) A plant located within the marketing area or in the State of Virginia that is operated by a cooperative association if pool plant status under this paragraph is requested for such plant by the cooperative association and during the month at least 60 percent of the producer milk of members of such cooperative association is delivered directly from farms to pool distributing plants or is transferred to such plants as a fluid milk product (excluding concentrated milk transferred to a distributing plant for an agreed-upon use other than Class I) from the cooperative's plant.

(e) Two or more plants operated by the same handler and that are located within the marketing area may qualify for pool status as a unit by meeting the total and in-area route disposition requirements specified in paragraph (a) of this section and the following additional requirements:

(1) At least one of the plants in the unit must qualify as a pool plant pursuant to paragraph (a) of this section;

(2) Other plants in the unit must process only Class I or Class II products and must be located in a pricing zone providing the same or a lower Class I price than the price applicable at the distributing plant included in the unit pursuant to paragraph (e)(1) of this section; and

(3) A written request to form a unit, or to add or remove plants from a unit, must be filed with the market administrator prior to the first day of the month for which it is to be effective.

(f) The applicable shipping percentages of paragraphs (c) and (d) of this section may be increased or decreased by the market administrator if the market administrator finds that such adjustment is necessary to encourage needed shipments or to prevent uneconomic shipments. Before making such a finding, the market administrator

shall investigate the need for adjustment either on the market administrator's own initiative or at the request of interested parties if the request is made in writing at least 15 days prior to the date for which the requested revision is desired effective. If the investigation shows that an adjustment of the shipping percentages might be appropriate, the market administrator shall issue a notice stating that an adjustment is being considered and invite data, views and arguments. Any decision to revise an applicable shipping percentage must be issued in writing at least one day before the effective date.

(g) Any distributing plant other than a plant qualified as a pool plant pursuant to paragraph § 1005.(7)(a) or paragraph (b) of this section or § _____.7(b) of any other Federal milk order or § 1005.(7)(e) or § 1000.(8)(a) or § 1000.(8)(e); located within the marketing area as described on May 1, 2006, in § 1005.2, from which there is route disposition and/or transfers of packaged fluid milk products in any non-Federally regulated marketing area(s) located within one or more States that require handlers to pay minimum prices for raw milk provided that 25 percent or more of the total quantity of fluid milk products physically received at such plant (excluding concentrated milk received from another plant by agreement for other than Class I use) is disposed of as route disposition and/or is transferred in the form of packaged fluid milk products to other plants. At least 25 percent of such route disposition and/or transfers, in aggregate, are in any non-Federally regulated marketing area(s) located within one or more States that require handlers to pay minimum prices for raw milk. Subject to the following exclusion:

(1) The plant is subject to the pricing provisions of a State-operated milk pricing plan which provides for the payment of minimum class prices for raw milk;

(2) A producer-handler described in § 1005.10 with less than three million pounds during the month of route disposition and/or transfers of packaged fluid milk products to other plants.

(h) The term pool plant shall not apply to the following plants:

(1) A producer-handler plant;

(2) An exempt plant as defined in § 1000.8(e);

(3) A plant qualified pursuant to paragraph (a) of this section which is not located within any Federal order marketing area, meets the pooling requirements of another Federal order, and has had greater route disposition in such other Federal order marketing area for 3 consecutive months;

(4) A plant qualified pursuant to paragraph (a) of this section which is located in another Federal order marketing area, meets the pooling standards of the other Federal order, and has not had a majority of its route disposition in this marketing area for 3 consecutive months or is locked into pool status under such other Federal order without regard to its route disposition in any other Federal order marketing area;

(5) A plant qualified pursuant to paragraph (c) of this section which also meets the pooling requirements of another Federal order and from which greater qualifying shipments are made to plants regulated under such other order than are made to plants regulated under the order in this part, or such plant has automatic pooling status under such other order; and

(6) That portion of a pool plant designated as a "nonpool plant" that is physically separate and operated separately from the pool portion of such plant. The designation of a portion of a regulated plant as a nonpool plant must be requested in writing by the handler and must be approved by the market administrator.

[64 FR 47960, Sept. 1, 1999, as amended at 71 FR 25497, May 1, 2006; 71 FR 28249, May 16, 2006]

§ 1005.8 Nonpool plant.

See § 1000.8.

§ 1005.9 Handler.

See § 1000.9.

§ 1005.10 Producer-handler.

Producer-handler means a person who:

(a) Operates a dairy farm and a distributing plant from which there is route disposition in the marketing area, and from which total route disposition and packaged sales of fluid milk products to other plants during the month does not exceed 3 million pounds;

(b) Receives no fluid milk products, and acquires no fluid milk products for route disposition, from sources other than own farm production;

(c) Disposes of no other source milk as Class I milk except by increasing the nonfat milk solids content of the fluid milk products received from own farm production; and

(d) Provides proof satisfactory to the market administrator that the care and management of the dairy animals and other resources necessary to produce all Class I milk handled, and the processing and packaging operations are the producer-handler's own enterprise and are operated at the producer-handler's own risk.

(e) Any producer-handler with Class I route dispositions and/or transfers of packaged fluid milk products in the marketing area described in § 1131.2 of this chapter shall be subject to payments into the Order 1131 producer settlement fund on such dispositions pursuant to § 1000.76(a) and payments into the Order 1131 administrative fund provided such dispositions are less than three million pounds in the current month and such producer-handler had total Class I route dispositions and/or transfers of packaged fluid milk products from own farm production of three million pounds or more the previous month. If the producer-handler has Class I route dispositions and/or transfers of packaged fluid milk products into the marketing area described in § 1131.2 of this chapter of three million pounds or more during the current month, such producer-handler shall be subject to the provisions described in § 1131.7 of this chapter or § 1000.76(a).

[64 FR 47960, Sept. 1, 1999, as amended at 71 FR 25498, May 1, 2006; 75 FR 21160, Apr. 23, 2010]

§ 1005.11 [Reserved]

§ 1005.12 Producer.

(a) Except as provided in paragraph (b) of this section, *producer* means any person who produces milk approved by a duly constituted regulatory agency for fluid consumption as Grade A milk and whose milk (or components of milk) is:

(1) Received at a pool plant directly from the producer or diverted by the plant operator in accordance with § 1005.13; or

(2) Received by a handler described in § 1000.9(c).

(b) Producer shall not include:

(1) A producer-handler as defined in any Federal order;

(2) A dairy farmer whose milk is received at an exempt plant, excluding producer milk diverted to the exempt plant pursuant to § 1005.13(d);

(3) A dairy farmer whose milk is received by diversion at a pool plant from a handler regulated under another Federal order if the other Federal order designates the dairy farmer as a producer under that order and that milk is allocated by request to a utilization other than Class I; and

(4) A dairy farmer whose milk is reported as diverted to a plant fully regulated under another order with respect to that portion of the milk so diverted that is assigned to Class I under the provisions of such other order.

§ 1005.13 Producer milk.

Except as provided for in paragraph (e) of this section, *Producer milk* means the skim milk (or the skim equivalent of components of skim milk) and butterfat contained in milk of a producer that is:

(a) Received by the operator of a pool plant directly from a producer or a handler described in § 1000.9(c). All milk received pursuant to this paragraph shall be priced at the location of the plant where it is first physically received;

(b) Received by a handler described in § 1000.9(c) in excess of the quantity delivered to pool plants;

(c) Diverted by a pool plant operator to another pool plant. Milk so diverted shall be priced at the location of the plant to which diverted; or

(d) Diverted by the operator of a pool plant or a handler described in § 1000.9(c) to a nonpool plant, subject to the following conditions:

(1) In any month of July through December, not less than 1 days' production of the producer whose milk is diverted is physically received at a pool plant during the month;

(2) In any month of January through June, not less than 1 days' production of the producer whose milk is diverted is physically received at a pool plant during the month;

(3) The total quantity of milk so diverted during the month by a cooperative association shall not exceed 25 percent during the months of July through November, January, and February, and 35 percent during the months of December and March through June, of the producer milk that the cooperative association caused to be delivered to, and physically received at, pool plants during the month, excluding the total pounds of bulk milk received directly from producers meeting the conditions as described in § 1005.82(c)(2)(ii) and (iii), and for which a transportation credit is requested;

(4) The operator of a pool plant that is not a cooperative association may divert any milk that is not under the control of a cooperative association that diverts milk during the month pursuant to paragraph (d) of this section. The total quantity of milk so diverted during the month shall not exceed 25 percent during the months of July through November, January, and February, and 35 percent during the months of December and March through June, of the producer milk physically received at such plant (or such unit of plants in the case of plants that pool as a unit pursuant to § 1005.7(e)) during the month, excluding the quantity of producer milk received from a handler described in § 1000.9(c) of this chapter and excluding the total pounds of bulk milk received directly from producers meeting the conditions as described in § 1005.82(c)(2)(ii) and (iii), and for which a transportation credit is requested;

(5) Any milk diverted in excess of the limits prescribed in paragraphs (d)(3) and (4) of this section shall not be producer milk. If the diverting handler or cooperative association fails to designate the dairy farmers' deliveries that will not be producer milk, no milk diverted by the handler or cooperative association shall be producer milk;

(6) Diverted milk shall be priced at the location of the plant to which diverted; and

(7) The delivery day requirements and the diversion percentages in paragraphs (d)(1) through (4) of this section may be increased or decreased by the market administrator if the market administrator finds that such revision is necessary to assure orderly marketing and efficient handling of milk in the marketing area. Before making such a finding, the market administrator shall investigate the need for the revision either on the market administrator's own initiative or at the request of interested persons. If the investigation shows that a revision might be appropriate, the market administrator shall issue a notice stating that the revision is being considered and inviting written data, views, and arguments. Any decision to revise an applicable percentage must be issued in writing at least one day before the effective date.

(e) Producer milk shall not include milk of a producer that is subject to inclusion and participation in a marketwide equalization pool under a milk classification and pricing program imposed under the authority of a State government maintaining marketwide pooling of returns.

[64 FR 47960, Sept. 1, 1999, as amended at 70 FR 59223, Oct. 12, 2005; 71 FR 62378; Oct. 25, 2006; 73 FR 14156, Mar. 17, 2008; 73 FR 26315, May 9, 2008; 79 FR 25005, May 2, 2014; 79 FR 26591, May 9, 2014]

§ 1005.14   Other source milk.

See § 1000.14.

§ 1005.15   Fluid milk product.

See § 1000.15.

§ 1005.16   Fluid cream product.

See § 1000.16.

§ 1005.17   [Reserved]

§ 1005.18   Cooperative association.

See § 1000.18.

§ 1005.19   Commercial food processing establishment.

See § 1000.19.

HANDLER REPORTS

§ 1005.30   Reports of receipts and utilization.

Each handler shall report monthly so that the market administrator's office receives the report on or before the 7th day after the end of the month, in the detail and on prescribed forms, as follows:

(a) With respect to each of its pool plants, the quantities of skim milk and butterfat contained in or represented by:

(1) Receipts of producer milk, including producer milk diverted by the reporting handler, from sources other than handlers described in § 1000.9(c);

(2) Receipts of milk from handlers described in § 1000.9(c);

(3) Receipts of fluid milk products and bulk fluid cream products from other pool plants;

(4) Receipts of other source milk;

(5) Receipts of bulk milk from a plant regulated under another Federal order, except Federal Order 1007, for which a transportation credit is requested pursuant to § 1005.82;

(6) Receipts of producer milk described in § 1005.82(c)(2), including the identity of the individual producers whose milk is eligible for the transportation credit pursuant to that paragraph and the date that such milk was received;

(7) For handlers submitting transportation credit requests, transfers of bulk milk to nonpool plants, including the dates that such milk was transferred;

(8) Inventories at the beginning and end of the month of fluid milk products and bulk fluid cream products; and

(9) The utilization or disposition of all milk and milk products required to be reported pursuant to this paragraph.

(b) Each handler operating a partially regulated distributing plant shall report with respect to such plant in the same manner as prescribed for reports required by paragraph (a) of this section. Receipts of milk that would have been producer milk if the plant had been fully regulated shall be reported in lieu of producer milk. The report shall show also the quantity of any reconstituted skim milk in route disposition in the marketing area.

(c) Each handler described in § 1000.9(c) shall report:

(1) The quantities of all skim milk and butterfat contained in receipts of milk from producers;

(2) The utilization or disposition of all such receipts; and

(3) With respect to milk for which a cooperative association is requesting a transportation credit pursuant to § 1005.82, all of the information required in paragraphs (a)(5), (a)(6), and (a)(7) of this section.

(d) Each handler not specified in paragraphs (a) through (c) of this section shall report with respect to its receipts and utilization of milk and milk products in such manner as the market administrator may prescribe.

§ 1005.31 Payroll reports.

(a) On or before the 20th day after the end of each month, each handler that operates a pool plant pursuant to § 1005.7 and each handler described in § 1000.9(c) shall report to the market administrator its producer payroll for the month, in detail prescribed by the market administrator, showing for each producer the information specified in § 1005.73(e).

(b) Each handler operating a partially regulated distributing plant who elects to make payment pursuant to § 1000.76(b) shall report for each dairy farmer who would have been a producer if the plant had been fully regulated in the same manner as prescribed for reports required by paragraph (a) of this section.

§ 1005.32 Other reports.

(a) On or before the 20th day after the end of each month, each handler described in § 1000.9(a) and (c) shall report to the market administrator any adjustments to transportation credit requests as reported pursuant to § 1005.30(a)(5), (6), and (7).

(b) In addition to the reports required pursuant to §§ 1005.30, 1005.31, and 1005.32(a), each handler shall report any information the market administrator deems necessary to verify or establish each handler's obligation under the order.

CLASSIFICATION OF MILK

§ 1005.40 Classes of utilization.

See § 1000.40.

§ 1005.41 [Reserved]

§ 1005.42 Classification of transfers and diversions.

See § 1000.42.

§ 1005.43 General classification rules.

See § 1000.43.

§ 1005.44 Classification of producer milk.

See § 1000.44.

§ 1005.45 Market administrator's reports and announcements concerning classification.

See § 1000.45.

CLASS PRICES

§ 1005.50 Class prices, component prices, and advanced pricing factors.

See § 1000.50.

§ 1005.51 Class I differential, adjustments to Class I prices, and Class I price.

(a) The Class I differential shall be the differential established for Mecklenburg County, North Carolina, which is reported in § 1000.52. The Class I price shall be the price computed pursuant to § 1005.50(a) for Mecklenburg County, North Carolina.

(b) Adjustment to Class I prices. Class I prices shall be established pursuant to § 1000.50(a), (b) and (c) using the following adjustments:

State	County/parish	FIPS	Class I price adjustment
GA	CATOOSA	13047	0.60
GA	CHATTOOGA	13055	0.60
GA	DADE	13083	0.60
GA	FANNIN	13111	0.60
GA	MURRAY	13213	0.60
GA	WALKER	13295	0.60
GA	WHITFIELD	13313	0.60
IN	CLARK	18019	0.10
IN	CRAWFORD	18025	0.10
IN	DAVIESS	18027	0.10
IN	DUBOIS	18037	0.10
IN	FLOYD	18043	0.10
IN	GIBSON	18051	0.10
IN	GREENE	18055	0.10
IN	HARRISON	18061	0.10
IN	KNOX	18083	0.10

State	County/parish	FIPS	Class I price ad- justment	State	County/parish	FIPS	Class I price ad- justment
IN	MARTIN	18101	0.10	KY	OWSLEY	21189	0.70
IN	ORANGE	18117	0.10	KY	PERRY	21193	0.50
IN	PERRY	18123	0.10	KY	POWELL	21197	0.40
IN	PIKE	18125	0.10	KY	PULASKI	21199	0.50
IN	POSEY	18129	0.10	KY	ROCKCASTLE	21203	0.70
IN	SCOTT	18143	0.10	KY	ROWAN	21205	0.40
IN	SPENCER	18147	0.10	KY	RUSSELL	21207	0.50
IN	SULLIVAN	18153	0.10	KY	SCOTT	21209	0.10
IN	VANDERBURGH	18163	0.10	KY	SHELBY	21211	0.10
IN	WARRICK	18173	0.10	KY	SPENCER	21215	0.10
IN	WASHINGTON	18175	0.10	KY	TAYLOR	21217	0.20
KY	ADAIR	21001	0.20	KY	TRIMBLE	21223	0.10
KY	ANDERSON	21005	0.40	KY	UNION	21225	0.10
KY	BATH	21011	0.40	KY	WASHINGTON	21229	0.40
KY	BELL	21013	0.50	KY	WAYNE	21231	0.50
KY	BOURBON	21017	0.40	KY	WEBSTER	21233	0.20
KY	BOYLE	21021	0.40	KY	WHITLEY	21235	0.50
KY	BREATHITT	21025	0.70	KY	WOLFE	21237	0.40
KY	BRECKINRIDGE	21027	0.10	KY	WOODFORD	21239	0.40
KY	BULLITT	21029	0.10	NC	ALAMANCE	37001	0.30
KY	BUTLER	21031	0.20	NC	ALEXANDER	37003	0.45
KY	CARROLL	21041	0.10	NC	ALLEGHANY	37005	0.45
KY	CARTER	21043	0.40	NC	ANSON	37007	0.50
KY	CASEY	21045	0.20	NC	ASHE	37009	0.45
KY	CLARK	21049	0.40	NC	AVERY	37011	0.45
KY	CLAY	21051	0.50	NC	BEAUFORT	37013	0.40
KY	CLINTON	21053	0.50	NC	BERTIE	37015	0.20
KY	CUMBERLAND	21057	0.50	NC	BLADEN	37017	0.70
KY	DAVIESS	21059	0.10	NC	BRUNSWICK	37019	0.70
KY	EDMONSON	21061	0.20	NC	BUNCOMBE	37021	0.45
KY	ELLIOTT	21063	0.40	NC	BURKE	37023	0.45
KY	ESTILL	21065	0.40	NC	CABARRUS	37025	0.30
KY	FAYETTE	21067	0.40	NC	CALDWELL	37027	0.45
KY	FLEMING	21069	0.40	NC	CAMDEN	37029	0.20
KY	FRANKLIN	21073	0.10	NC	CARTERET	37031	0.40
KY	GALLATIN	21077	0.10	NC	CASWELL	37033	0.30
KY	GARRARD	21079	0.40	NC	CATAWBA	37035	0.30
KY	GRAYSON	21085	0.20	NC	CHATHAM	37037	0.30
KY	GREEN	21087	0.20	NC	CHEROKEE	37039	0.45
KY	HANCOCK	21091	0.10	NC	CHOWAN	37041	0.20
KY	HARDIN	21093	0.10	NC	CLAY	37043	0.45
KY	HARLAN	21095	0.50	NC	CLEVELAND	37045	0.30
KY	HART	21099	0.20	NC	COLUMBUS	37047	0.70
KY	HENDERSON	21101	0.10	NC	CRAVEN	37049	0.40
KY	HENRY	21103	0.10	NC	CUMBERLAND	37051	0.30
KY	HOPKINS	21107	0.20	NC	CURRITUCK	37053	0.20
KY	JACKSON	21109	0.70	NC	DARE	37055	0.40
KY	JEFFERSON	21111	0.10	NC	DAVIDSON	37057	0.30
KY	JESSAMINE	21113	0.40	NC	DAVIE	37059	0.30
KY	KNOTT	21119	0.50	NC	DUPLIN	37061	0.30
KY	KNOX	21121	0.50	NC	DURHAM	37063	0.30
KY	LARUE	21123	0.40	NC	EDGECOMBE	37065	0.20
KY	LAUREL	21125	0.50	NC	FORSYTH	37067	0.30
KY	LEE	21129	0.40	NC	FRANKLIN	37069	0.30
KY	LESLIE	21131	0.50	NC	GASTON	37071	0.30
KY	LETCHER	21133	0.50	NC	GATES	37073	0.20
KY	LINCOLN	21137	0.40	NC	GRAHAM	37075	0.45
KY	MCCREARY	21147	0.50	NC	GRANVILLE	37077	0.30
KY	MCLEAN	21149	0.40	NC	GREENE	37079	0.40
KY	MADISON	21151	0.40	NC	GUILFORD	37081	0.30
KY	MARION	21155	0.40	NC	HALIFAX	37083	0.30
KY	MEADE	21163	0.10	NC	HARNETT	37085	0.10
KY	MENIFEE	21165	0.40	NC	HAYWOOD	37087	0.45
KY	MERCER	21167	0.40	NC	HENDERSON	37089	0.45
KY	MONTGOMERY	21173	0.40	NC	HERTFORD	37091	0.20
KY	MORGAN	21175	0.40	NC	HOKE	37093	0.30
KY	MUHLENBURG	21177	0.20	NC	HYDE	37095	0.40
KY	NELSON	21179	0.10	NC	IREDELL	37097	0.30
KY	NICHOLAS	21181	0.40	NC	JACKSON	37099	0.45
KY	OHIO	21183	0.20	NC	JOHNSTON	37101	0.20
KY	OLDHAM	21185	0.10	NC	JONES	37103	0.40
KY	OWEN	21187	0.10	NC	LEE	37105	0.30

State	County/parish	FIPS	Class I price adjustment	State	County/parish	FIPS	Class I price adjustment
NC	LENOIR	37107	0.40	SC	HORRY	45051	0.70
NC	LINCOLN	37109	0.30	SC	JASPER	45053	1.00
NC	MCDOWELL	37111	0.45	SC	KERSHAW	45055	0.30
NC	MACON	37113	0.45	SC	LANCASTER	45057	0.50
NC	MADISON	37115	0.45	SC	LAURENS	45059	0.50
NC	MARTIN	37117	0.40	SC	LEE	45061	0.70
NC	MECKLENBURG	37119	0.30	SC	LEXINGTON	45063	0.70
NC	MITCHELL	37121	0.45	SC	MCCORMICK	45065	0.50
NC	MONTGOMERY	37123	0.30	SC	MARION	45067	0.70
NC	MOORE	37125	0.30	SC	MARLBORO	45069	0.70
NC	NASH	37127	0.30	SC	NEWBERRY	45071	0.30
NC	NEW HANOVER	37129	0.70	SC	OCONEE	45073	0.50
NC	NORTHAMPTON	37131	0.30	SC	ORANGEBURG	45075	0.70
NC	ONSLOW	37133	0.30	SC	PICKENS	45077	0.50
NC	ORANGE	37135	0.30	SC	RICHLAND	45079	0.70
NC	PAMLICO	37137	0.40	SC	SALUDA	45081	0.30
NC	PASQUOTANK	37139	0.20	SC	SPARTANBURG	45083	0.50
NC	PENDER	37141	0.70	SC	SUMTER	45085	0.70
NC	PERQUIMANS	37143	0.20	SC	UNION	45087	0.50
NC	PERSON	37145	0.30	SC	WILLIAMSBURG	45089	0.70
NC	PITT	37147	0.40	SC	YORK	45091	0.50
NC	POLK	37149	0.30	TN	ANDERSON	47001	0.40
NC	RANDOLPH	37151	0.30	TN	BLOUNT	47009	0.40
NC	RICHMOND	37153	0.50	TN	BRADLEY	47011	0.60
NC	ROBESON	37155	0.70	TN	CAMPBELL	47013	0.40
NC	ROCKINGHAM	37157	0.45	TN	CARTER	47019	0.40
NC	ROWAN	37159	0.30	TN	CLAIBORNE	47025	0.40
NC	RUTHERFORD	37161	0.30	TN	COCKE	47029	0.40
NC	SAMPSON	37163	0.30	TN	CUMBERLAND	47035	0.40
NC	SCOTLAND	37165	0.30	TN	GRAINGER	47057	0.40
NC	STANLY	37167	0.30	TN	GREENE	47059	0.40
NC	STOKES	37169	0.45	TN	HAMBLEN	47063	0.40
NC	SURRY	37171	0.45	TN	HAMILTON	47065	0.60
NC	SWAIN	37173	0.45	TN	HANCOCK	47067	0.40
NC	TRANSYLVANIA	37175	0.45	TN	HAWKINS	47073	0.40
NC	TYRRELL	37177	0.40	TN	JEFFERSON	47089	0.40
NC	UNION	37179	0.50	TN	JOHNSON	47091	0.40
NC	VANCE	37181	0.30	TN	KNOX	47093	0.40
NC	WAKE	37183	0.30	TN	LOUDON	47105	0.40
NC	WARREN	37185	0.30	TN	MCMINN	47107	0.60
NC	WASHINGTON	37187	0.40	TN	MARION	47115	0.60
NC	WATAUGA	37189	0.45	TN	MEIGS	47121	0.60
NC	WAYNE	37191	0.40	TN	MONROE	47123	0.60
NC	WILKES	37193	0.45	TN	MORGAN	47129	0.40
NC	WILSON	37195	0.20	TN	POLK	47139	0.60
NC	YADKIN	37197	0.30	TN	RHEA	47143	0.40
NC	YANCEY	37199	0.45	TN	ROANE	47145	0.40
SC	ABBEVILLE	45001	0.50	TN	SCOTT	47151	0.10
SC	AIKEN	45003	0.70	TN	SEQUATCHIE	47153	0.40
SC	ALLENDALE	45005	1.00	TN	SEVIER	47155	0.40
SC	ANDERSON	45007	0.50	TN	SULLIVAN	47163	0.40
SC	BAMBERG	45009	0.70	TN	UNICOI	47171	0.40
SC	BARNWELL	45011	0.70	TN	UNION	47173	0.40
SC	BEAUFORT	45013	1.00	TN	WASHINGTON	47179	0.40
SC	BERKELEY	45015	1.00	VA	ALLEGHANY	51005	0.10
SC	CALHOUN	45017	0.70	VA	AMHERST	51009	0.40
SC	CHARLESTON	45019	1.00	VA	AUGUSTA	51015	0.10
SC	CHEROKEE	45021	0.50	VA	BATH	51017	0.10
SC	CHESTER	45023	0.50	VA	BEDFORD	51019	0.40
SC	CHESTERFIELD	45025	0.30	VA	BLAND	51021	0.40
SC	CLARENDON	45027	0.70	VA	BOTETOURT	51023	0.10
SC	COLLETON	45029	1.00	VA	BUCHANAN	51027	0.10
SC	DARLINGTON	45031	0.70	VA	CAMPBELL	51031	0.40
SC	DILLON	45033	0.70	VA	CARROLL	51035	0.40
SC	DORCHESTER	45035	1.00	VA	CRAIG	51045	0.10
SC	EDGEFIELD	45037	0.30	VA	DICKENSON	51051	0.40
SC	FAIRFIELD	45039	0.30	VA	FLOYD	51063	0.40
SC	FLORENCE	45041	0.70	VA	FRANKLIN	51067	0.40
SC	GEORGETOWN	45043	0.70	VA	GILES	51071	0.10
SC	GREENVILLE	45045	0.50	VA	GRAYSON	51077	0.40
SC	GREENWOOD	45047	0.50	VA	HENRY	51089	0.40
SC	HAMPTON	45049	1.00	VA	HIGHLAND	51091	0.10

State	County/parish	FIPS	Class I price adjustment	State	County/parish	FIPS	Class I price adjustment
VA	LEE	51105	0.40	VA	CLIFTON FORGE CITY	51560	0.10
VA	MONTGOMERY	51121	0.40	VA	COVINGTON CITY	51580	0.10
VA	PATRICK	51141	0.40	VA	DANVILLE CITY	51590	0.40
VA	PITTSYLVANIA	51143	0.40	VA	GALAX CITY	51640	0.40
VA	PULASKI	51155	0.40	VA	HARRISONBURG CITY	51660	0.10
VA	ROANOKE	51161	0.40	VA	LEXINGTON CITY	51678	0.10
VA	ROCKBRIDGE	51163	0.10	VA	LYNCHBURG CITY	51680	0.40
VA	ROCKINGHAM	51165	0.10	VA	MARTINSVILLE CITY	51690	0.40
VA	RUSSELL	51167	0.40	VA	NORTON CITY	51720	0.40
VA	SCOTT	51169	0.40	VA	RADFORD CITY	51750	0.40
VA	SMYTH	51173	0.40	VA	ROANOKE CITY	51770	0.40
VA	TAZEWELL	51185	0.40	VA	SALEM CITY	51775	0.40
VA	WASHINGTON	51191	0.40	VA	STAUNTON CITY	51790	0.10
VA	WISE	51195	0.40	VA	WAYNESBORO CITY	51820	0.10
VA	WYTHE	51197	0.40	WV	MCDOWELL	54047	0.10
VA	BEDFORD CITY	51515	0.40	WV	MERCER	54055	0.10
VA	BRISTOL CITY	51520	0.40				
VA	BUENA VISTA CITY	51530	0.10				

[73 FR 14156, Mar. 17, 2008]

§ 1005.52 Adjusted Class I differentials.

See § 1000.52.

§ 1005.53 Announcement of class prices, component prices, and advanced pricing factors.

See § 1000.53.

§ 1005.54 Equivalent price.

See § 1000.54.

UNIFORM PRICES

§ 1005.60 Handler's value of milk.

For the purpose of computing a handler's obligation for producer milk, the market administrator shall determine for each month the value of milk of each handler with respect to each of the handler's pool plants and of each handler described in § 1000.9(c) with respect to milk that was not received at a pool plant by adding the amounts computed in paragraphs (a) through (e) of this section and subtracting from that total amount the value computed in paragraph (f) of this section. Receipts of nonfluid milk products that are distributed as labeled reconstituted milk for which payments are made to the producer-settlement fund of another Federal order under § 1000.76(a)(4) or (d) shall be excluded from pricing under this section.

(a) Multiply the pounds of skim milk and butterfat in producer milk that were classified in each class pursuant to § 1000.44(c) by the applicable skim milk and butterfat prices, and add the resulting amounts; except that for the months of January 2005 through March 2005, the Class I skim milk price for this purpose shall be the Class I skim milk price as determined in § 1000.50(b) plus $0.04 per hundredweight, and the Class I butterfat price for this purpose shall be the Class I butterfat price as determined in § 1000.50(c) plus $0.0004 per pound. The adjustments to the Class I skim milk and butterfat prices provided herein may be reduced by the market administrator for any month if the market administrator determines that the payments yet unpaid computed pursuant to paragraphs (g)(1) through (5) and paragraph (g)(7) of this section will be less than the amount computed pursuant to paragraph (g)(6) of this section. The adjustments to the Class I skim milk and butterfat prices provided herein during the months of January 2005 through March 2005 shall be announced along with the prices announced in § 1000.53(b);

(b) Multiply the pounds of skim milk and butterfat overage assigned to each class pursuant to § 1000.44(a)(11) by the respective skim milk and butterfat prices applicable at the location of the pool plant;

(c) Multiply the difference between the Class IV price for the preceding month and the current month's Class I, II, or III price, as the case may be, by the hundredweight of skim milk and

69

butterfat subtracted from Class I, II, or III, respectively, pursuant to §1000.44(a)(7) and the corresponding step of §1000.44(b);

(d) Multiply the difference between the Class I price applicable at the location of the pool plant and the Class IV price by the hundredweight of skim milk and butterfat assigned to Class I pursuant to §1000.43(d) and the hundredweight of skim milk and butterfat subtracted from Class I pursuant to §1000.44(a)(3)(i) through (vi) and the corresponding step of §1000.44(b), excluding receipts of bulk fluid cream products from a plant regulated under other Federal orders and bulk concentrated fluid milk products from pool plants, plants regulated under other Federal orders, and unregulated supply plants;

(e) Multiply the Class I skim milk and Class I butterfat prices applicable at the location of the nearest unregulated supply plants from which an equivalent volume was received by the pounds of skim milk and butterfat in receipts of concentrated fluid milk products assigned to Class I pursuant to §1000.43(d) and §1000.44(a)(3)(i) and the corresponding step of §1000.44(b) and the pounds of skim milk and butterfat subtracted from Class I pursuant to §1000.44(a)(8) and the corresponding step of §1000.44(b), excluding such skim milk and butterfat in receipts of fluid milk products from an unregulated supply plant to the extent that an equivalent amount of skim milk or butterfat disposed of to such plant by handlers fully regulated under any Federal milk order is classified and priced as Class I milk and is not used as an offset for any other payment obligation under any order.

(f) For reconstituted milk made from receipts of nonfluid milk products, multiply $1.00 (but not more than the difference between the Class I price applicable at the location of the pool plant and the Class IV price) by the hundredweight of skim milk and butterfat contained in receipts of nonfluid milk products that are allocated to Class I use pursuant to §1000.43(d).

(g) For the months of January 2005 through March 2005 for handlers who have submitted proof satisfactory to the market administrator to determine eligibility for reimbursement of transportation costs, subtract an amount equal to:

(1) The cost of transportation on loads of producer milk delivered or rerouted to a pool distributing plant which were delivered as a result of hurricanes Charley, Frances, Ivan, and Jeanne;

(2) The cost of transportation on loads of producer milk delivered or rerouted to a pool supply plant that was then transferred to a pool distributing plant which were delivered as a result of hurricanes Charley, Frances, Ivan, and Jeanne;

(3) The cost of transportation on loads of bulk milk delivered or rerouted to a pool distributing plant from a pool supply plant which were delivered as a result of hurricanes Charley, Frances, Ivan, and Jeanne;

(4) The cost of transportation on loads of bulk milk delivered or rerouted to a pool distributing plant from another order plant which were delivered as a result of hurricanes Charley, Frances, Ivan, and Jeanne; and

(5) The cost of transportation on loads of bulk milk transferred or diverted to a plant regulated under another Federal order or to other nonpool plants which were delivered as a result of hurricanes Charley, Frances, Ivan, and Jeanne.

(6) The total amount of payment to all handlers under this section shall be limited for each month to an amount determined by multiplying the total Class I producer milk for all handlers pursuant to §1000.44(c) times $0.04 per hundredweight.

(7) If the cost of transportation computed pursuant to paragraphs (g)(1) through (5) of this section exceeds the amount computed pursuant to paragraph (g)(6) of this section, the market administrator shall prorate such payments to each handler based on the handler's proportion of transportation costs submitted pursuant to paragraphs (g)(1) through (5) of this section. Transportation costs submitted pursuant to paragraphs (g)(1) through (5) of this section which are not paid as a result of such a proration shall be included in each subsequent month's

transportation costs submitted pursuant to paragraphs (g)(1) through (5) of this section until paid, or until the time period for such payments is concluded.

(8) The reimbursement of transportation costs pursuant to this section shall be the actual demonstrated cost of such transportation of bulk milk delivered or rerouted as described in paragraphs (g)(1) through (5) of this section, or the miles of transportation on loads of bulk milk delivered or rerouted as described in paragraphs (g)(1) through (5) of this section multiplied by $2.25 per loaded mile, whichever is less.

(9) For each handler, the reimbursement of transportation costs pursuant to paragraph (g) of this section for bulk milk delivered or rerouted as described in paragraphs (g)(1) through (5) of this section shall be reduced by the amount of payments received for such milk movements from the transportation credit balancing fund pursuant to §1005.82.

[64 FR 47960, Sept. 1, 1999, as amended at 65 FR 82835, Dec. 28, 2000; 69 FR 71699, Dec. 10, 2004]

§1005.61 Computation of uniform prices.

On or before the 11th day of each month, the market administrator shall compute a uniform butterfat price, a uniform skim milk price, and a uniform price for producer milk receipts reported for the prior month. The report of any handler who has not made payments required pursuant to §1005.71 for the preceding month shall not be included in the computation of these prices, and such handler's report shall not be included in the computation for succeeding months until the handler has made full payment of outstanding monthly obligations.

(a) *Uniform butterfat price.* The uniform butterfat price per pound, rounded to the nearest one-hundredth cent, shall be computed by:

(1) Multiplying the pounds of butterfat in producer milk allocated to each class pursuant to §1000.44(b) by the respective class butterfat prices;

(2) Adding the butterfat value calculated in §1005.60(e) for other source milk allocated to Class I pursuant to

§1000.43(d) and the steps of §1000.44(b) that correspond to §1000.44(a)(3)(i) and §1000.44(a)(8) by the Class I price; and

(3) Dividing the sum of paragraphs (a)(1) and (a)(2) of this section by the sum of the pounds of butterfat in producer milk and other source milk used to calculate the values in paragraphs (a)(1) and (a)(2) of this section.

(b) *Uniform skim milk price.* The uniform skim milk price per hundredweight, rounded to the nearest cent, shall be computed as follows:

(1) Combine into one total the values computed pursuant to §1005.60 for all handlers;

(2) Add an amount equal to the minus location adjustments and subtract an amount equal to the plus location adjustments computed pursuant to §1005.75;

(3) Add an amount equal to not less than one-half of the unobligated balance in the producer-settlement fund;

(4) Subtract the value of the total pounds of butterfat for all handlers. The butterfat value shall be computed by multiplying the sum of the pounds of butterfat in producer milk and other source milk used to calculate the values in paragraphs (a)(1) and (a)(2) of this section by the butterfat price computed in paragraph (a) of this section;

(5) Divide the resulting amount by the sum of the following for all handlers included in these computations:

(i) The total skim pounds of producer milk; and

(ii) The total skim pounds for which a value is computed pursuant to §1005.60(e); and

(6) Subtract not less than 4 cents and not more than 5 cents.

(c) *Uniform price.* The uniform price per hundredweight, rounded to the nearest cent, shall be the sum of the following:

(1) Multiply the uniform butterfat price for the month pursuant to paragraph (a) of this section times 3.5 pounds of butterfat; and

(2) Multiply the uniform skim milk price for the month pursuant to paragraph (b) of this section times 96.5 pounds of skim milk.

[64 FR 47960, Sept. 1, 1999, as amended at 65 FR 82835, Dec. 28, 2000]

§ 1005.62 Announcement of uniform prices.

On or before the 11th day after the end of the month, the market administrator shall announce the uniform prices for the month computed pursuant to § 1005.61.

PAYMENTS FOR MILK

§ 1005.70 Producer-settlement fund.

See § 1000.70.

§ 1005.71 Payments to the producer-settlement fund.

Each handler shall make a payment to the producer-settlement fund in a manner that provides receipt of the funds by the market administrator no later than the 12th day after the end of the month (except as provided in § 1000.90). Payment shall be the amount, if any, by which the amount specified in paragraph (a) of this section exceeds the amount specified in paragraph (b) of this section:

(a) The total value of milk of the handler for the month as determined pursuant to § 1005.60; and

(b) The sum of the value at the uniform prices for skim milk and butterfat, adjusted for plant location, of the handler's receipts of producer milk; and the value at the uniform price, as adjusted pursuant to § 1005.75, applicable at the location of the plant from which received of other source milk for which a value is computed pursuant to § 1005.60(e).

§ 1005.72 Payments from the producer-settlement fund.

No later than one day after the date of payment receipt required under § 1005.71, the market administrator shall pay to each handler the amount, if any, by which the amount computed pursuant to § 1005.71(b) exceeds the amount computed pursuant to § 1005.71(a). If, at such time, the balance in the producer-settlement fund is insufficient to make all payments pursuant to this section, the market administrator shall reduce uniformly such payments and shall complete the payments as soon as the funds are available.

§ 1005.73 Payments to producers and to cooperative associations.

(a) Each handler that is not paying a cooperative association for producer milk shall pay each producer as follows:

(1) *Partial payment.* For each producer who has not discontinued shipments as of the 23rd day of the month, payment shall be made so that it is received by the producer on or before the 26th day of the month (except as provided in § 1000.90) for milk received during the first 15 days of the month at not less than 90 percent of the preceding month's uniform price, adjusted for plant location pursuant to § 1005.75 and proper deductions authorized in writing by the producer.

(2) *Final payment.* For milk received during the month, a payment computed as follows shall be made so that it is received by each producer one day after the payment date required in § 1005.72:

(i) Multiply the hundredweight of producer skim milk received times the uniform skim milk price for the month;

(ii) Multiply the pounds of butterfat received times the uniform butterfat price for the month;

(iii) Multiply the hundredweight of producer milk received times the plant location adjustment pursuant to § 1005.75; and

(iv) Add the amounts computed in paragraph (a)(2)(i), (ii), and (iii) of this section, and from that sum:

(A) Subtract the partial payment made pursuant to paragraph (a)(1) of this section;

(B) Subtract the deduction for marketing services pursuant to § 1000.86;

(C) Add or subtract for errors made in previous payments to the producer; and

(D) Subtract proper deductions authorized in writing by the producer.

(b) One day before partial and final payments are due pursuant to paragraph (a) of this section, each handler shall pay a cooperative association for milk received as follows:

(1) *Partial payment to a cooperative association for bulk milk received directly from producers' farms.* For bulk milk (including the milk of producers who are not members of such association

and who the market administrator determines have authorized the cooperative association to collect payment for their milk) received during the first 15 days of the month from a cooperative association in any capacity, except as the operator of a pool plant, the payment shall be equal to the hundredweight of milk received multiplied by 90 percent of the preceding month's uniform price, adjusted for plant location pursuant to §1005.75.

(2) *Partial payment to a cooperative association for milk transferred from its pool plant.* For bulk fluid milk products and bulk fluid cream products received during the first 15 days of the month from a cooperative association in its capacity as the operator of a pool plant, the partial payment shall be at the pool plant operator's estimated use value of the milk using the most recent class prices available for skim milk and butterfat at the receiving plant's location.

(3) *Final payment to a cooperative association for milk transferred from its pool plant.* For bulk fluid milk products and bulk fluid cream products received during the month from a cooperative association in its capacity as the operator of a pool plant, the final payment shall be the classified value of such milk as determined by multiplying the pounds of skim milk and butterfat assigned to each class pursuant to §1000.44 by the class prices for the month at the receiving plant's location, and subtracting from this sum the partial payment made pursuant to paragraph (b)(2) of this section.

(4) *Final payment to a cooperative association for bulk milk received directly from producers' farms.* For bulk milk received from a cooperative association during the month, including the milk of producers who are not members of such association and who the market administrator determines have authorized the cooperative association to collect payment for their milk, the final payment for such milk shall be an amount equal to the sum of the individual payments otherwise payable for such milk pursuant to paragraph (a)(2) of this section.

(c) If a handler has not received full payment from the market administrator pursuant to §1005.72 by the payment date specified in paragraph (a) or (b) of this section, the handler may reduce payments pursuant to paragraphs (a) and (b) of this section, but by not more than the amount of the underpayment. The payments shall be completed on the next scheduled payment date after receipt of the balance due from the market administrator.

(d) If a handler claims that a required payment to a producer cannot be made because the producer is deceased or cannot be located, or because the cooperative association or its lawful successor or assignee is no longer in existence, the payment shall be made to the producer-settlement fund, and in the event that the handler subsequently locates and pays the producer or a lawful claimant, or in the event that the handler no longer exists and a lawful claim is later established, the market administrator shall make the required payment from the producer-settlement fund to the handler or to the lawful claimant as the case may be.

(e) In making payments to producers pursuant to this section, each pool plant operator shall furnish each producer, except a producer whose milk was received from a cooperative association described in §1000.9(a) or (c), a supporting statement in such form that it may be retained by the recipient which shall show:

(1) The name, address, Grade A identifier assigned by a duly constituted regulatory agency, and the payroll number of the producer;

(2) The month and dates that milk was received from the producer, including the daily and total pounds of milk received;

(3) The total pounds of butterfat in the producer's milk;

(4) The minimum rate or rates at which payment to the producer is required pursuant to the order in this part;

(5) The rate used in making payment if the rate is other than the applicable minimum rate;

(6) The amount, or rate per hundredweight, and nature of each deduction claimed by the handler; and

(7) The net amount of payment to the producer or cooperative association.

[64 FR 47960, Sept. 1, 1999, as amended at 65 FR 32010, May 22, 2000]

§ 1005.74 [Reserved]

§ 1005.75 Plant location adjustments for producer milk and nonpool milk.

For purposes of making payments for producer milk and nonpool milk, a plant location adjustment shall be determined by subtracting the Class I price specified in § 1005.51 from the Class I price at the plant's location. The difference, plus or minus as the case may be, shall be used to adjust the payments required pursuant to §§ 1005.73 and 1000.76.

§ 1005.76 Payments by a handler operating a partially regulated distributing plant.

See § 1000.76.

§ 1005.77 Adjustment of accounts.

See § 1000.77.

§ 1005.78 Charges on overdue accounts.

See § 1000.78.

MARKETWIDE SERVICE PAYMENTS

§ 1005.80 Transportation credit balancing fund.

The market administrator shall maintain a separate fund known as the *Transportation Credit Balancing Fund* into which shall be deposited the payments made by handlers pursuant to § 1005.81 and out of which shall be made the payments due handlers pursuant to § 1005.82. Payments due a handler shall be offset against payments due from the handler.

§ 1005.81 Payments to the transportation credit balancing fund.

(a) On or before the 12th day after the end of the month (except as provided in § 1000.90 of this chapter), each handler operating a pool plant and each handler specified in § 1000.9(c) shall pay to the market administrator a transportation credit balancing fund assessment determined by multiplying the pounds of Class I producer milk assigned pursuant to § 1005.44 by $0.15 per hundredweight or such lesser amount as the market administrator deems necessary to maintain a balance in the fund equal to the total transportation credits disbursed during the prior June-February period. In the event that during any month of the June-February period the fund balance is insufficient to cover the amount of credits that are due, the assessment should be based upon the amount of credits that would had been disbursed had the fund balance been sufficient.

(b) The market administrator shall announce publicly on or before the 23rd day of the month (except as provided in § 1000.90) the assessment pursuant to paragraph (a) of this section for the following month.

[79 FR 25005, May 2, 2014; 79 FR 26591, May 9, 2014]

§ 1005.82 Payments from the transportation credit balancing fund.

(a) Payments from the transportation credit balancing fund to handlers and cooperative associations requesting transportation credits shall be made as follows:

(1) On or before the 13th day (except as provided in § 1000.90) after the end of each of the months of January, February and July through December and any other month in which transportation credits are in effect pursuant to paragraph (b) of this section, the market administrator shall pay to each handler that received, and reported pursuant to § 1005.30(a)(5), bulk milk transferred from a plant fully regulated under another Federal order as described in paragraph (c)(1) of this section or that received, and reported pursuant to § 1005.30(a)(6), milk directly from producers'' farms as specified in paragraph (c)(2) of this section, a preliminary amount determined pursuant to paragraph (d) of this section to the extent that funds are available in the transportation credit balancing fund. If an insufficient balance exists to pay all of the credits computed pursuant to this section, the market administrator shall distribute the balance available in the transportation credit balancing fund by reducing payments prorata using the percentage derived by dividing the balance in the fund by the total credits that are due for the month. The amount of credits resulting from this initial proration shall be subject to audit adjustment pursuant to paragraph (a)(2) of this section.

(2) The market administrator shall accept adjusted requests for transportation credits on or before the 20th day of the month following the month for which such credits were requested pursuant to §1005.32(a). After such date, a preliminary audit will be conducted by the market administrator, who will recalculate any necessary proration of transportation credit payments for the preceding month pursuant to paragraph (a) of this section. Handlers will be promptly notified of an overpayment of credits based upon this final computation and remedial payments to or from the transportation credit balancing fund will be made on or before the next payment date for the following month.

(3) Transportation credits paid pursuant to paragraphs (a)(1) and (2) of this section shall be subject to final verification by the market administrator pursuant to §1000.77. Adjusted payments to or from the transportation credit balancing fund will remain subject to the final proration established pursuant to paragraph (a)(2) of this section.

(4) In the event that a qualified cooperative association is the responsible party for whose account such milk is received and written documentation of this fact is provided to the market administrator pursuant to §1005.30(c)(3) prior to the date payment is due, the transportation credits for such milk computed pursuant to this section shall be made to such cooperative association rather than to the operator of the pool plant at which the milk was received.

(b) The Market Administrator may extend the period during which transportation credits are in effect (i.e., the transportation credit period) to the month of June if a written request to do so is received 15 days prior to the beginning of the month for which the request is made and, after conducting an independent investigation, finds that such extension is necessary to assure the market of an adequate supply of milk for fluid use. Before making such a finding, the Market Administrator shall notify the Deputy Administrator of the Dairy Programs and all handlers in the market that an extension is being considered and invite written data, views, and arguments. Any decision to extend the transportation credit period must be issued in writing prior to the first day of the month for which the extension is to be effective.

(c) Transportation credits shall apply to the following milk:

(1) Bulk milk received at a pool distributing plant from a plant regulated under another Federal order, except Federal Order 1007; and

(2) Bulk milk received directly from the farms of dairy farmers at pool distributing plants subject to the following conditions:

(i) The dairy farmer was not a "producer" under this order for more than 45 days during the immediately preceding months of March through May, or not more than 50 percent of the production of the dairy farmer during those 3 months, in aggregate, was received as producer milk under this order during those 3 months; and

(ii) The farm on which the milk was produced is not located within the specified marketing area of the order in this part or the marketing area of Federal Order 1007 (7 CFR part 1007).

(iii) The market administrator may increase or decrease the milk production standard specified in paragraph (c)(2)(i) of this section if the market administrator finds that such revision is necessary to assure orderly marketing and efficient handling of milk in the marketing area. Before making such a finding, the market administrator shall investigate the need for the revision either on the market administrator's own initiative or at the request of interested persons. If the investigation shows that a revision might be appropriate, the market administrator shall issue a notice stating that the revision is being considered and inviting written data, views, and arguments. Any decision to revise an applicable percentage must be issued in writing at least one day before the effective date.

(d) Transportation credits shall be computed as follows:

(1) The market administrator shall subtract from the pounds of milk described in paragraphs (c)(1) and (2) of this section the pounds of bulk milk

transferred from the pool plant receiving the supplemental milk if milk was transferred to a nonpool plant on the same calendar day that the supplemental milk was received. For this purpose, the transferred milk shall be subtracted from the most distant load of supplemental milk received, and then in sequence with the next most distant load until all of the transfers have been offset.

(2) With respect to the pounds of milk described in paragraph (c)(1) of this section that remain after the computations described in paragraph (d)(1) of this section, the market administrator shall:

(i) Determine the shortest hard-surface highway distance between the shipping plant and the receiving plant;

(ii) Multiply the number of miles so determined by the mileage rate for the month computed pursuant to §1005.83(a)(6);

(iii) Subtract the applicable Class I price specified in §1000.50(a) for the county in which the shipping plant is located from the Class I price applicable for the county in which the receiving plant is located;

(iv) Subtract any positive difference computed in paragraph (d)(2)(iii) of this section from the amount computed in paragraph (d)(2)(ii) of this section; and

(v) Multiply the remainder computed in paragraph (d)(2)(iv) of this section by the hundredweight of milk described in paragraph (d)(2) of this section.

(3) For the remaining milk described in paragraph (c)(2) of this section after computations described in paragraph (d)(1) of this section, the market administrator shall:

(i) Determine an origination point for each load of milk by locating the nearest city to the last producer's farm from which milk was picked up for delivery to the receiving pool plant;

(ii) Determine the shortest hard-surface highway distance between the receiving pool plant and the origination point;

(iii) Subtract 85 miles from the mileage so determined;

(iv) Multiply the remaining miles so computed by the mileage rate for the month computed pursuant to §1005.83(a)(6);

(v) Subtract the Class I price specified in §1000.50(a) applicable for the county in which the origination point is located from the Class I price applicable at the receiving pool plant's location;

(vi) Subtract any positive difference computed in paragraph (d)(3)(v) of this section from the amount computed in paragraph (d)(3)(iv) of this section; and

(vii) Multiply the remainder computed in paragraph (d)(3)(vi) of this section by the hundredweight of milk described in paragraph (d)(3) of this section.

[64 FR 47960, Sept. 1, 1999, as amended at 70 FR 59223, Oct. 12, 2005; 71 FR 62379, Oct. 25, 2006; 73 FR 14161, Mar. 17, 2008; 79 FR 25005, May 2, 2014; 79 FR 26591, May 9, 2014]

§ 1005.83   Mileage rate for the transportation credit balancing fund.

(a) The market administrator shall compute a mileage rate each month as follows:

(1) Compute the simple average rounded to three decimal places for the most recent four (4) weeks of the Diesel Price per Gallon as reported by the Energy Information Administration of the United States Department of Energy for the Lower Atlantic and Gulf Coast Districts combined.

(2) From the result in paragraph (a)(1) in this section subtract $1.42 per gallon;

(3) Divide the result in paragraph (a)(2) of this section by 5.5, and round down to three decimal places to compute the fuel cost adjustment factor;

(4) Add the result in paragraph (a)(3) of this section to $1.91;

(5) Divide the result in paragraph (a)(4) of this section by 480;

(6) Round the result in paragraph (a)(5) of this section down to five decimal places to compute the mileage rate.

(b) The market administrator shall announce publicly on or before the 23rd day of the month (except as provided in §1000.90 of this chapter) the mileage rate pursuant to paragraph (a) of this section for the following month.

[79 FR 25005, May 2, 2014; 79 FR 26591, May 9, 2014]

ADMINISTRATIVE ASSESSMENT AND
MARKETING SERVICE DEDUCTION

§ 1005.85 Assessment for order administration.

On or before the payment receipt date specified under § 1005.71, each handler shall pay to the market administrator its *pro rata* share of the expense of administration to the order at a rate specified by the market administrator that is no more than $.08 per hundredweight with respect to:

(a) Receipts of producer milk (including the handler's own production) other than such receipts by a handler described in § 1000.9(c) of this chapter that were delivered to pool plants of other handlers;

(b) Receipts from a handler described in § 1000.9(c) of this chapter;

(c) Receipts of concentrated fluid milk products from unregulated supply plants and receipts of nonfluid milk products assigned to Class I use pursuant to § 1000.43(d) of this chapter and other source milk allocated to Class I pursuant to § 1000.44(a)(3) and (8) of this chapter and the corresponding steps of § 1000.44(b) of this chapter, except other source milk that is excluded from the computations pursuant to § 1005.60(d) and (e); and

(d) Route disposition in the marketing area from a partially regulated distributing plant that exceeds the skim milk and butterfat subtracted pursuant to § 1000.76(a)(1)(i) and (ii) of this chapter.

[79 FR 25002, May 2, 2014; 79 FR 26591, May 9, 2014]

§ 1005.86 Deduction for marketing services.

See § 1000.86.

PART 1006—MILK IN THE FLORIDA MARKETING AREA

Subpart A—Order Regulating Handling

GENERAL PROVISIONS

ADMINISTRATIVE ASSESSMENT AND MARKETING
SERVICE DEDUCTION

1006.85　Assessment for order administration.

1006.86　Deduction for marketing services.

AUTHORITY: 7 U.S.C. 601–674, and 7253.

SOURCE: 64 FR 47966, Sept. 1, 1999, unless otherwise noted.

Subpart A—Order Regulating Handling

GENERAL PROVISIONS

§ 1006.1　General provisions.

The terms, definitions, and provisions in part 1000 of this chapter apply to this part 1006. In this part 1006, all references to sections in part 1000 refer to part 1000 of this chapter.

DEFINITIONS

§ 1006.2　Florida marketing area.

The marketing area means all the territory within the State of Florida, except the counties of Escambia, Okaloosa, Santa Rosa, and Walton, including all piers, docks and wharves connected therewith and all craft moored thereat, and all territory occupied by government (municipal, State or Federal) reservations, installations, institutions, or other similar establishments if any part thereof is within any of the listed states or political subdivisions.

§ 1006.3　Route disposition.

See § 1000.3.

§ 1006.4　Plant.

See § 1000.4.

§ 1006.5　Distributing plant.

See § 1000.5.

§ 1006.6　Supply plant.

See § 1000.6.

§ 1006.7　Pool plant.

Pool plant means a plant specified in paragraphs (a) through (d) of this section, a unit of plants as specified in paragraph (e) of this section, or a plant specified in paragraph (h) of this section, but excluding a plant specified in paragraph (g) of this section. The pool-

ing standards described in paragraphs (c) and (d) of this section are subject to modification pursuant to paragraph (f) of this section:

(a) A distributing plant, other than a plant qualified as a pool plant pursuant to paragraph (b) of this section or §_____.7(b) of any other Federal milk order, from which during the month 50 percent or more of the fluid milk products physically received at such plant (excluding concentrated milk received from another plant by agreement for other than Class I use) are disposed of as route disposition or are transferred in the form of packaged fluid milk products to other distributing plants. At least 25 percent of such route disposition and transfers must be to outlets in the marketing area.

(b) Any distributing plant located in the marketing area which during the month processed at least 50 percent of the total quantity of fluid milk products physically received at the plant (excluding concentrated milk received from another plant by agreement for other than Class I use) into ultra-pasteurized or aseptically-processed fluid milk products.

(c) A supply plant from which 60 percent or more of the total quantity of milk that is physically received during the month from dairy farmers and handlers described in § 1000.9(c), including milk that is diverted from the plant, is transferred to pool distributing plants. Concentrated milk transferred from the supply plant to a distributing plant for an agreed-upon use other than Class I shall be excluded from the supply plant's shipments in computing the plant's shipping percentage.

(d) A plant located within the marketing area that is operated by a cooperative association if pool plant status under this paragraph is requested for such plant by the cooperative association and during the month 60 percent of the producer milk of members of such cooperative association is delivered directly from farms to pool distributing plants or is transferred to such plants as a fluid milk product (excluding concentrated milk transferred to a distributing plant for an agreed-upon use other than Class I) from the cooperative's plant.

(e) Two or more plants operated by the same handler and that are located within the marketing area may qualify for pool status as a unit by meeting the total and in-area route disposition requirements specified in paragraph (a) of this section and the following additional requirements:

(1) At least one of the plants in the unit must qualify as a pool plant pursuant to paragraph (a) of this section;

(2) Other plants in the unit must process only Class I or Class II products and must be located in a pricing zone providing the same or a lower Class I price than the price applicable at the distributing plant included in the unit pursuant to paragraph (e)(1) of this section; and

(3) A written request to form a unit, or to add or remove plants from a unit, must be filed with the market administrator prior to the first day of the month for which it is to be effective.

(f) The applicable shipping percentages of paragraphs (c) and (d) of this section may be increased or decreased by the market administrator if the market administrator finds that such adjustment is necessary to encourage needed shipments or to prevent uneconomic shipments. Before making such a finding, the market administrator shall investigate the need for adjustment either on the market administrator's own initiative or at the request of interested parties if the request is made in writing at least 15 days prior to the date for which the requested revision is desired effective. If the investigation shows that an adjustment of the shipping percentages might be appropriate, the market administrator shall issue a notice stating that an adjustment is being considered and invite data, views and arguments. Any decision to revise an applicable shipping percentage must be issued in writing at least one day before the effective date.

(g) The term pool plant shall not apply to the following plants:

(1) A producer-handler plant;

(2) An exempt plant as defined in § 1000.8(e);

(3) A plant qualified pursuant to paragraph (a) of this section which is not located within any Federal order marketing area, meets the pooling requirements of another Federal order, and has had greater route disposition in such other Federal order marketing area for 3 consecutive months;

(4) A plant qualified pursuant to paragraph (a) of this section which is located in another Federal order marketing area, meets the pooling standards of the other Federal order, and has not had a majority of its route disposition in this marketing area for 3 consecutive months or is locked into pool status under such other Federal order without regard to its route disposition in any other Federal order marketing area; and

(5) A plant qualified pursuant to paragraph (c) of this section which also meets the pooling requirements of another Federal order and from which greater qualifying shipments are made to plants regulated under such other order than are made to plants regulated under the order in this part, or such plant has automatic pooling status under such other order.

(h) Any distributing plant, located within the marketing area as described on May 1, 2006, in § 1006.2;

(1) From which there is route disposition and/or transfers of packaged fluid milk products in any non-Federally regulated marketing area(s) located within one or more States that require handlers to pay minimum prices for raw milk provided that 25 percent or more of the total quantity of fluid milk products physically received at such plant (excluding concentrated milk received from another plant by agreement for other than Class I use) is disposed of as route disposition and/or is transferred in the form of packaged fluid milk products to other plants. At least 25 percent of such route disposition and/or transfers, in aggregate, are in any non-Federally regulated marketing area(s) located within one or more States that require handlers to pay minimum prices for raw milk. Subject to the following exclusions:

(i) The plant is described in § 1006.7(a), (b), or (e);

(ii) The plant is subject to the pricing provisions of a State-operated milk pricing plan which provides for the payment of minimum class prices for raw milk;

(iii) The plant is described in § 1000.8(a) or (e); or

(iv) A producer-handler described in § 1006.10 with less than three million pounds during the month of route disposition and/or transfers of packaged fluid milk products to other plants.

(2) [Reserved]

[64 FR 47966, Sept. 1, 1999, as amended at 71 FR 25498, May 1, 2006; 71 FR 28249, May 16, 2006]

§ 1006.8 Nonpool plant.

See § 1000.8.

§ 1006.9 Handler.

See § 1000.9.

§ 1006.10 Producer-handler.

Producer-handler means a person who:

(a) Operates a dairy farm and a distributing plant from which there is route disposition in the marketing area, and from which total route disposition and packaged sales of fluid milk products to other plants during the month does not exceed 3 million pounds;

(b) Receives no fluid milk products, and acquires no fluid milk products for route disposition, from sources other than own farm production;

(c) Disposes of no other source milk as Class I milk except by increasing the nonfat milk solids content of the fluid milk products received from own farm production; and

(d) Provides proof satisfactory to the market administrator that the care and management of the dairy animals and other resources necessary to produce all Class I milk handled, and the processing and packaging operations, are the producer-handler's own enterprise and are operated at the producer-handler's own risk.

(e) Any producer-handler with Class I route dispositions and/or transfers of packaged fluid milk products in the marketing area described in § 1131.2 of this chapter shall be subject to payments into the Order 1131 producer settlement fund on such dispositions pursuant to § 1000.76(a) and payments into the Order 1131 administrative fund provided such dispositions are less than three million pounds in the current month and such producer-handler had total Class I route dispositions and/or transfers of packaged fluid milk products from own farm production of three million pounds or more the previous month. If the producer-handler has Class I route dispositions and/or transfers of packaged fluid milk products into the marketing area described in § 1131.2 of this chapter of three million pounds or more during the current month, such producer-handler shall be subject to the provisions described in § 1131.7 of this chapter or § 1000.76(a).

[64 FR 47966, Sept. 1, 1999, as amended at 71 FR 25498, May 1, 2006; 75 FR 21160, Apr. 23, 2010]

§ 1006.11 [Reserved]

§ 1006.12 Producer.

(a) Except as provided in paragraph (b) of this section, *producer* means any person who produces milk approved by a duly constituted regulatory agency for fluid consumption as Grade A milk and whose milk (or components of milk) is:

(1) Received at a pool plant directly from the producer or diverted by the plant operator in accordance with § 1006.13; or

(2) Received by a handler described in § 1000.9(c).

(b) Producer shall not include:

(1) A producer-handler as defined in any Federal order;

(2) A dairy farmer whose milk is received at an exempt plant, excluding producer milk diverted to the exempt plant pursuant to § 1006.13(d);

(3) A dairy farmer whose milk is received by diversion at a pool plant from a handler regulated under another Federal order if the other Federal order designates the dairy farmer as a producer under that order and that milk is allocated by request to a utilization other than Class I; and

(4) A dairy farmer whose milk is reported as diverted to a plant fully regulated under another Federal order with respect to that portion of the milk so diverted that is assigned to Class I under the provisions of such other order.

§ 1006.13 Producer milk.

Producer milk means the skim milk (or the skim equivalent of components of skim milk) and butterfat contained in milk of a producer that is:

(a) Received by the operator of a pool plant directly from a producer or a handler described in §1000.9(c). All milk received pursuant to this paragraph shall be priced at the location of the plant where it is first physically received;

(b) Received by a handler described in §1000.9(c) in excess of the quantity delivered to pool plants;

(c) Diverted by a pool plant operator to another pool plant. Milk so diverted shall be priced at the location of the plant to which diverted; or

(d) Diverted by the operator of a pool plant or a handler described in §1000.9(c) to a nonpool plant, subject to the following conditions:

(1) In any month, not less than 10 days' production of the producer whose milk is diverted is physically received at a pool plant during the month;

(2) The total quantity of milk so diverted during the month by a cooperative association shall not exceed 20 percent during the months of July through November, 25 percent during the months of December through February, and 40 percent during all other months, of the producer milk that the cooperative association caused to be delivered to, and physically received at, pool plants during the month;

(3) The operator of a pool plant that is not a cooperative association may divert any milk that is not under the control of a cooperative association that diverts milk during the month pursuant to paragraph (d) of this section. The total quantity of milk so diverted during the month shall not exceed 20 percent during the months of July through November, 25 percent during the months of December through February, and 40 percent during all other months, of the producer milk physically received at such plant (or such unit of plants in the case of plants that pool as a unit pursuant to §1006.7(d)) during the month, excluding the quantity of producer milk received from a handler described in §1000.9(c);

(4) Any milk diverted in excess of the limits prescribed in paragraphs (d) (3) and (4) of this section shall not be producer milk. If the diverting handler or cooperative association fails to designate the dairy farmers' deliveries that will not be producer milk, no milk

diverted by the handler or cooperative association shall be producer milk;

(5) Diverted milk shall be priced at the location of the plant to which diverted; and

(6) The delivery day requirements and the diversion percentages in paragraphs (d) (1) through (3) of this section may be increased or decreased by the market administrator if the market administrator finds that such revision is necessary to assure orderly marketing and efficient handling of milk in the marketing area. Before making such a finding, the market administrator shall investigate the need for the revision either on the market administrator's own initiative or at the request of interested persons. If the investigation shows that a revision might be appropriate, the market administrator shall issue a notice stating that the revision is being considered and inviting written data, views, and arguments. Any decision to revise an applicable percentage must be issued in writing at least one day before the effective date.

§1006.14 Other source milk.

See §1000.14.

§1006.15 Fluid milk product.

See §1000.15.

§1006.16 Fluid cream product.

See §1000.16.

§1006.17 [Reserved]

§1006.18 Cooperative association.

See §1000.18.

§1006.19 Commercial food processing establishment.

See §1000.19.

HANDLER REPORTS

§1006.30 Reports of receipts and utilization.

Each handler shall report monthly so that the market administrator's office receives the report on or before the 7th day after the end of the month, in the detail and on prescribed forms, as follows:

(a) With respect to each of its pool plants, the quantities of skim milk and butterfat contained in or represented by:

(1) Receipts of producer milk, including producer milk diverted by the reporting handler, from sources other than handlers described in § 1000.9(c);

(2) Receipts of milk from handlers described in § 1000.9(c);

(3) Receipts of fluid milk products and bulk fluid cream products from other pool plants;

(4) Receipts of other source milk;

(5) Inventories at the beginning and end of the month of fluid milk products and bulk fluid cream products; and

(6) The utilization or disposition of all milk and milk products required to be reported pursuant to this paragraph.

(b) Each handler operating a partially regulated distributing plant shall report with respect to such plant in the same manner as prescribed for reports required by paragraph (a) of this section. Receipts of milk that would have been producer milk if the plant had been fully regulated shall be reported in lieu of producer milk. The report shall show also the quantity of any reconstituted skim milk in route disposition in the marketing area.

(c) Each handler described in § 1000.9(c) shall report:

(1) The quantities of all skim milk and butterfat contained in receipts of milk from producers; and

(2) The utilization or disposition of all such receipts.

(d) Each handler not specified in paragraphs (a) through (c) of this section shall report with respect to its receipts and utilization of milk and milk products in such manner as the market administrator may prescribe.

§ 1006.31   Payroll reports.

(a) On or before the 20th day after the end of each month, each handler that operates a pool plant pursuant to § 1006.7 and each handler described in § 1000.9(c) shall report to the market administrator its producer payroll for the month, in detail prescribed by the market administrator, showing for each producer the information specified in § 1006.73(e).

(b) Each handler operating a partially regulated distributing plant who elects to make payment pursuant to § 1000.76(b) shall report for each dairy farmer who would have been a producer if the plant had been fully regulated in the same manner as prescribed for reports required by paragraph (a) of this section.

§ 1006.32   Other reports.

In addition to the reports required pursuant to §§ 1006.30 and 1006.31, each handler shall report any information the market administrator deems necessary to verify or establish each handler's obligation under the order.

CLASSIFICATION OF MILK

§ 1006.40   Classes of utilization.

See § 1000.40.

§ 1006.41   [Reserved]

§ 1006.42   Classification of transfers and diversions.

See § 1000.42.

§ 1006.43   General classification rules.

See § 1000.43.

§ 1006.44   Classification of producer milk.

See § 1000.44.

§ 1006.45   Market administrator's reports and announcements concerning classification.

See § 1000.45.

CLASS PRICES

§ 1006.50   Class prices, component prices, and advanced pricing factors.

See § 1000.50.

§ 1006.51   Class I differential, adjustments to Class I prices, and Class I price.

(a) The Class I differential shall be the differential established for Hillsborough County, Florida, which is reported in § 1000.52. The Class I price shall be the price computed pursuant to § 1006.50(a) for Hillsborough County, Florida.

(b) *Adjustment to Class I prices.* Class I prices shall be established pursuant to § 1000.50(a), (b) and (c) using the following adjustments:

State	County/parish	FIPS	Class I price adjustment	State	County/parish	FIPS	Class I price adjustment
FL	ALACHUA	12001	1.30	FL	LAFAYETTE	12067	1.30
FL	BAKER	12003	1.30	FL	LAKE	12069	1.40
FL	BAY	12005	0.60	FL	LEE	12071	1.70
FL	BRADFORD	12007	1.30	FL	LEON	12073	0.90
FL	BREVARD	12009	1.40	FL	LEVY	12075	1.00
FL	BROWARD	12011	1.70	FL	LIBERTY	12077	0.90
FL	CALHOUN	12013	0.60	FL	MADISON	12079	1.30
FL	CHARLOTTE	12015	1.50	FL	MANATEE	12081	1.80
FL	CITRUS	12017	1.40	FL	MARION	12083	1.00
FL	CLAY	12019	1.30	FL	MARTIN	12085	1.50
FL	COLLIER	12021	1.70	FL	MONROE	12087	1.70
FL	COLUMBIA	12023	1.30	FL	NASSAU	12089	1.30
FL	DADE	12025	1.70	FL	OKEECHOBEE	12093	1.80
FL	DE SOTO	12027	1.80	FL	ORANGE	12095	1.40
FL	DIXIE	12029	1.30	FL	OSCEOLA	12097	1.40
FL	DUVAL	12031	1.30	FL	PALM BEACH	12099	1.70
FL	FLAGLER	12035	1.00	FL	PASCO	12101	1.40
FL	FRANKLIN	12037	0.90	FL	PINELLAS	12103	1.40
FL	GADSDEN	12039	0.90	FL	POLK	12105	1.40
FL	GILCHRIST	12041	1.30	FL	PUTNAM	12107	1.30
FL	GLADES	12043	1.50	FL	SAINT JOHNS	12109	1.30
FL	GULF	12045	0.90	FL	SAINT LUCIE	12111	1.80
FL	HAMILTON	12047	1.30	FL	SARASOTA	12115	1.80
FL	HARDEE	12049	1.80	FL	SEMINOLE	12117	1.40
FL	HENDRY	12051	1.70	FL	SUMTER	12119	1.40
FL	HERNANDO	12053	1.40	FL	SUWANNEE	12121	1.30
FL	HIGHLANDS	12055	1.80	FL	TAYLOR	12123	1.30
FL	HILLSBOROUGH	12057	1.40	FL	UNION	12125	1.30
FL	HOLMES	12059	0.60	FL	VOLUSIA	12127	1.40
FL	INDIAN RIVER	12061	1.80	FL	WAKULLA	12129	0.90
FL	JACKSON	12063	0.60	FL	WASHINGTON	12133	0.60
FL	JEFFERSON	12065	0.90				

[73 FR 14161, Mar. 17, 2008]

§ 1006.52 Adjusted Class I differentials.

See § 1000.52.

§ 1006.53 Announcement of class prices, component prices, and advanced pricing factors.

See § 1000.53.

§ 1006.54 Equivalent price.

See § 1000.54.

UNIFORM PRICES

§ 1006.60 Handler's value of milk.

For the purpose of computing a handler's obligation for producer milk, the market administrator shall determine for each month the value of milk of each handler with respect to each of the handler's pool plants and of each handler described in § 1000.9(c) with respect to milk that was not received at a pool plant by adding the amounts computed in paragraphs (a) through (e) of this section and subtracting from that total amount the value computed in paragraph (f) of this section. Receipts of nonfluid milk products that are distributed as labeled reconstituted milk for which payments are made to the producer-settlement fund of another Federal order under § 1000.76(a)(4) or (d) shall be excluded from pricing under this section.

(a) Multiply the pounds of skim milk and butterfat in producer milk that were classified in each class pursuant to § 1000.44(c) of this chapter by the applicable skim milk and butterfat prices, and add the resulting amounts; except that for the months of July 2018 through January 2019, the Class I skim milk price for this purpose shall be the Class I skim milk price as determined in § 1000.50(b) of this chapter plus $0.09 per hundredweight, and the Class I butterfat price for this purpose shall be the Class I butterfat price as determined in § 1000.50(c) of this chapter plus $0.0009 per pound. The adjustments to the Class I skim milk and butterfat prices provided herein may be reduced

83

by the market administrator for any month if the market administrator determines that the payments yet unpaid computed pursuant to paragraphs (g)(1) through (g)(6) of this section will be less than the amount computed pursuant to paragraph (h) of this section. The adjustments to the Class I skim milk and butterfat prices provided herein during the months of July 2018 through January 2019 shall be announced along with the prices announced in §1000.53(b) of this chapter.

(b) Multiply the pounds of skim milk and butterfat overage assigned to each class pursuant to §1000.44(a)(11) by the respective skim milk and butterfat prices applicable at the location of the pool plant;

(c) Multiply the difference between the Class IV price for the preceding month and the current month's Class I, II, or III price, as the case may be, by the hundredweight of skim milk and butterfat subtracted from Class I, II, or III, respectively, pursuant to §1000.44(a)(7) and the corresponding step of §1000.44(b);

(d) Multiply the difference between the Class I price applicable at the location of the pool plant and the Class IV price by the hundredweight of skim milk and butterfat assigned to Class I pursuant to §1000.43(d) and the hundredweight of skim milk and butterfat subtracted from Class I pursuant to §1000.44(a)(3)(i) through (vi) and the corresponding step of §1000.44(b), excluding receipts of bulk fluid cream products from a plant regulated under other Federal orders and bulk concentrated fluid milk products from pool plants, plants regulated under other Federal orders, and unregulated supply plants;

(e) Multiply the Class I skim milk and Class I butterfat prices applicable at the location of the nearest unregulated supply plants from which an equivalent volume was received by the pounds of skim milk and butterfat in receipts of concentrated fluid milk products assigned to Class I pursuant to §1000.43(d) and §1000.44(a)(3)(i) and the corresponding step of §1000.44(b) and the pounds of skim milk and butterfat subtracted from Class I pursuant to §1000.44(a)(8) and the corresponding step of §1000.44(b), excluding such skim

milk and butterfat in receipts of fluid milk products from an unregulated supply plant to the extent that an equivalent amount of skim milk or butterfat disposed of to such plant by handlers fully regulated under any Federal milk order is classified and priced as Class I milk and is not used as an offset for any other payment obligation under any order; and

(f) For reconstituted milk made from receipts of nonfluid milk products, multiply $1.00 (but not more than the difference between the Class I price applicable at the location of the pool plant and the Class IV price) by the hundredweight of skim milk and butterfat contained in receipts of nonfluid milk products that are allocated to Class I use pursuant to §1000.43(d).

(g) For transactions occurring during the period of September 6, 2017, through September 15, 2017, for handlers who have submitted proof satisfactory to the market administrator no later than August 1, 2018, to determine eligibility for reimbursement of hurricane-imposed costs, subtract an amount equal to:

(1) The additional cost of transportation on loads of milk rerouted from pool distributing plants to plants outside the state of Florida as a result of Hurricane Irma, and the additional cost of transportation on loads of milk moved and then dumped. The reimbursement of transportation costs pursuant to this section shall be the actual demonstrated cost of such transportation of bulk milk or the miles of transportation on such loads of bulk milk multiplied by $3.75 per loaded mile, whichever is less;

(2) The lost location value on loads of milk rerouted to plants outside the state of Florida as a result of Hurricane Irma. The lost location value shall be the difference per hundredweight between the value specified in §1000.52 of this chapter, adjusted by §1006.51(b), at the location of the plant where the milk would have normally been received and the value specified in §1000.52, as adjusted by §1005.51(b) and §1007.51(b) of this chapter, at the location of the plant to which the milk was rerouted;

(3) The value per hundredweight at the lowest classified price for the

month of September 2017 for milk dumped at the farm and classified as other use milk pursuant to § 1000.40(e) of this chapter as a result of Hurricane Irma;

(4) The value per hundredweight at the lowest classified price for the month of September 2017 for milk dumped from milk tankers after being moved off-farm and classified as other use milk pursuant to § 1000.40(e) of this chapter as a result of Hurricane Irma;

(5) The value per hundredweight at the lowest classified price for the month of September 2017 for skim portion of milk dumped and classified as other use milk pursuant to § 1000.40(e) of this chapter as a result of Hurricane Irma; and

(6) The difference between the announced class price applicable to the milk as classified by the market administrator for the month of September 2017 and the actual price received for milk delivered to nonpool plants outside the state of Florida as a result of Hurricane Irma.

(h) The total amount of payment to all handlers under paragraph (g) of this section shall be limited for each month to an amount determined by multiplying the total Class I producer milk for all handlers pursuant to § 1000.44(c) of this chapter times $0.09 per hundredweight.

(i) If the cost of payments computed pursuant to paragraphs (g)(1) through (g)(6) of this section exceeds the amount computed pursuant to paragraph (h) of this section, the market administrator shall prorate such payments to each handler based on each handler's proportion of transportation and other use milk costs submitted pursuant to paragraphs (g)(1) through (g)(6). Costs submitted pursuant to paragraphs (g)(1) through (g)(6) which are not paid as a result of such a proration shall be paid in subsequent months until all costs incurred and documented through (g)(1) through (g)(6) have been paid.

[64 FR 47966, Sept. 1, 1999, as amended at 65 FR 82835, Dec. 28, 2000; 69 FR 71700, Dec. 10, 2004; 83 FR 21845, May 11, 2018]

§ 1006.61 Computation of uniform prices.

On or before the 11th day of each month, the market administrator shall compute a uniform butterfat price, a uniform skim milk price, and a uniform price for producer milk receipts reported for the prior month. The report of any handler who has not made payments required pursuant to § 1006.71 for the preceding month shall not be included in the computation of these prices, and such handler's report shall not be included in the computation for succeeding months until the handler has made full payment of outstanding monthly obligations.

(a) *Uniform butterfat price.* The uniform butterfat price per pound, rounded to the nearest one-hundredth cent, shall be computed by:

(1) Multiplying the pounds of butterfat in producer milk allocated to each class pursuant to § 1000.44(b) by the respective class butterfat prices;

(2) Adding the butterfat value calculated in § 1006.60(e) for other source milk allocated to Class I pursuant to § 1000.43(d) and the steps of § 1000.44(b) that correspond to § 1000.44(a)(3)(i) and § 1000.44(a)(8) by the Class I price; and

(3) Dividing the sum of paragraphs (a)(1) and (a)(2) of this section by the sum of the pounds of butterfat in producer milk and other source milk used to calculate the values in paragraphs (a)(1) and (a)(2) of this section.

(b) *Uniform skim milk price.* The uniform skim milk price per hundredweight, rounded to the nearest cent, shall be computed as follows:

(1) Combine into one total the values computed pursuant to § 1006.60 for all handlers;

(2) Add an amount equal to the minus location adjustments and subtract an amount equal to the plus location adjustments computed pursuant to § 1006.75;

(3) Add an amount equal to not less than one-half of the unobligated balance in the producer-settlement fund;

(4) Subtract the value of the total pounds of butterfat for all handlers. The butterfat value shall be computed by multiplying the sum of the pounds of butterfat in producer milk and other source milk used to calculate the values in paragraphs (a)(1) and (a)(2) of

this section by the butterfat price computed in paragraph (a) of this section;

(5) Divide the resulting amount by the sum of the following for all handlers included in these computations:

(i) The total skim pounds of producer milk; and

(ii) The total skim pounds for which a value is computed pursuant to § 1006.60(e); and

(6) Subtract not less than 4 cents and not more than 5 cents.

(c) *Uniform price.* The uniform price per hundredweight, rounded to the nearest cent, shall be the sum of the following:

(1) Multiply the uniform butterfat price for the month pursuant to paragraph (a) of this section times 3.5 pounds of butterfat; and

(2) Multiply the uniform skim milk price for the month pursuant to paragraph (b) of this section times 96.5 pounds of skim milk.

[64 FR 47966, Sept. 1, 1999, as amended at 65 FR 82835, Dec. 28, 2000]

§ 1006.62 Announcement of uniform prices.

On or before the 11th day after the end of the month, the market administrator shall announce the uniform prices for the month computed pursuant to § 1006.61.

PAYMENTS FOR MILK

§ 1006.70 Producer-settlement fund.

See § 1000.70.

§ 1006.71 Payments to the producer-settlement fund.

Each handler shall make a payment to the producer-settlement fund in a manner that provides receipt of the funds by the market administrator no later than the 12th day after the end of the month (except as provided in § 1000.90). Payment shall be the amount, if any, by which the amount specified in paragraph (a) of this section exceeds the amount specified in paragraph (b) of this section:

(a) The total value of milk of the handler for the month as determined pursuant to § 1006.60; and

(b) The sum of the value at the uniform prices for skim milk and butterfat, adjusted for plant location, of the handler's receipts of producer milk; and the value at the uniform price, as adjusted pursuant to § 1006.75, applicable at the location of the plant from which received of other source milk for which a value is computed pursuant to § 1006.60(e).

§ 1006.72 Payments from the producer-settlement fund.

No later than one day after the date of payment receipt required under § 1006.71, the market administrator shall pay to each handler the amount, if any, by which the amount computed pursuant to § 1006.71(b) exceeds the amount computed pursuant to § 1006.71(a). If, at such time, the balance in the producer-settlement fund is insufficient to make all payments pursuant to this section, the market administrator shall reduce uniformly such payments and shall complete the payments as soon as the funds are available.

§ 1006.73 Payments to producers and to cooperative associations.

(a) Each handler that is not paying a cooperative association for producer milk shall pay each producer as follows:

(1) *Partial payments.* (i) For each producer who has not discontinued shipments as of the 15th day of the month, payment shall be made so that it is received by the producer on or before the 20th day of the month (except as provided in § 1000.90) for milk received during the first 15 days of the month at not less than 85 percent of the preceding month's uniform price, adjusted for plant location pursuant to § 1006.75 and proper deductions authorized in writing by the producer; and

(ii) For each producer who has not discontinued shipments as of the last day of the month, payment shall be made so that it is received by the producer on or before the 5th day of the following month (except as provided in § 1000.90) for milk received from the 16th to the last day of the month at not less than 85 percent of the preceding month's uniform price, adjusted for plant location pursuant to § 1006.75 and proper deductions authorized in writing by the producer.

(2) *Final payment.* For milk received during the month, a payment computed as follows shall be made so that it is received by each producer one day after the payment date required in § 1006.72:

(i) Multiply the hundredweight of producer skim milk received times the uniform skim milk price for the month;

(ii) Multiply the pounds of butterfat received times the uniform butterfat price for the month;

(iii) Multiply the hundredweight of producer milk received times the plant location adjustment pursuant to § 1006.75; and

(iv) Add the amounts computed in paragraphs (a)(2)(i), (ii), and (iii) of this section, and from that sum:

(A) Subtract the partial payments made pursuant to paragraph (a)(1) of this section;

(B) Subtract the deduction for marketing services pursuant to § 1000.86;

(C) Add or subtract for errors made in previous payments to the producer; and

(D) Subtract proper deductions authorized in writing by the producer.

(b) One day before partial and final payments are due pursuant to paragraph (a) of this section, each handler shall pay a cooperative association for milk received as follows:

(1) *Partial payment to a cooperative association for bulk milk received directly from producers' farms.* For bulk milk (including the milk of producers who are not members of such association and who the market administrator determines have authorized the cooperative association to collect payment for their milk) received from a cooperative association in any capacity, except as the operator of a pool plant, the payment shall be equal to the hundredweight of milk received multiplied by 90 percent of the preceding month's uniform price, adjusted for plant location pursuant to § 1006.75.

(2) *Partial payment to a cooperative association for milk transferred from its pool plant.* For bulk fluid milk products and bulk fluid cream products received during the first 15 days of the month from a cooperative association in its capacity as the operator of a pool plant, the partial payment shall be at the pool plant operator's estimated use value of the milk using the most recent class prices available for skim milk and butterfat at the receiving plant's location.

(3) *Final payment to a cooperative association for milk transferred from its pool plant.* For bulk fluid milk products and bulk fluid cream products received during the month from a cooperative association in its capacity as the operator of a pool plant, the final payment shall be the classified value of such milk as determined by multiplying the pounds of skim milk and butterfat assigned to each class pursuant to § 1000.44 by the class prices for the month at the receiving plant's location, and subtracting from this sum the partial payment made pursuant to paragraph (b)(2) of this section.

(4) *Final payment to a cooperative association for bulk milk received directly from producers' farms.* For bulk milk received from a cooperative association during the month, including the milk of producers who are not members of such association and who the market administrator determines have authorized the cooperative association to collect payment for their milk, the final payment shall be an amount equal to the sum of the individual payments otherwise payable for such milk pursuant to paragraph (a)(2) of this section.

(c) If a handler has not received full payment from the market administrator pursuant to § 1006.72 by the payment date specified in paragraph (a) or (b) of this section, the handler may reduce payments pursuant to paragraphs (a) and (b) of this section, but by not more than the amount of the underpayment. The payments shall be completed on the next scheduled payment date after receipt of the balance due from the market administrator.

(d) If a handler claims that a required payment to a producer cannot be made because the producer is deceased or cannot be located, or because the cooperative association or its lawful successor or assignee is no longer in existence, the payment shall be made to the producer-settlement fund, and in the event that the handler subsequently locates and pays the producer or a lawful claimant, or in the event that the handler no longer exists and a lawful claim

87

is later established, the market administrator shall make the required payment from the producer-settlement fund to the handler or to the lawful claimant as the case may be.

(e) In making payments to producers pursuant to this section, each pool plant operator shall furnish each producer, except a producer whose milk was received from a cooperative association described in § 1000.9(a) or (c), a supporting statement in such form that it may be retained by the recipient which shall show:

(1) The name, address, Grade A identifier assigned by a duly constituted regulatory agency, and the payroll number of the producer;

(2) The month and dates that milk was received from the producer, including the daily and total pounds of milk received;

(3) The total pounds of butterfat in the producer's milk;

(4) The minimum rate or rates at which payment to the producer is required pursuant to the order in this part;

(5) The rate used in making payment if the rate is other than the applicable minimum rate;

(6) The amount, or rate per hundredweight, and nature of each deduction claimed by the handler; and

(7) The net amount of payment to the producer or cooperative association.

[64 FR 47966, Sept. 1, 1999, as amended at 65 FR 32010, May 22, 2000]

§ 1006.74 [Reserved]

§ 1006.75 Plant location adjustments for producer milk and nonpool milk.

For purposes of making payments for producer milk and nonpool milk, a plant location adjustment shall be determined by subtracting the Class I price specified in § 1006.51 from the Class I price at the plant's location. The difference, plus or minus as the case may be, shall be used to adjust the payments required pursuant to §§ 1006.73 and 1000.76.

§ 1006.76 Payments by a handler operating a partially regulated distributing plant.

See § 1000.76.

§ 1006.77 Adjustment of accounts.

See § 1000.77.

§ 1006.78 Charges on overdue accounts.

See § 1000.78.

ADMINISTRATIVE ASSESSMENT AND MARKETING SERVICE DEDUCTION

§ 1006.85 Assessment for order administration.

On or before the payment receipt date specified under § 1006.71, each handler shall pay to the market administrator its *pro rata* share of the expense of administration of the order at a rate specified by the market administrator that is no more than $.08 per hundredweight with respect to:

(a) Receipts of producer milk (including the handler's own production) other than such receipts by a handler described in § 1000.9(c) of this chapter that were delivered to pool plants of other handlers;

(b) Receipts from a handler described in § 1000.9(c) of this chapter;

(c) Receipts of concentrated fluid milk products from unregulated supply plants and receipts of nonfluid milk products assigned to Class I use pursuant to § 1000.43(d) of this chapter and other source milk allocated to Class I pursuant to § 1000.44(a)(3) and (8) chapter and the corresponding steps of § 1000.44(b) of this chapter, except other source milk that is excluded from the computations pursuant to § 1006.60(d) and (e); and

(d) Route disposition in the marketing area from a partially regulated distributing plant that exceeds the skim milk and butterfat subtracted pursuant to § 1000.76(a)(1)(i) and (ii) of this chapter.

[79 FR 25002, May 2, 2014; 79 FR 26591, May 9, 2014]

§ 1006.86 Deduction for marketing services.

See § 1000.86.

PART 1007—MILK IN THE SOUTHEAST MARKETING AREA

Subpart—Order Regulating Handling

AUTHORITY: 7 U.S.C. 601–674, and 7253.

SOURCE: 64 FR 47971, Sept. 1, 1999, unless otherwise noted.

Subpart—Order Regulating Handling

GENERAL PROVISIONS

§ 1007.1 General provisions.

The terms, definitions, and provisions in part 1000 of this chapter apply to this part 1007. In this part 1007, all references to sections in part 1000 refer to part 1000 of this chapter.

DEFINITIONS

§ 1007.2 Southeast marketing area.

The marketing area means all territory within the bounds of the following states and political subdivisions, including all piers, docks and wharves connected therewith and all craft moored thereat, and all territory occupied by government (municipal, State or Federal) reservations, installations, institutions, or other similar establishments if any part thereof is within any of the listed states or political subdivisions:

ALABAMA, ARKANSAS, LOUISIANA, AND
MISSISSIPPI

All of the States of Alabama, Arkansas,
Louisiana, and Mississippi.

Florida Counties

Escambia, Okaloosa, Santa Rosa, and Walton.

Georgia Counties

All of the State of Georgia except for the
counties of Catoosa, Chattooga, Dade,
Fannin, Murray, Walker, and Whitfield.

Kentucky Counties

Allen, Ballard, Barren, Caldwell, Calloway,
Carlisle, Christian, Crittenden, Fulton,
Graves, Hickman, Livingston, Logan, Lyon,
Marshall, McCracken, Metcalfe, Monroe,
Simpson, Todd, Trigg, and Warren.

Missouri Counties

Barry, Barton, Bollinger, Butler, Cape
Girardeau, Carter, Cedar, Christian,
Crawford, Dade, Dallas, Dent, Douglas,
Dunklin, Greene, Howell, Iron, Jasper,
Laclede, Lawrence, Madison, McDonald, Mississippi, New Madrid, Newton, Oregon,
Ozark, Pemiscot, Perry, Polk, Reynolds, Ripley, Scott, Shannon, St. Francois, Stoddard,
Stone, Taney, Texas, Vernon, Washington,
Wayne, Webster, and Wright.

Tennessee Counties

All of the State of Tennessee except for the
counties of Anderson, Blount, Bradley,
Campbell, Carter, Claiborne, Cocke, Cumberland, Grainger, Greene, Hamblen, Hamilton, Hancock, Hawkins, Jefferson, Johnson,
Knox, Loudon, Marion, McMinn, Meigs, Monroe, Morgan, Polk, Rhea, Roane, Scott,
Sequatchie, Sevier, Sullivan, Unicoi, Union,
and Washington.

§ 1007.3 Route disposition.

See § 1000.3.

§ 1007.4 Plant.

See § 1000.4.

§ 1007.5 Distributing plant.

See § 1000.5.

§ 1007.6 Supply plant.

See § 1000.6.

§ 1007.7 Pool plant.

Pool plant means a plant specified in
paragraphs (a) through (d) of this section, a unit of plants as specified in
paragraph (e) of this section, or a plant
specified in paragraph (h) of this section, but excluding a plant specified in
paragraph (g) of this section. The pooling standards described in paragraphs
(c) and (d) of this section are subject to
modification pursuant to paragraph (f)
of this section:

(a) A distributing plant, other than a
plant qualified as a pool plant pursuant
to paragraph (b) of this section or
§ _____.7(b) of any other Federal milk
order, from which during the month 50
percent or more of the fluid milk products physically received at such plant
(excluding concentrated milk received
from another plant by agreement for
other than Class I use) are disposed of
as route disposition or are transferred
in the form of packaged fluid milk
products to other distributing plants.
At least 25 percent of such route disposition and transfers must be to outlets in the marketing area.

(b) Any distributing plant located in
the marketing area which during the
month processed at least 50 percent of
the total quantity of fluid milk products physically received at the plant
(excluding concentrated milk received
from another plant by agreement for
other than Class I use) into ultra-pasteurized or aseptically-processed fluid
milk products.

(c) A supply plant from which 50 percent or more of the total quantity of
milk that is physically received during
the month from dairy farmers and handlers described in § 1000.9(c), including
milk that is diverted from the plant, is
transferred to pool distributing plants.
Concentrated milk transferred from
the supply plant to a distributing plant
for an agreed-upon use other than Class
I shall be excluded from the supply
plant's shipments in computing the
plant's shipping percentage.

(d) A plant located within the marketing area that is operated by a cooperative association if pool plant status
under this paragraph is requested for
such plant by the cooperative association and during the month at least 60
percent of the producer milk of members of such cooperative association is
delivered directly from farms to pool
distributing plants or is transferred to
such plants as a fluid milk product (excluding concentrated milk transferred
to a distributing plant for an agreed-

upon use other than Class I) from the cooperative's plant.

(e) Two or more plants operated by the same handler and located within the marketing area may qualify for pool status as a unit by meeting the total and in-area route disposition requirements specified in paragraph (a) of this section and the following additional requirements:

(1) At least one of the plants in the unit must qualify as a pool plant pursuant to paragraph (a) of this section;

(2) Other plants in the unit must process only Class I or Class II products and must be located in a pricing zone providing the same or a lower Class I price than the price applicable at the distributing plant included in the unit pursuant to paragraph (e)(1) of this section; and

(3) A written request to form a unit, or to add or remove plants from a unit, must be filed with the market administrator prior to the first day of the month for which it is to be effective.

(f) The applicable shipping percentages of paragraphs (c) and (d) of this section may be increased or decreased by the market administrator if the market administrator finds that such adjustment is necessary to encourage needed shipments or to prevent uneconomic shipments. Before making such a finding, the market administrator shall investigate the need for adjustment either on the market administrator's own initiative or at the request of interested parties if the request is made in writing at least 15 days prior to the date for which the requested revision is desired effective. If the investigation shows that an adjustment of the shipping percentages might be appropriate, the market administrator shall issue a notice stating that an adjustment is being considered and invite data, views and arguments. Any decision to revise an applicable shipping percentage must be issued in writing at least one day before the effective date.

(g) The term pool plant shall not apply to the following plants:

(1) A producer-handler plant;

(2) An exempt plant as defined in § 1000.8(e);

(3) A plant qualified pursuant to paragraph (a) of this section which is not located within any Federal order

marketing area, meets the pooling requirements of another Federal order, and has had greater route disposition in such other Federal order marketing area for 3 consecutive months;

(4) A plant qualified pursuant to paragraph (a) of this section which is located in another Federal order marketing area, meets the pooling standards of the other Federal order, and has not had a majority of its route disposition in this marketing area for 3 consecutive months or is locked into pool status under such other Federal order without regard to its route disposition in any other Federal order marketing area; and

(5) A plant qualified pursuant to paragraph (c) of this section which also meets the pooling requirements of another Federal order and from which greater qualifying shipments are made to plants regulated under such other order than are made to plants regulated under the order in this part, or such plant has automatic pooling status under such other order.

(h) Any distributing plant, located within the marketing area as described on May 1, 2006, in § 1007.2;

(1) From which there is route disposition and/or transfers of packaged fluid milk products in any non-Federally regulated marketing area(s) located within one or more States that require handlers to pay minimum prices for raw milk provided that 25 percent or more of the total quantity of fluid milk products physically received at such plant (excluding concentrated milk received from another plant by agreement for other than Class I use) is disposed of as route disposition and/or is transferred in the form of packaged fluid milk products to other plants. At least 25 percent of such route disposition and/or transfers, in aggregate, are in any non-Federally regulated marketing area(s) located within one or more States that require handlers to pay minimum prices for raw milk. Subject to the following exclusions:

(i) The plant is described in § 1007.7(a), (b), or (e);

(ii) The plant is subject to the pricing provisions of a State-operated milk pricing plan which provides for the payment of minimum class prices for raw milk;

(iii) The plant is described in § 1000.8(a) or (e); or

(iv) A producer-handler described in § 1007.10 with less than three million pounds during the month of route disposition and/or transfers of packaged fluid milk products to other plants.

[64 FR 47971, Sept. 1, 1999, as amended at 71 FR 25498, May 1, 2006; 71 FR 28249, May 16, 2006]

§ 1007.8 Nonpool plant.

See § 1000.8.

§ 1007.9 Handler.

See § 1000.9.

§ 1007.10 Producer-handler.

Producer-handler means a person who:

(a) Operates a dairy farm and a distributing plant from which there is route disposition in the marketing area, and from which total route disposition and packaged sales of fluid milk products to other plants during the month does not exceed 3 million pounds;

(b) Receives no fluid milk products, and acquires no fluid milk products for route disposition, from sources other than own farm production;

(c) Disposes of no other source milk as Class I milk except by increasing the nonfat milk solids content of the fluid milk products received from own farm production; and

(d) Provides proof satisfactory to the market administrator that the care and management of the dairy animals and other resources necessary to produce all Class I milk handled, and the processing and packaging operations, are the producer-handler's own enterprise and are operated at the producer-handler's own risk.

(e) Any producer-handler with Class I route dispositions and/or transfers of packaged fluid milk products in the marketing area described in § 1131.2 of this chapter shall be subject to payments into the Order 1131 producer settlement fund on such dispositions pursuant to § 1000.76(a) and payments into the Order 1131 administrative fund provided such dispositions are less than three million pounds in the current month and such producer-handler had total Class I route dispositions and/or transfers of packaged fluid milk prod-

ucts from own farm production of three million pounds or more the previous month. If the producer-handler has Class I route dispositions and/or transfers of packaged fluid milk products into the marketing area described in § 1131.2 of this chapter of three million pounds or more during the current month, such producer-handler shall be subject to the provisions described in § 1131.7 of this chapter or § 1000.76(a).

[64 FR 47971, Sept. 1, 1999, as amended at 71 FR 25499, May 1, 2006; 75 FR 21160, Apr. 23, 2010]

§ 1007.11 [Reserved]

§ 1007.12 Producer.

(a) Except as provided in paragraph (b) of this section, *producer* means any person who produces milk approved by a duly constituted regulatory agency for fluid consumption as Grade A milk and whose milk (or components of milk) is:

(1) Received at a pool plant directly from the producer or diverted by the plant operator in accordance with § 1007.13; or

(2) Received by a handler described in § 1000.9(c).

(b) Producer shall not include:

(1) A producer-handler as defined in any Federal order;

(2) A dairy farmer whose milk is received at an exempt plant, excluding producer milk diverted to the exempt plant pursuant to § 1007.13(d);

(3) A dairy farmer whose milk is received by diversion at a pool plant from a handler regulated under another Federal order if the other Federal order designates the dairy farmer as a producer under that order and that milk is allocated by request to a utilization other than Class I; and

(4) A dairy farmer whose milk is reported as diverted to a plant fully regulated under another Federal order with respect to that portion of the milk so diverted that is assigned to Class I under the provisions of such other order.

§ 1007.13 Producer milk.

Except as provided for in paragraph (e) of this section, *Producer milk* means the skim milk (or the skim equivalent of components of skim milk) and butterfat contained in milk of a producer that is:

(a) Received by the operator of a pool plant directly from a producer or a handler described in §1000.9(c). All milk received pursuant to this paragraph shall be priced at the location of the plant where it is first physically received;

(b) Received by a handler described in §1000.9(c) in excess of the quantity delivered to pool plants;

(c) Diverted by a pool plant operator to another pool plant. Milk so diverted shall be priced at the location of the plant to which diverted; or

(d) Diverted by the operator of a pool plant or a handler described in §1000.9(c) to a nonpool plant, subject to the following conditions:

(1) In any month of January through June, not less than 1 days' production of the producer whose milk is diverted is physically received at a pool plant during the month;

(2) In any month of July through December, not less than 1 days' production of the producer whose milk is diverted is physically received at a pool plant during the month;

(3) The total quantity of milk so diverted during the month by a cooperative association shall not exceed 25 percent during the months of July through November, January, and February, and 35 percent during the months of December and March through June, of the producer milk that the cooperative association caused to be delivered to, and physically received at, pool plants during the month, excluding the total pounds of bulk milk received directly from producers meeting the conditions as described in §1005.82(c)(2)(ii) and (iii), and for which a transportation credit is requested;

(4) The operator of a pool plant that is not a cooperative association may divert any milk that is not under the control of a cooperative association that diverts milk during the month pursuant to paragraph (d) of this section. The total quantity of milk so diverted during the month shall not exceed 25 percent during the months of July through November, January, and February, and 35 percent during the

months of December and March through June, of the producer milk physically received at such plant (or such unit of plants in the case of plants that pool as a unit pursuant to §1005.7(e)) during the month, excluding the quantity of producer milk received from a handler described in §1000.9(c) of this chapter and excluding the total pounds of bulk milk received directly from producers meeting the conditions as described in §1005.82(c)(2)(ii) and (iii), and for which a transportation credit is requested;

(5) Any milk diverted in excess of the limits prescribed in paragraphs (d)(3) and (4) of this section shall not be producer milk. If the diverting handler or cooperative association fails to designate the dairy farmers' deliveries that will not be producer milk, no milk diverted by the handler or cooperative association shall be producer milk;

(6) Diverted milk shall be priced at the location of the plant to which diverted; and

(7) The delivery day requirements and the diversion percentages in paragraphs (d)(1) through (4) of this section may be increased or decreased by the market administrator if the market administrator finds that such revision is necessary to assure orderly marketing and efficient handling of milk in the marketing area. Before making such a finding, the market administrator shall investigate the need for the revision either on the market administrator's own initiative or at the request of interested persons. If the investigation shows that a revision might be appropriate, the market administrator shall issue a notice stating that the revision is being considered and inviting written data, views, and arguments. Any decision to revise an applicable percentage must be issued in writing at least one day before the effective date.

(e) Producer milk shall not include milk of a producer that is subject to inclusion and participation in a marketwide equalization pool under a milk classification and pricing program imposed under the authority of a

State government maintaining marketwide pooling of returns.

[64 FR 47971, Sept. 1, 1999, as amended at 70 FR 59223, Oct. 12, 2005; 71 FR 62379, Oct. 25, 2006; 73 FR 14162, Mar. 17, 2008; 73 FR 26315, May 9, 2008; 79 FR 25005, May 2, 2014; 79 FR 26591, May 9, 2014]

§ 1007.14 Other source milk.

See § 1000.14.

§ 1007.15 Fluid milk product.

See § 1000.15.

§ 1007.16 Fluid cream product.

See § 1000.16.

§ 1007.17 [Reserved]

§ 1007.18 Cooperative association.

See § 1000.18.

§ 1007.19 Commercial food processing establishment.

See § 1000.19.

HANDLER REPORTS

§ 1007.30 Reports of receipts and utilization.

Each handler shall report monthly so that the market administrator's office receives the report on or before the 7th day after the end of the month, in the detail and on prescribed forms, as follows:

(a) With respect to each of its pool plants, the quantities of skim milk and butterfat contained in or represented by:

(1) Receipts of producer milk, including producer milk diverted by the reporting handler, from sources other than handlers described in § 1000.9(c);

(2) Receipts of milk from handlers described in § 1000.9(c);

(3) Receipts of fluid milk products and bulk fluid cream products from other pool plants;

(4) Receipts of other source milk;

(5) Receipts of bulk milk from a plant regulated under another Federal order, except Federal Order 1005, for which a transportation credit is requested pursuant to § 1007.82;

(6) Receipts of producer milk described in § 1007.82(c)(2), including the identity of the individual producers whose milk is eligible for the transportation credit pursuant to that paragraph and the date that such milk was received;

(7) For handlers submitting transportation credit requests, transfers of bulk milk to nonpool plants, including the dates that such milk was transferred;

(8) Inventories at the beginning and end of the month of fluid milk products and bulk fluid cream products; and

(9) The utilization or disposition of all milk and milk products required to be reported pursuant to this paragraph.

(b) Each handler operating a partially regulated distributing plant shall report with respect to such plant in the same manner as prescribed for reports required by paragraphs (a)(1), (a)(2), (a)(3), (a)(4), and (a)(8) of this section. Receipts of milk that would have been producer milk if the plant had been fully regulated shall be reported in lieu of producer milk. The report shall show also the quantity of any reconstituted skim milk in route disposition in the marketing area.

(c) Each handler described in § 1000.9(c) shall report:

(1) The quantities of all skim milk and butterfat contained in receipts of milk from producers;

(2) The utilization or disposition of all such receipts; and

(3) With respect to milk for which a cooperative association is requesting a transportation credit pursuant to § 1007.82, all of the information required in paragraphs (a)(5), (a)(6), and (a)(7) of this section.

(d) Each handler not specified in paragraphs (a) through (c) of this section shall report with respect to its receipts and utilization of milk and milk products in such manner as the market administrator may prescribe.

§ 1007.31 Payroll reports.

(a) On or before the 20th day after the end of each month, each handler that operates a pool plant pursuant to § 1007.7 and each handler described in § 1000.9(c) shall report to the market administrator its producer payroll for the month, in detail prescribed by the market administrator, showing for each producer the information specified in § 1007.73(e).

(b) Each handler operating a partially regulated distributing plant who elects to make payment pursuant to § 1000.76(b) shall report for each dairy farmer who would have been a producer if the plant had been fully regulated in the same manner as prescribed for reports required by paragraph (a) of this section.

§ 1007.32 Other reports.

(a) On or before the 20th day after the end of each month, each handler described in § 1000.9(a) and (c) shall report to the market administrator any adjustments to transportation credit requests as reported pursuant to § 1007.30(a)(5), (6), and (7).

(b) In addition to the reports required pursuant to §§ 1007.30, 31, and 32(a), each handler shall report any information the market administrator deems necessary to verify or establish each handler's obligation under the order.

CLASSIFICATION OF MILK

§ 1007.40 Classes of utilization.

See § 1000.40.

§ 1007.41 [Reserved]

§ 1007.42 Classification of transfers and diversions.

See § 1000.42.

§ 1007.43 General classification rules.

See § 1000.43.

§ 1007.44 Classification of producer milk.

See § 1000.44.

§ 1007.45 Market administrator's reports and announcements concerning classification.

See § 1000.45.

CLASS PRICES

§ 1007.50 Class prices, component prices, and advanced pricing factors.

See § 1000.50.

§ 1007.51 Class I differential, adjustments to Class I prices, and Class I price.

(a) The Class I differential shall be the differential established for Fulton County, Georgia, which is reported in § 1000.52. The Class I price shall be the price computed pursuant to § 1007.50(a) for Fulton County, Georgia.

(b) Adjustment to Class I prices. Class I prices shall be established pursuant to § 1000.50(a), (b) and (c) using the following adjustments:

State	Country/parish	FIPS	Class I price adjustment
AL	AUTAUGA	01001	0.50
AL	BALDWIN	01003	0.50
AL	BARBOUR	01005	0.55
AL	BIBB	01007	0.30
AL	BLOUNT	01009	0.20
AL	BULLOCK	01011	0.70
AL	BUTLER	01013	0.55
AL	CALHOUN	01015	0.30
AL	CHAMBERS	01017	0.70
AL	CHEROKEE	01019	0.30
AL	CHILTON	01021	0.70
AL	CHOCTAW	01023	0.50
AL	CLARKE	01025	0.35
AL	CLAY	01027	0.70
AL	CLEBURNE	01029	0.70
AL	COFFEE	01031	0.85
AL	COLBERT	01033	0.30
AL	CONECUH	01035	0.55
AL	COOSA	01037	0.70
AL	COVINGTON	01039	0.55
AL	CRENSHAW	01041	0.55
AL	CULLMAN	01043	0.20
AL	DALE	01045	0.85
AL	DALLAS	01047	0.50
AL	DE KALB	01049	0.40
AL	ELMORE	01051	0.50
AL	ESCAMBIA	01053	0.55
AL	ETOWAH	01055	0.30
AL	FAYETTE	01057	0.20
AL	FRANKLIN	01059	0.30
AL	GENEVA	01061	0.85
AL	GREENE	01063	0.30
AL	HALE	01065	0.30
AL	HENRY	01067	0.85
AL	HOUSTON	01069	0.85
AL	JACKSON	01071	0.40
AL	JEFFERSON	01073	0.30
AL	LAMAR	01075	0.20
AL	LAUDERDALE	01077	0.30
AL	LAWRENCE	01079	0.30
AL	LEE	01081	0.70
AL	LIMESTONE	01083	0.30
AL	LOWNDES	01085	0.70
AL	MACON	01087	0.70
AL	MADISON	01089	0.30
AL	MARENGO	01091	0.50
AL	MARION	01093	0.20
AL	MARSHALL	01095	0.40
AL	MOBILE	01097	0.50
AL	MONROE	01099	0.35
AL	MONTGOMERY	01101	0.70
AL	MORGAN	01103	0.30
AL	PERRY	01105	0.30
AL	PICKENS	01107	0.30
AL	PIKE	01109	0.55
AL	RANDOLPH	01111	0.70
AL	RUSSELL	01113	0.70
AL	SAINT CLAIR	01115	0.30
AL	SHELBY	01117	0.30
AL	SUMTER	01119	0.30
AL	TALLADEGA	01121	0.30
AL	TALLAPOOSA	01123	0.70
AL	TUSCALOOSA	01125	0.30
AL	WALKER	01127	0.20

State	Country/parish	FIPS	Class I price ad-justment	State	Country/parish	FIPS	Class I price ad-justment
AL	WASHINGTON	01129	0.35	AR	UNION	05139	0.10
AL	WILCOX	01131	0.50	AR	VAN BUREN	05141	0.10
AL	WINSTON	01133	0.20	AR	WASHINGTON	05143	0.10
AR	ARKANSAS	05001	0.00	AR	WHITE	05145	0.10
AR	ASHLEY	05003	0.10	AR	WOODRUFF	05147	0.10
AR	BAXTER	05005	0.10	AR	YELL	05149	0.10
AR	BENTON	05007	0.10	FL	ESCAMBIA	12033	0.55
AR	BOONE	05009	0.10	FL	OKALOOSA	12091	0.55
AR	BRADLEY	05011	0.30	FL	SANTA ROSA	12113	0.55
AR	CALHOUN	05013	0.30	FL	WALTON	12131	0.55
AR	CARROLL	05015	0.10	GA	APPLING	13001	1.15
AR	CHICOT	05017	0.10	GA	ATKINSON	13003	1.15
AR	CLARK	05019	0.00	GA	BACON	13005	1.15
AR	CLAY	05021	0.10	GA	BAKER	13007	0.85
AR	CLEBURNE	05023	0.10	GA	BALDWIN	13009	0.70
AR	CLEVELAND	05025	0.30	GA	BANKS	13011	0.70
AR	COLUMBIA	05027	0.10	GA	BARROW	13013	0.70
AR	CONWAY	05029	0.10	GA	BARTOW	13015	0.30
AR	CRAIGHEAD	05031	0.10	GA	BEN HILL	13017	1.15
AR	CRAWFORD	05033	0.10	GA	BERRIEN	13019	1.15
AR	CRITTENDEN	05035	0.10	GA	BIBB	13021	0.70
AR	CROSS	05037	0.10	GA	BLECKLEY	13023	1.00
AR	DALLAS	05039	0.00	GA	BRANTLEY	13025	1.15
AR	DESHA	05041	0.30	GA	BROOKS	13027	1.15
AR	DREW	05043	0.30	GA	BRYAN	13029	1.15
AR	FAULKNER	05045	0.10	GA	BULLOCH	13031	1.00
AR	FRANKLIN	05047	0.10	GA	BURKE	13033	0.70
AR	FULTON	05049	0.10	GA	BUTTS	13035	0.70
AR	GARLAND	05051	0.10	GA	CALHOUN	13037	0.85
AR	GRANT	05053	0.00	GA	CAMDEN	13039	1.15
AR	GREENE	05055	0.10	GA	CANDLER	13043	1.00
AR	HEMPSTEAD	05057	0.30	GA	CARROLL	13045	0.70
AR	HOT SPRING	05059	0.00	GA	CHARLTON	13049	1.15
AR	HOWARD	05061	0.00	GA	CHATHAM	13051	1.15
AR	INDEPENDENCE	05063	0.10	GA	CHATTAHOOCHEE	13053	0.70
AR	IZARD	05065	0.10	GA	CHEROKEE	13057	0.30
AR	JACKSON	05067	0.10	GA	CLARKE	13059	0.70
AR	JEFFERSON	05069	0.00	GA	CLAY	13061	0.85
AR	JOHNSON	05071	0.10	GA	CLAYTON	13063	0.70
AR	LAFAYETTE	05073	0.10	GA	CLINCH	13065	1.15
AR	LAWRENCE	05075	0.10	GA	COBB	13067	0.70
AR	LEE	05077	0.10	GA	COFFEE	13069	1.15
AR	LINCOLN	05079	0.30	GA	COLQUITT	13071	1.15
AR	LITTLE RIVER	05081	0.30	GA	COLUMBIA	13073	0.70
AR	LOGAN	05083	0.10	GA	COOK	13075	1.15
AR	LONOKE	05085	0.10	GA	COWETA	13077	0.70
AR	MADISON	05087	0.10	GA	CRAWFORD	13079	0.70
AR	MARION	05089	0.10	GA	CRISP	13081	0.85
AR	MILLER	05091	0.10	GA	DAWSON	13085	0.30
AR	MISSISSIPPI	05093	0.30	GA	DECATUR	13087	1.15
AR	MONROE	05095	0.10	GA	DE KALB	13089	0.70
AR	MONTGOMERY	05097	0.10	GA	DODGE	13091	0.85
AR	NEVADA	05099	0.30	GA	DOOLY	13093	0.85
AR	NEWTON	05101	0.10	GA	DOUGHERTY	13095	0.85
AR	OUACHITA	05103	0.30	GA	DOUGLAS	13097	0.70
AR	PERRY	05105	0.10	GA	EARLY	13099	0.85
AR	PHILLIPS	05107	0.00	GA	ECHOLS	13101	1.15
AR	PIKE	05109	0.00	GA	EFFINGHAM	13103	1.00
AR	POINSETT	05111	0.30	GA	ELBERT	13105	0.70
AR	POLK	05113	0.10	GA	EMANUEL	13107	1.00
AR	POPE	05115	0.10	GA	EVANS	13109	1.15
AR	PRAIRIE	05117	0.10	GA	FAYETTE	13113	0.70
AR	PULASKI	05119	0.10	GA	FLOYD	13115	0.30
AR	RANDOLPH	05121	0.10	GA	FORSYTH	13117	0.70
AR	SAINT FRANCIS	05123	0.10	GA	FRANKLIN	13119	0.70
AR	SALINE	05125	0.10	GA	FULTON	13121	0.70
AR	SCOTT	05127	0.10	GA	GILMER	13123	0.30
AR	SEARCY	05129	0.10	GA	GLASCOCK	13125	0.90
AR	SEBASTIAN	05131	0.10	GA	GLYNN	13127	1.15
AR	SEVIER	05133	0.00	GA	GORDON	13129	0.30
AR	SHARP	05135	0.10	GA	GRADY	13131	1.15
AR	STONE	05137	0.10	GA	GREENE	13133	0.70

State	Country/parish	FIPS	Class I price adjustment	State	Country/parish	FIPS	Class I price adjustment
GA	GWINNETT	13135	0.70	GA	TREUTLEN	13283	1.00
GA	HABERSHAM	13137	0.30	GA	TROUP	13285	0.70
GA	HALL	13139	0.70	GA	TURNER	13287	0.85
GA	HANCOCK	13141	0.70	GA	TWIGGS	13289	0.70
GA	HARALSON	13143	0.70	GA	UNION	13291	0.30
GA	HARRIS	13145	0.70	GA	UPSON	13293	0.70
GA	HART	13147	0.70	GA	WALTON	13297	0.70
GA	HEARD	13149	0.70	GA	WARE	13299	1.15
GA	HENRY	13151	0.70	GA	WARREN	13301	0.70
GA	HOUSTON	13153	0.70	GA	WASHINGTON	13303	0.70
GA	IRWIN	13155	1.15	GA	WAYNE	13305	1.15
GA	JACKSON	13157	0.70	GA	WEBSTER	13307	0.55
GA	JASPER	13159	0.70	GA	WHEELER	13309	1.15
GA	JEFF DAVIS	13161	1.15	GA	WHITE	13311	0.30
GA	JEFFERSON	13163	0.70	GA	WILCOX	13315	0.85
GA	JENKINS	13165	1.00	GA	WILKES	13317	0.70
GA	JOHNSON	13167	1.00	GA	WILKINSON	13319	0.70
GA	JONES	13169	0.70	GA	WORTH	13321	0.85
GA	LAMAR	13171	0.70	KY	ALLEN	21003	0.20
GA	LANIER	13173	1.15	KY	BALLARD	21007	0.30
GA	LAURENS	13175	1.00	KY	BARREN	21009	0.20
GA	LEE	13177	0.85	KY	CALDWELL	21033	0.20
GA	LIBERTY	13179	1.15	KY	CALLOWAY	21035	0.30
GA	LINCOLN	13181	0.70	KY	CARLISLE	21039	0.30
GA	LONG	13183	1.15	KY	CHRISTIAN	21047	0.20
GA	LOWNDES	13185	1.15	KY	CRITTENDEN	21055	0.20
GA	LUMPKIN	13187	0.30	KY	FULTON	21075	0.30
GA	MCDUFFIE	13189	0.70	KY	GRAVES	21083	0.30
GA	MCINTOSH	13191	1.15	KY	HICKMAN	21105	0.30
GA	MACON	13193	0.70	KY	LIVINGSTON	21139	0.30
GA	MADISON	13195	0.70	KY	LOGAN	21141	0.20
GA	MARION	13197	0.70	KY	LYON	21143	0.20
GA	MERIWETHER	13199	0.70	KY	MCCRACKEN	21145	0.30
GA	MILLER	13201	0.85	KY	MARSHALL	21157	0.30
GA	MITCHELL	13205	1.15	KY	METCALFE	21169	0.20
GA	MONROE	13207	0.70	KY	MONROE	21171	0.50
GA	MONTGOMERY	13209	1.15	KY	SIMPSON	21213	0.20
GA	MORGAN	13211	0.70	KY	TODD	21219	0.20
GA	MUSCOGEE	13215	0.70	KY	TRIGG	21221	0.20
GA	NEWTON	13217	0.70	KY	WARREN	21227	0.20
GA	OCONEE	13219	0.70	LA	ACADIA	22001	0.30
GA	OGLETHORPE	13221	0.70	LA	ALLEN	22003	0.30
GA	PAULDING	13223	0.70	LA	ASCENSION	22005	0.20
GA	PEACH	13225	0.70	LA	ASSUMPTION	22007	0.20
GA	PICKENS	13227	0.30	LA	AVOYELLES	22009	0.00
GA	PIERCE	13229	1.15	LA	BEAUREGARD	22011	0.30
GA	PIKE	13231	0.70	LA	BIENVILLE	22013	0.00
GA	POLK	13233	0.70	LA	BOSSIER	22015	0.10
GA	PULASKI	13235	0.85	LA	CADDO	22017	0.10
GA	PUTNAM	13237	0.70	LA	CALCASIEU	22019	0.30
GA	QUITMAN	13239	0.85	LA	CALDWELL	22021	0.00
GA	RABUN	13241	0.30	LA	CAMERON	22023	0.20
GA	RANDOLPH	13243	0.85	LA	CATAHOULA	22025	0.00
GA	RICHMOND	13245	0.70	LA	CLAIBORNE	22027	0.10
GA	ROCKDALE	13247	0.70	LA	CONCORDIA	22029	0.00
GA	SCHLEY	13249	0.70	LA	DE SOTO	22031	0.00
GA	SCREVEN	13251	1.00	LA	EAST BATON ROUGE	22033	0.20
GA	SEMINOLE	13253	1.15	LA	EAST CARROLL	22035	0.20
GA	SPALDING	13255	0.70	LA	EAST FELICIANA	22037	0.30
GA	STEPHENS	13257	0.30	LA	EVANGELINE	22039	0.30
GA	STEWART	13259	0.55	LA	FRANKLIN	22041	0.00
GA	SUMTER	13261	0.85	LA	GRANT	22043	0.00
GA	TALBOT	13263	0.70	LA	IBERIA	22045	0.20
GA	TALIAFERRO	13265	0.70	LA	IBERVILLE	22047	0.20
GA	TATTNALL	13267	1.15	LA	JACKSON	22049	0.00
GA	TAYLOR	13269	0.70	LA	JEFFERSON	22051	0.20
GA	TELFAIR	13271	1.15	LA	JEFFERSON DAVIS	22053	0.30
GA	TERRELL	13273	0.85	LA	LAFAYETTE	22055	0.20
GA	THOMAS	13275	1.15	LA	LAFOURCHE	22057	0.20
GA	TIFT	13277	1.15	LA	LA SALLE	22059	0.00
GA	TOOMBS	13279	1.15	LA	LINCOLN	22061	0.10
GA	TOWNS	13281	0.30	LA	LIVINGSTON	22063	0.20

State	Country/parish	FIPS	Class I price adjustment	State	Country/parish	FIPS	Class I price adjustment
LA	MADISON	22065	0.00	MS	LEE	28081	0.30
LA	MOREHOUSE	22067	0.10	MS	LEFLORE	28083	0.10
LA	NATCHITOCHES	22069	0.00	MS	LINCOLN	28085	0.00
LA	ORLEANS	22071	0.20	MS	LOWNDES	28087	0.20
LA	OUACHITA	22073	0.10	MS	MADISON	28089	0.20
LA	PLAQUEMINES	22075	0.20	MS	MARION	28091	0.40
LA	POINTE COUPEE	22077	0.30	MS	MARSHALL	28093	0.00
LA	RAPIDES	22079	0.00	MS	MONROE	28095	0.20
LA	RED RIVER	22081	0.00	MS	MONTGOMERY	28097	0.20
LA	RICHLAND	22083	0.20	MS	NESHOBA	28099	0.20
LA	SABINE	22085	0.00	MS	NEWTON	28101	0.10
LA	SAINT BERNARD	22087	0.20	MS	NOXUBEE	28103	0.30
LA	SAINT CHARLES	22089	0.20	MS	OKTIBBEHA	28105	0.20
LA	SAINT HELENA	22091	0.30	MS	PANOLA	28107	0.30
LA	SAINT JAMES	22093	0.20	MS	PEARL RIVER	28109	0.40
LA	SAINT JOHN THE BAPTIST	22095	0.20	MS	PERRY	28111	0.40
LA	SAINT LANDRY	22097	0.30	MS	PIKE	28113	0.40
LA	SAINT MARTIN	22099	0.20	MS	PONTOTOC	28115	0.30
LA	SAINT MARY	22101	0.20	MS	PRENTISS	28117	0.30
LA	SAINT TAMMANY	22103	0.30	MS	QUITMAN	28119	0.30
LA	TANGIPAHOA	22105	0.20	MS	RANKIN	28121	0.10
LA	TENSAS	22107	0.00	MS	SCOTT	28123	0.10
LA	TERREBONNE	22109	0.20	MS	SHARKEY	28125	0.20
LA	UNION	22111	0.10	MS	SIMPSON	28127	0.10
LA	VERMILION	22113	0.20	MS	SMITH	28129	0.10
LA	VERNON	22115	0.00	MS	STONE	28131	0.40
LA	WASHINGTON	22117	0.30	MS	SUNFLOWER	28133	0.10
LA	WEBSTER	22119	0.10	MS	TALLAHATCHIE	28135	0.10
LA	WEST BATON ROUGE	22121	0.20	MS	TATE	28137	0.00
LA	WEST CARROLL	22123	0.10	MS	TIPPAH	28139	0.30
LA	WEST FELICIANA	22125	0.30	MS	TISHOMINGO	28141	0.30
LA	WINN	22127	0.00	MS	TUNICA	28143	0.00
MS	ADAMS	28001	0.00	MS	UNION	28145	0.30
MS	ALCORN	28003	0.30	MS	WALTHALL	28147	0.40
MS	AMITE	28005	0.40	MS	WARREN	28149	0.00
MS	ATTALA	28007	0.20	MS	WASHINGTON	28151	0.10
MS	BENTON	28009	0.30	MS	WAYNE	28153	0.40
MS	BOLIVAR	28011	0.10	MS	WEBSTER	28155	0.20
MS	CALHOUN	28013	0.10	MS	WILKINSON	28157	0.40
MS	CARROLL	28015	0.20	MS	WINSTON	28159	0.20
MS	CHICKASAW	28017	0.10	MS	YALOBUSHA	28161	0.10
MS	CHOCTAW	28019	0.20	MS	YAZOO	28163	0.20
MS	CLAIBORNE	28021	0.10	MO	BARRY	29009	0.20
MS	CLARKE	28023	0.50	MO	BARTON	29011	0.20
MS	CLAY	28025	0.20	MO	BOLLINGER	29017	0.20
MS	COAHOMA	28027	0.30	MO	BUTLER	29023	0.20
MS	COPIAH	28029	0.10	MO	CAPE GIRARDEAU	29031	0.20
MS	COVINGTON	28031	0.00	MO	CARTER	29035	0.20
MS	DE SOTO	28033	0.00	MO	CEDAR	29039	0.20
MS	FORREST	28035	0.40	MO	CHRISTIAN	29043	0.20
MS	FRANKLIN	28037	0.00	MO	CRAWFORD	29055	0.40
MS	GEORGE	28039	0.40	MO	DADE	29057	0.20
MS	GREENE	28041	0.40	MO	DALLAS	29059	0.20
MS	GRENADA	28043	0.10	MO	DENT	29065	0.40
MS	HANCOCK	28045	0.30	MO	DOUGLAS	29067	0.20
MS	HARRISON	28047	0.30	MO	DUNKLIN	29069	0.50
MS	HINDS	28049	0.00	MO	GREENE	29077	0.20
MS	HOLMES	28051	0.20	MO	HOWELL	29091	0.20
MS	HUMPHREYS	28053	0.20	MO	IRON	29093	0.40
MS	ISSAQUENA	28055	0.20	MO	JASPER	29097	0.20
MS	ITAWAMBA	28057	0.30	MO	LACLEDE	29105	0.20
MS	JACKSON	28059	0.30	MO	LAWRENCE	29109	0.20
MS	JASPER	28061	0.10	MO	MCDONALD	29119	0.20
MS	JEFFERSON	28063	0.00	MO	MADISON	29123	0.20
MS	JEFFERSON DAVIS	28065	0.00	MO	MISSISSIPPI	29133	0.50
MS	JONES	28067	0.40	MO	NEW MADRID	29143	0.50
MS	KEMPER	28069	0.30	MO	NEWTON	29145	0.20
MS	LAFAYETTE	28071	0.30	MO	OREGON	29149	0.20
MS	LAMAR	28073	0.40	MO	OZARK	29153	0.20
MS	LAUDERDALE	28075	0.10	MO	PEMISCOT	29155	0.50
MS	LAWRENCE	28077	0.00	MO	PERRY	29157	0.20
MS	LEAKE	28079	0.20	MO	POLK	29167	0.20

State	Country/parish	FIPS	Class I price adjustment	State	Country/parish	FIPS	Class I price adjustment
MO	REYNOLDS	29179	0.20	TN	HENRY	47079	0.10
MO	RIPLEY	29181	0.20	TN	HICKMAN	47081	0.30
MO	SAINT FRANCOIS	29187	0.40	TN	HOUSTON	47083	0.30
MO	SCOTT	29201	0.20	TN	HUMPHREYS	47085	0.30
MO	SHANNON	29203	0.20	TN	JACKSON	47087	0.30
MO	STODDARD	29207	0.20	TN	LAKE	47095	0.10
MO	STONE	29209	0.20	TN	LAUDERDALE	47097	0.30
MO	TANEY	29213	0.20	TN	LAWRENCE	47099	0.40
MO	TEXAS	29215	0.20	TN	LEWIS	47101	0.30
MO	VERNON	29217	0.20	TN	LINCOLN	47103	0.40
MO	WASHINGTON	29221	0.40	TN	MCNAIRY	47109	0.10
MO	WAYNE	29223	0.20	TN	MACON	47111	0.30
MO	WEBSTER	29225	0.20	TN	MADISON	47113	0.30
MO	WRIGHT	29229	0.20	TN	MARSHALL	47117	0.30
TN	BEDFORD	47003	0.30	TN	MAURY	47119	0.30
TN	BENTON	47005	0.30	TN	MONTGOMERY	47125	0.30
TN	BLEDSOE	47007	0.60	TN	MOORE	47127	0.40
TN	CANNON	47015	0.30	TN	OBION	47131	0.10
TN	CARROLL	47017	0.10	TN	OVERTON	47133	0.30
TN	CHEATHAM	47021	0.30	TN	PERRY	47135	0.30
TN	CHESTER	47023	0.10	TN	PICKETT	47137	0.30
TN	CLAY	47027	0.30	TN	PUTNAM	47141	0.30
TN	COFFEE	47031	0.60	TN	ROBERTSON	47147	0.30
TN	CROCKETT	47033	0.30	TN	RUTHERFORD	47149	0.30
TN	DAVIDSON	47037	0.30	TN	SHELBY	47157	0.10
TN	DECATUR	47039	0.30	TN	SMITH	47159	0.30
TN	DE KALB	47041	0.30	TN	STEWART	47161	0.30
TN	DICKSON	47043	0.30	TN	SUMNER	47165	0.30
TN	DYER	47045	0.10	TN	TIPTON	47167	0.10
TN	FAYETTE	47047	0.10	TN	TROUSDALE	47169	0.30
TN	FENTRESS	47049	0.30	TN	VAN BUREN	47175	0.60
TN	FRANKLIN	47051	0.40	TN	WARREN	47177	0.60
TN	GIBSON	47053	0.10	TN	WAYNE	47181	0.40
TN	GILES	47055	0.40	TN	WEAKLEY	47183	0.10
TN	GRUNDY	47061	0.60	TN	WHITE	47185	0.30
TN	HARDEMAN	47069	0.10	TN	WILLIAMSON	47187	0.30
TN	HARDIN	47071	0.10	TN	WILSON	47189	0.30
TN	HAYWOOD	47075	0.30				
TN	HENDERSON	47077	0.30				

[73 FR 14163, Mar. 17, 2008]

§ 1007.52 Adjusted Class I differentials.

See § 1000.52.

§ 1007.53 Announcement of class prices, component prices, and advanced pricing factors.

See § 1000.53.

§ 1007.54 Equivalent price.

See § 1000.54.

UNIFORM PRICES

§ 1007.60 Handler's value of milk.

For the purpose of computing a handler's obligation for producer milk, the market administrator shall determine for each month the value of milk of each handler with respect to each of the handler's pool plants and of each handler described in § 1000.9(c) with respect to milk that was not received at a pool plant by adding the amounts computed in paragraphs (a) through (e) of this section and subtracting from that total amount the value computed in paragraph (f) of this section. Receipts of nonfluid milk products that are distributed as labeled reconstituted milk for which payments are made to the producer-settlement fund of another Federal order under § 1000.76(a)(4) or (d) shall be excluded from pricing under this section.

(a) Multiply the pounds of skim milk and butterfat in producer milk that were classified in each class pursuant to § 1000.44(c) by the applicable skim milk and butterfat prices, and add the resulting amounts; except that for the months of January 2005 through March 2005, the Class I skim milk price for this purpose shall be the Class I skim milk price as determined in § 1000.50(b) plus $0.04 per hundredweight, and the Class I butterfat price for this purpose shall be the Class I butterfat price as determined in § 1000.50(c) plus $0.0004 per pound. The adjustments to the Class I skim milk and butterfat prices provided herein may be reduced by the market administrator for any month if the market administrator determines that the payments yet unpaid computed pursuant to paragraphs (g)(1) through (5) and paragraph (g)(7) of this section will be less than the amount computed pursuant to paragraph (g)(6) of this section. The adjustments to the Class I skim milk and butterfat prices provided herein during the months of January 2005 through March 2005 shall be announced along with the prices announced in § 1000.53(b);

(b) Multiply the pounds of skim milk and butterfat overage assigned to each class pursuant to § 1000.44(a)(11) by the respective skim milk and butterfat prices applicable at the location of the pool plant;

(c) Multiply the difference between the Class IV price for the preceding month and the current month's Class I, II, or III price, as the case may be, by the hundredweight of skim milk and butterfat subtracted from Class I, II, or III, respectively, pursuant to § 1000.44(a)(7) and the corresponding step of § 1000.44(b);

(d) Multiply the difference between the Class I price applicable at the location of the pool plant and the Class IV price by the hundredweight of skim milk and butterfat assigned to Class I pursuant to § 1000.43(d) and the hundredweight of skim milk and butterfat subtracted from Class I pursuant to § 1000.44(a)(3)(i) through (vi) and the corresponding step of § 1000.44(b), excluding receipts of bulk fluid cream products from a plant regulated under other Federal orders and bulk concentrated fluid milk products from pool plants, plants regulated under other Federal orders, and unregulated supply plants;

(e) Multiply the Class I skim milk and Class I butterfat prices applicable at the location of the nearest unregulated supply plants from which an equivalent volume was received by the pounds of skim milk and butterfat in receipts of concentrated fluid milk products assigned to Class I pursuant to § 1000.43(d) and § 1000.44(a)(3)(i) and the corresponding step of § 1000.44(b) and the pounds of skim milk and butterfat subtracted from Class I pursuant to § 1000.44(a)(8) and the corresponding step of § 1000.44(b), excluding such skim milk and butterfat in receipts of fluid milk products from an unregulated supply plant to the extent that an equivalent amount of skim milk or butterfat disposed of to such plant by handlers fully regulated under any Federal milk order is classified and priced as Class I milk and is not used as an offset for any other payment obligation under any order; and

(f) For reconstituted milk made from receipts of nonfluid milk products, multiply $1.00 (but not more than the difference between the Class I price applicable at the location of the pool plant and the Class IV price) by the hundredweight of skim milk and butterfat contained in receipts of nonfluid milk products that are allocated to Class I use pursuant to § 1000.43(d).

(g) For the months of January 2005 through March 2005 for handlers who have submitted proof satisfactory to the market administrator to determine eligibility for reimbursement of transportation costs, subtract an amount equal to:

(1) The cost of transportation on loads of producer milk delivered or rerouted to a pool distributing plant which were delivered as a result of hurricanes Charley, Frances, Ivan, and Jeanne;

(2) The cost of transportation on loads of producer milk delivered or rerouted to a pool supply plant that was then transferred to a pool distributing plant which were delivered as a result of hurricanes Charley, Frances, Ivan, and Jeanne;

(3) The cost of transportation on loads of bulk milk delivered or rerouted to a pool distributing plant from a pool supply plant which were delivered as a result of hurricanes Charley, Frances, Ivan, and Jeanne;

(4) The cost of transportation on loads of bulk milk delivered or rerouted to a pool distributing plant from another order plant which were

delivered as a result of hurricanes Charley, Frances, Ivan, and Jeanne; and

(5) The cost of transportation on loads of bulk milk transferred or diverted to a plant regulated under another Federal order or to other nonpool plants which were delivered as a result of hurricanes Charley, Frances, Ivan, and Jeanne.

(6) The total amount of payment to all handlers under this section shall be limited for each month to an amount determined by multiplying the total Class I producer milk for all handlers pursuant to §1000.44(c) times $0.04 per hundredweight.

(7) If the cost of transportation computed pursuant to paragraphs (g)(1) through (5) of this section exceeds the amount computed pursuant to paragraph (g)(6) of this section, the market administrator shall prorate such payments to each handler based on each handler's proportion of transportation costs submitted pursuant to paragraphs (g)(1) through (5) of this section. Transportation costs submitted pursuant to paragraphs (g)(1) through (5) of this section which are not paid as a result of such a proration shall be included in each subsequent month's transportation costs submitted pursuant to paragraphs (g)(1) through (5) of this section until paid, or until the time period for such payments has concluded.

(8) The reimbursement of transportation costs pursuant to this section shall be the actual demonstrated cost of such transportation of bulk milk delivered or rerouted as described in paragraphs (g)(1) through (5) of this section, or the miles of transportation on loads of bulk milk delivered or rerouted as described in paragraphs (g)(1) through (5) of this section multiplied by $2.25 per loaded mile, whichever is less.

(9) For each handler, the reimbursement of transportation costs pursuant to paragraph (g) of this section for bulk milk delivered or rerouted as described in paragraphs (g)(1) through (5) of this section shall be reduced by the amount of payments received for such milk movements from the transportation credit balancing fund pursuant to §1007.82.

[64 FR 47966, Sept. 1, 1999, as amended at 65 FR 82835, Dec. 28, 2000; 69 FR 71700, Dec. 10, 2004]

§1007.61 Computation of uniform prices.

On or before the 11th day of each month, the market administrator shall compute a uniform butterfat price, a uniform skim milk price, and a uniform price for producer milk receipts reported for the prior month. The report of any handler who has not made payments required pursuant to §1007.71 for the preceding month shall not be included in the computation of these prices, and such handler's report shall not be included in the computation for succeeding months until the handler has made full payment of outstanding monthly obligations.

(a) *Uniform butterfat price.* The uniform butterfat price per pound, rounded to the nearest one-hundredth cent, shall be computed by:

(1) Multiplying the pounds of butterfat in producer milk allocated to each class pursuant to §1000.44(b) by the respective class butterfat prices;

(2) Adding the butterfat value calculated in §1007.60(e) for other source milk allocated to Class I pursuant to §1000.43(d) and the steps of §1000.44(b) that correspond to §1000.44(a)(3)(i) and §1000.44(a)(8) by the Class I price; and

(3) Dividing the sum of paragraphs (a)(1) and (a)(2) of this section by the sum of the pounds of butterfat in producer milk and other source milk used to calculate the values in paragraphs (a)(1) and (a)(2) of this section.

(b) *Uniform skim milk price.* The uniform skim milk price per hundredweight, rounded to the nearest cent, shall be computed as follows:

(1) Combine into one total the values computed pursuant to §1007.60 for all handlers;

(2) Add an amount equal to the minus location adjustments and subtract an amount equal to the plus location adjustments computed pursuant to §1007.75;

(3) Add an amount equal to not less than one-half of the unobligated balance in the producer-settlement fund;

(4) Subtract the value of the total pounds of butterfat for all handlers. The butterfat value shall be computed by multiplying the sum of the pounds of butterfat in producer milk and other source milk used to calculate the values in paragraphs (a)(1) and (a)(2) of this section by the butterfat price computed in paragraph (a) of this section;

(5) Divide the resulting amount by the sum of the following for all handlers included in these computations:

(i) The total skim pounds of producer milk; and

(ii) The total skim pounds for which a value is computed pursuant to § 1007.60(e); and

(6) Subtract not less than 4 cents and not more than 5 cents.

(c) *Uniform price.* The uniform price per hundredweight, rounded to the nearest cent, shall be the sum of the following:

(1) Multiply the uniform butterfat price for the month pursuant to paragraph (a) of this section times 3.5 pounds of butterfat; and

(2) Multiply the uniform skim milk price for the month pursuant to paragraph (b) of this section times 96.5 pounds of skim milk.

[64 FR 47966, Sept. 1, 1999, as amended at 65 FR 82835, Dec. 28, 2000]

§ 1007.62 Announcement of uniform prices.

On or before the 11th day after the end of the month, the market administrator shall announce the uniform prices for the month computed pursuant to § 1007.61.

PAYMENTS FOR MILK

§ 1007.70 Producer-settlement fund.

See § 1000.70.

§ 1007.71 Payments to the producer-settlement fund.

Each handler shall make a payment to the producer-settlement fund in a manner that provides receipt of the funds by the market administrator no later than the 12th day after the end of the month (except as provided in § 1000.90). Payment shall be the amount, if any, by which the amount specified in paragraph (a) of this sec-tion exceeds the amount specified in paragraph (b) of this section:

(a) The total value of milk of the handler for the month as determined pursuant to § 1007.60; and

(b) The sum of the value at the uniform prices for skim milk and butterfat, adjusted for plant location, of the handler's receipts of producer milk; and the value at the uniform price, as adjusted pursuant to § 1007.75, applicable at the location of the plant from which received of other source milk for which a value is computed pursuant to § 1007.60(e).

§ 1007.72 Payments from the producer-settlement fund.

No later than one day after the date of payment receipt required under § 1007.71, the market administrator shall pay to each handler the amount, if any, by which the amount computed pursuant to § 1007.71(b) exceeds the amount computed pursuant to § 1007.71(a). If, at such time, the balance in the producer-settlement fund is insufficient to make all payments pursuant to this section, the market administrator shall reduce uniformly such payments and shall complete the payments as soon as the funds are available.

§ 1007.73 Payments to producers and to cooperative associations.

(a) Each handler that is not paying a cooperative association for producer milk shall pay each producer as follows:

(1) *Partial payment.* For each producer who has not discontinued shipments as of the 23rd day of the month, payment shall be made so that it is received by the producer on or before the 26th day of the month (except as provided in § 1000.90) for milk received during the first 15 days of the month at not less than 90 percent of the preceding month's uniform price, adjusted for plant location pursuant to § 1007.75 and proper deductions authorized in writing by the producer.

(2) *Final payment.* For milk received during the month, a payment computed as follows shall be made so that it is received by each producer one day after the payment date required in § 1007.72:

(i) Multiply the hundredweight of producer skim milk received times the uniform skim milk price for the month;

(ii) Multiply the pounds of butterfat received times the uniform butterfat price for the month;

(iii) Multiply the hundredweight of producer milk received times the plant location adjustment pursuant to §1007.75; and

(iv) Add the amounts computed in paragraph (a)(2)(i), (ii), and (iii) of this section, and from that sum:

(A) Subtract the partial payment made pursuant to paragraph (a)(1) of this section;

(B) Subtract the deduction for marketing services pursuant to §1000.86;

(C) Add or subtract for errors made in previous payments to the producer; and

(D) Subtract proper deductions authorized in writing by the producer.

(b) One day before partial and final payments are due pursuant to paragraph (a) of this section, each handler shall pay a cooperative association for milk received as follows:

(1) *Partial payment to a cooperative association for bulk milk received directly from producers' farms.* For bulk milk (including the milk of producers who are not members of such association and who the market administrator determines have authorized the cooperative association to collect payment for their milk) received during the first 15 days of the month from a cooperative association in any capacity, except as the operator of a pool plant, the payment shall be equal to the hundredweight of milk received multiplied by 90 percent of the preceding month's uniform price, adjusted for plant location pursuant to §1007.75.

(2) *Partial payment to a cooperative association for milk transferred from its pool plant.* For bulk fluid milk products and bulk fluid cream products received during the first 15 days of the month from a cooperative association in its capacity as the operator of a pool plant, the partial payment shall be at the pool plant operator's estimated use value of the milk using the most recent class prices available for skim milk and butterfat at the receiving plant's location.

(3) *Final payment to a cooperative association for milk transferred from its pool plant.* For bulk fluid milk products and bulk fluid cream products received during the month from a cooperative association in its capacity as the operator of a pool plant, the final payment shall be the classified value of such milk as determined by multiplying the pounds of skim milk and butterfat assigned to each class pursuant to §1000.44 by the class prices for the month at the receiving plant's location, and subtracting from this sum the partial payment made pursuant to paragraph (b)(2) of this section.

(4) *Final payment to a cooperative association for bulk milk received directly from producers' farms.* For bulk milk received from a cooperative association during the month, including the milk of producers who are not members of such association and who the market administrator determines have authorized the cooperative association to collect payment for their milk, the final payment for such milk shall be an amount equal to the sum of the individual payments otherwise payable for such milk pursuant to paragraph (a)(2) of this section.

(c) If a handler has not received full payment from the market administrator pursuant to §1007.72 by the payment date specified in paragraph (a) or (b) of this section, the handler may reduce payments pursuant to paragraphs (a) and (b) of this section, but by not more than the amount of the underpayment. The payments shall be completed on the next scheduled payment date after receipt of the balance due from the market administrator.

(d) If a handler claims that a required payment to a producer cannot be made because the producer is deceased or cannot be located, or because the cooperative association or its lawful successor or assignee is no longer in existence, the payment shall be made to the producer-settlement fund, and in the event that the handler subsequently locates and pays the producer or a lawful claimant, or in the event that the handler no longer exists and a lawful claim is later established, the market administrator shall make the required payment from the producer-settlement fund to the handler or to the lawful claimant as the case may be.

(e) In making payments to producers pursuant to this section, each pool plant operator shall furnish each producer, except a producer whose milk was received from a cooperative association described in § 1000.9(a) or (c), a supporting statement in such form that it may be retained by the recipient which shall show:

(1) The name, address, Grade A identifier assigned by a duly constituted regulatory agency, and the payroll number of the producer;

(2) The month and dates that milk was received from the producer, including the daily and total pounds of milk received;

(3) The total pounds of butterfat in the producer's milk;

(4) The minimum rate or rates at which payment to the producer is required pursuant to this order;

(5) The rate used in making payment if the rate is other than the applicable minimum rate;

(6) The amount, or rate per hundredweight, and nature of each deduction claimed by the handler; and

(7) The net amount of payment to the producer or cooperative association.

[64 FR 47971, Sept. 1, 1999, as amended at 65 FR 32010, May 22, 2000]

§ 1007.74 [Reserved]

§ 1007.75 Plant location adjustments for producer milk and nonpool milk.

For purposes of making payments for producer milk and nonpool milk, a plant location adjustment shall be determined by subtracting the Class I price specified in § 1007.51 from the Class I price at the plant's location. The difference, plus or minus as the case may be, shall be used to adjust the payments required pursuant to §§ 1007.73 and 1000.76.

§ 1007.76 Payments by a handler operating a partially regulated distributing plant.

See § 1000.76.

§ 1007.77 Adjustment of accounts.

See § 1000.77.

§ 1007.78 Charges on overdue accounts.

See § 1000.78.

MARKETWIDE SERVICE PAYMENTS

§ 1007.80 Transportation credit balancing fund.

The market administrator shall maintain a separate fund known as the *Transportation Credit Balancing Fund* into which shall be deposited the payments made by handlers pursuant to § 1007.81 and out of which shall be made the payments due handlers pursuant to § 1007.82. Payments due a handler shall be offset against payments due from the handler.

§ 1007.81 Payments to the transportation credit balancing fund.

(a) On or before the 12th day after the end of the month (except as provided in § 1000.90), each handler operating a pool plant and each handler specified in § 1000.9 (c) shall pay to the market administrator a transportation credit balancing fund assessment determined by multiplying the pounds of Class I producer milk assigned pursuant to § 1007.44 by $0.30 per hundredweight or such lesser amount as the market administrator deems necessary to maintain a balance in the fund equal to the total transportation credits disbursed during the prior June–February period to reflect any changes in the current mileage rate versus the mileage rate(s) in effect during the prior June–February period. In the event that during any month of the June–February period the fund balance is insufficient to cover the amount of credits that are due, the assessment should be based upon the amount of credits that would have been disbursed had the fund balance been sufficient.

(b) The market administrator shall announce publicly on or before the 23rd day of the month (except as provided in § 1000.90 of this chapter) the assessment pursuant to paragraph (a) of this section for the following month.

[71 FR 62379, Oct. 25, 2006, as amended at 73 FR 14171, Mar. 17, 2008; 79 FR 25006, May 2, 2014; 79 FR 26591, May 9, 2014]

§ 1007.82 Payments from the transportation credit balancing fund.

(a) Payments from the transportation credit balancing fund to handlers and cooperative associations requesting transportation credits shall be made as follows:

(1) On or before the 13th day (except as provided in § 1000.90) after the end of each of the months of January, February and July through December and any other month in which transportation credits are in effect pursuant to paragraph (b) of this section, the market administrator shall pay to each handler that received, and reported pursuant to § 1007.30(a)(5), bulk milk transferred from a plant fully regulated under another Federal order as described in paragraph (c)(1) of this section or that received, and reported pursuant to § 1007.30(a)(6), milk directly from producers' farms as specified in paragraph (c)(2) of this section, a preliminary amount determined pursuant to paragraph (d) of this section to the extent that funds are available in the transportation credit balancing fund. If an insufficient balance exists to pay all of the credits computed pursuant to this section, the market administrator shall distribute the balance available in the transportation credit balancing fund by reducing payments pro rata using the percentage derived by dividing the balance in the fund by the total credits that are due for the month. The amount of credits resulting from this initial proration shall be subject to audit adjustment pursuant to paragraph (a)(2) of this section.

(2) The market administrator shall accept adjusted requests for transportation credits on or before the 20th day of the month following the month for which such credits were requested pursuant to § 1007.32(a). After such date, a preliminary audit will be conducted by the market administrator, who will recalculate any necessary proration of transportation credit payments for the preceding month pursuant to paragraph (a) of this section. Handlers will be promptly notified of an overpayment of credits based upon this final computation and remedial payments to or from the transportation credit balancing fund will be made on or before the next payment date for the following month;

(3) Transportation credits paid pursuant to paragraphs (a)(1) and (2) of this section shall be subject to final verification by the market administrator pursuant to § 1000.77. Adjusted payments to or from the transportation credit balancing fund will remain subject to the final proration established pursuant to paragraph (a)(2) of this section; and

(4) In the event that a qualified cooperative association is the responsible party for whose account such milk is received and written documentation of this fact is provided to the market administrator pursuant to § 1007.30(c)(3) prior to the date payment is due, the transportation credits for such milk computed pursuant to this section shall be made to such cooperative association rather than to the operator of the pool plant at which the milk was received.

(b) The market administrator may extend the period during which transportation credits are in effect (i.e., the transportation credit period) to the month of June if a written request to do so is received 15 days prior to the beginning of the month for which the request is made and, after conducting an independent investigation, finds that such extension is necessary to assure the market of an adequate supply of milk for fluid use. Before making such a finding, the market administrator shall notify the Deputy Administrator of Dairy Programs and all handlers in the market that an extension is being considered and invite written data, views, and arguments. Any decision to extend the transportation credit period must be issued in writing prior to the first day of the month for which the extension is to be effective.

(c) Transportation credits shall apply to the following milk:

(1) Bulk milk received at a pool distributing plant from a plant regulated under another Federal order, except Federal Order 1005; and

(2) Bulk milk received directly from the farms of dairy farmers at pool distributing plants subject to the following conditions:

(i) The dairy farmer was not a "producer" under this order for more than

45 days during the immediately preceding months of March through May, or not more than 50 percent of the production of the dairy farmer during those 3 months, in aggregate, was received as producer milk under this order during those 3 months; and

(ii) The farm on which the milk was produced is not located within the specified marketing area of the order in this part or the marketing area of Federal Order 1005 (7 CFR part 1005).

(iii) The market administrator may increase or decrease the milk production standard specified in paragraph (c)(2)(i) of this section if the market administrator finds that such revision is necessary to assure orderly marketing and efficient handling of milk in the marketing area. Before making such a finding, the market administrator shall investigate the need for the revision either on the market administrator's own initiative or at the request of interested persons. If the investigation shows that a revision might be appropriate, the market administrator shall issue a notice stating that the revision is being considered and inviting written data, views, and arguments. Any decision to revise an applicable percentage must be issued in writing at least one day before the effective date.

(d) Transportation credits shall be computed as follows:

(1) The market administrator shall subtract from the pounds of milk described in paragraphs (c)(1) and (2) of this section the pounds of bulk milk transferred from the pool plant receiving the supplemental milk if milk was transferred to a nonpool plant on the same calendar day that the supplemental milk was received. For this purpose, the transferred milk shall be subtracted from the most distant load of supplemental milk received, and then in sequence with the next most distant load until all of the transfers have been offset;

(2) With respect to the pounds of milk described in paragraph (c)(1) of this section that remain after the computations described in paragraph (d)(1) of this section, the market administrator shall:

(i) Determine the shortest hard-surface highway distance between the shipping plant and the receiving plant;

(ii) Multiply the number of miles so determined by the mileage rate for the month computed pursuant to § 1007.83(a)(6);

(iii) Subtract the applicable Class I price specified in § 1000.50(a) for the county in which the shipping plant is located from the Class I price applicable for the county in which the receiving plant is located;

(iv) Subtract any positive difference computed in paragraph (d)(2)(iii) of this section from the amount computed in paragraph (d)(2)(ii) of this section; and

(v) Multiply the remainder computed in paragraph (d)(2)(iv) of this section by the hundredweight of milk described in paragraph (d)(2) of this section.

(3) For the remaining milk described in paragraph (c)(2) of this section after computations described in paragraph (d)(1) of this section, the market administrator shall:

(i) Determine an origination point for each load of milk by locating the nearest city to the last producer's farm from which milk was picked up for delivery to the receiving pool plant;

(ii) Determine the shortest hard-surface highway distance between the receiving pool plant and the origination point;

(iii) Subtract 85 miles from the mileage so determined;

(iv) Multiply the remaining miles so computed by the mileage rate for the month computed pursuant to § 1007.83(a)(6);

(v) Subtract the Class I price specified in § 1000.50(a) applicable for the county in which the origination point is located from the Class I price applicable at the receiving pool plant's location;

(vi) Subtract any positive difference computed in paragraph (d)(3)(v) of this section from the amount computed in paragraph (d)(3)(iv) of this section; and

(vii) Multiply the remainder computed in paragraph (d)(3)(vi) of this section by the hundredweight of milk described in paragraph (d)(3) of this section.

[64 FR 47971, Sept. 1, 1999, as amended at 70 FR 59223, Oct. 12, 2005; 71 FR 62380, Oct. 25, 2006; 73 FR 14171, Mar. 17, 2008; 79 FR 25006, May 2, 2014; 79 FR 26591, May 9, 2014]

§ 1007.83 Mileage rate for the transportation credit balancing fund.

(a) The market administrator shall compute the mileage rate each month as follows:

(1) Compute the simple average rounded to three decimal places for the most recent 4 weeks of the Diesel Price per Gallon as reported by the Energy Information Administration of the United States Department of Energy for the Lower Atlantic and Gulf Coast Districts combined.

(2) From the result in paragraph (a)(1) in this section subtract $1.42 per gallon;

(3) Divide the result in paragraph (a)(2) of this section by 5.5, and round down to three decimal places to compute the fuel cost adjustment factor;

(4) Add the result in paragraph (a)(3) of this section to $1.91;

(5) Divide the result in paragraph (a)(4) of this section by 480;

(6) Round the result in paragraph (a)(5) of this section down to five decimal places to compute the mileage rate.

(b) The market administrator shall announce publicly on or before the 23rd day of the month (except as provided in § 1000.90 of this chapter) the mileage rate pursuant to paragraph (a) of this section for the following month.

[79 FR 25006, May 2, 2014; 79 FR 26591, May 9, 2014]

ADMINISTRATIVE ASSESSMENT AND MARKETING SERVICE DEDUCTION

§ 1007.85 Assessment for order administration.

On or before the payment receipt date specified under § 1007.71, each handler shall pay to the market administrator its *pro rata* share of the expense of administration of the order at a rate specified by the market administrator

that is no more than $.08 per hundredweight with respect to:

(a) Receipts of producer milk (including the handler's own production) other than such receipts by a handler described in § 1000.9(c) of this chapter that were delivered to pool plants of other handlers;

(b) Receipts from a handler described in § 1000.9(c) of this chapter;

(c) Receipts of concentrated fluid milk products from unregulated supply plants and receipts of nonfluid milk products assigned to Class I use pursuant to § 1000.43(d) of this chapter and other source milk allocated to Class I pursuant to § 1000.44(a)(3) and (8) of this chapter and the corresponding steps of § 1000.44(b) of this chapter, except other source milk that is excluded from the computations pursuant to § 1007.60(d) and (e); and

(d) Route disposition in the marketing area from a partially regulated distributing plant that exceeds the skim milk and butterfat subtracted pursuant to § 1000.76(a)(1)(i) and (ii) of this chapter.

[79 FR 25002, May 2, 2014; 79 FR 26591, May 9, 2014]

§ 1007.86 Deduction for marketing services.

See § 1000.86.

PARTS 1011–1013 [RESERVED]

PART 1030—MILK IN THE UPPER MIDWEST MARKETING AREA

Subpart—Order Regulating Handling

GENERAL PROVISIONS

AUTHORITY: 7 U.S.C. 601–674, and 7253.

SOURCE: 64 FR 47978, Sept. 1, 1999, unless otherwise noted.

Subpart—Order Regulating Handling

GENERAL PROVISIONS

§ 1030.1   General provisions.

The terms, definitions, and provisions in part 1000 of this chapter apply to this part 1030. In this part 1030, all references to sections in part 1000 refer to part 1000 of this chapter.

DEFINITIONS

§ 1030.2   Upper Midwest marketing area.

The marketing area means all territory within the bounds of the following states and political subdivisions, including all piers, docks, and wharves connected therewith and all craft moored thereat, and all territory occupied by government (municipal, State, or Federal) reservations, installations, institutions, or other similar establishments if any part thereof is within any of the listed states or political subdivisions:

ILLINOIS COUNTIES

Boone, Carroll, Cook, De Kalb, Du Page, Jo Daviess, Kane, Kendall, Lake, Lee, McHenry, Ogle, Stephenson, Will, and Winnebago.

IOWA COUNTIES

Howard, Kossuth, Mitchell, Winnebago, Winneshiek, and Worth.

MICHIGAN COUNTIES

Delta, Dickinson, Gogebic, Iron, Menominee, and Ontonagon.

MINNESOTA

All counties except Lincoln, Nobles, Pipestone, and Rock.

NORTH DAKOTA COUNTIES

Barnes, Cass, Cavalier, Dickey, Grand Forks, Griggs, La Moure, Nelson, Pembina, Ramsey, Ransom, Richland, Sargent, Steele, Traill, and Walsh.

SOUTH DAKOTA COUNTIES

Brown, Day, Edmunds, Grant, Marshall, McPherson, Roberts, and Walworth.

WISCONSIN COUNTIES

All counties except Crawford and Grant.

§ 1030.3   Route disposition.

See § 1000.3.

§1030.4 Plant.

See §1000.4.

§1030.5 Distributing plant.

See §1000.5.

§1030.6 Supply plant.

See §1000.6.

§1030.7 Pool plant.

Pool plant means a plant, unit of plants, or system of plants as specified in paragraphs (a) through (f) of this section, but excluding a plant specified in paragraph (h) of this section. The pooling standards described in paragraphs (c) and (f) of this section are subject to modification pursuant to paragraph (g) of this section:

(a) A distributing plant, other than a plant qualified as a pool plant pursuant to paragraph (b) of this section or (§_____.7b) of any other Federal milk order, from which during the month 15 percent or more of the total quantity of fluid milk products physically received at the plant (excluding concentrated milk received from another plant by agreement for other than Class I use) are disposed of as route disposition or are transferred in the form of packaged fluid milk products to other distributing plants. At least 25 percent of such route disposition and transfers must be to outlets in the marketing area.

(b) Any distributing plant located in the marketing area which during the month processed at least 15 percent of the total quantity of fluid milk products physically received at the plant (excluding concentrated milk received from another plant by agreement for other than Class I use) into ultra-pasteurized or aseptically-processed fluid milk products.

(c) A supply plant from which the quantity of bulk fluid milk products shipped to (and physically unloaded into) plants described in paragraph (c)(1) of this section is not less than 10 percent of the Grade A milk received from dairy farmers (except dairy farmers described in §1030.12(b)) and handlers described in §1000.9(c), including milk diverted pursuant to §1030.13, subject to the following conditions:

(1) Qualifying shipments may be made to plants described in paragraphs (c)(1)(i) through (iv) of this section, except that whenever shipping requirements are increased pursuant to paragraph (g) of this section, only shipments to pool plants described in paragraphs (a), (b), and (e) of this section shall count as qualifying shipments for the purpose of meeting the increased shipments:

(i) Pool plants described in §1030.7(a), (b), (d), and (e);

(ii) Plants of producer-handlers;

(iii) Partially regulated distributing plants, except that credit for such shipments shall be limited to the amount of such milk classified as Class I at the transferee plant; and

(iv) Distributing plants fully regulated under other Federal orders, except that credit for shipments to such plants shall be limited to the quantity shipped to pool distributing plants during the month and credits for shipments to other order plants shall not include any such shipments made on the basis of agreed-upon Class II, Class III, or Class IV utilization.

(2) The operator of a supply plant located within the States of Illinois, Iowa, Minnesota, North Dakota, South Dakota, Wisconsin and the Upper Peninsula of Michigan may include as qualifying shipments under this paragraph milk delivered directly from producers' farms pursuant to §§1000.9(c) or 1030.13(c) to plants described in paragraphs (a), (b), (d) and (e) of this section. Handlers may not use shipments pursuant to §1000.9(c) or §1030.13(c) to qualify plants located outside the area described above.

(3) Concentrated milk transferred from the supply plant to a distributing plant for an agreed-upon use other than Class I shall be excluded from the supply plant's shipments in computing the supply plant's shipping percentage.

(d) Any distributing plant, located within the marketing area as described on May 1, 2006, in §1030.2;

(1) From which there is route disposition and/or transfers of packaged fluid milk products in any non-federally regulated marketing area(s) located within one or more States that require handlers to pay minimum prices for raw milk provided that 25 percent or more

109

of the total quantity of fluid milk products physically received at such plant (excluding concentrated milk received from another plant by agreement for other than Class I use) is disposed of as route disposition and/or is transferred in the form of packaged fluid milk products to other plants. At least 25 percent of such route disposition and/or transfers, in aggregate, are in any non-federally regulated marketing area(s) located within one or more States that require handlers to pay minimum prices for raw milk. Subject to the following exclusions:

(i) The plant is described in § 1030.7(a), (b), or (e);

(ii) The plant is subject to the pricing provisions of a State-operated milk pricing plan which provides for the payment of minimum class prices for raw milk;

(iii) The plant is described in § 1000.8(a) or (e); or

(iv) A producer-handler described in § 1030.10 with less than three million pounds during the month of route disposition and/or transfers of packaged fluid milk products to other plants.

(2) [Reserved]

(e) Two or more plants operated by the same handler and located in the marketing area may qualify for pool status as a unit by meeting the total and in-area route disposition requirements of a pool distributing plant specified in paragraph (a) of this section and subject to the following additional requirements:

(1) At least one of the plants in the unit must qualify as a pool plant pursuant to paragraph (a) of this section;

(2) Other plants in the unit must process Class I or Class II products, using 50 percent or more of the total Grade A fluid milk products received in bulk form at such plant or diverted therefrom by the plant operator in Class I or Class II products; and

(3) The operator of the unit has filed a written request with the market administrator prior to the first day of the month for which such status is desired to be effective. The unit shall continue from month-to-month thereafter without further notification. The handler shall notify the market administrator in writing prior to the first day of any month for which termination or any change of the unit is desired.

(f) A system of 2 or more supply plants operated by one or more handlers may qualify for pooling by meeting the shipping requirements of paragraph (c) of this section in the same manner as a single plant subject to the following additional requirements:

(1) Each plant in the system is located within the marketing area or was a pool supply plant pursuant to § 1030.7(c) for each of the 3 months immediately preceding the applicability date of this paragraph so long as it continues to maintain pool status. Cooperative associations may not use shipments pursuant to § 1000.9(c) to qualify plants located outside the marketing area;

(2) The handler(s) establishing the system submits a written request to the market administrator on or before July 15 requesting that such plants qualify as a system for the period of August through July of the following year. Such request will contain a list of the plants participating in the system in the order, beginning with the last plant, in which the plants will be dropped from the system if the system fails to qualify. Each plant that qualifies as a pool plant within a system shall continue each month as a plant in the system through the following July unless the handler(s) establishing the system submits a written request to the market administrator that the plant be deleted from the system or that the system be discontinued. Any plant that has been so deleted from a system, or that has failed to qualify in any month, will not be part of any system for the remaining months through July. The handler(s) that established a system may add a plant operated by such handler(s) to a system if such plant has been a pool plant each of the 6 prior months and would otherwise be eligible to be in a system, upon written request to the market administrator no later than the 15th day of the prior month. In the event of an ownership change or the business failure of a handler that is a participant in a system, the system may be reorganized to reflect such changes if a written request to file a new marketing agreement is

submitted to the market administrator; and

(3) If a system fails to qualify under the requirements of this paragraph, the handler responsible for qualifying the system shall notify the market administrator which plant or plants will be deleted from the system so that the remaining plants may be pooled as a system. If the handler fails to do so, the market administrator shall exclude one or more plants, beginning at the bottom of the list of plants in the system and continuing up the list as necessary until the deliveries are sufficient to qualify the remaining plants in the system.

(g) The applicable shipping percentages of paragraphs (c) and (f) of this section and § 1030.13(d)(2), and (d)(3) may be increased or decreased, for all or part of the marketing area, by the market administrator if the market administrator finds that such adjustment is necessary to encourage needed shipments or to prevent uneconomic shipments. Before making such a finding, the market administrator shall investigate the need for adjustment either on the market administrator's own initiative or at the request of interested parties if the request is made in writing at least 15 days prior to the month for which the requested revision is desired effective. If the investigation shows that an adjustment of the shipping percentages might be appropriate, the market administrator shall issue a notice stating that an adjustment is being considered and invite data, views and arguments. Any decision to revise an applicable shipping or diversion percentage must be issued in writing at least one day before the effective date.

(h) The term pool plant shall not apply to the following plants:

(1) A producer-handler as defined under any Federal order;

(2) An exempt plant as defined in § 1000.8(e);

(3) A plant located within the marketing area and qualified pursuant to paragraph (a) of this section which meets the pooling requirements of another Federal order, and from which more than 50 percent of its route disposition has been in the other Federal order marketing area for 3 consecutive months;

(4) A plant located outside any Federal order marketing area and qualified pursuant to paragraph (a) of this section that meets the pooling requirements of another Federal order and has had greater route disposition in such other Federal order's marketing area for 3 consecutive months;

(5) A plant located in another Federal order marketing area and qualified pursuant to paragraph (a) of this section that meets the pooling requirements of such other Federal order and does not have a majority of its route distribution in this marketing area for 3 consecutive months or if the plant is required to be regulated under such other Federal order without regard to its route disposition in any other Federal order marketing area;

(6) A plant qualified pursuant to paragraph (c) of this section which also meets the pooling requirements of another Federal order and from which greater qualifying shipments are made to plants regulated under the other Federal order than are made to plants regulated under the order in this part, or the plant has automatic pooling status under the other Federal order; and

(7) That portion of a regulated plant designated as a nonpool plant that is physically separate and operated separately from the pool portion of such plant. The designation of a portion of a regulated plant as a nonpool plant must be requested in advance and in writing by the handler and must be approved by the market administrator.

(i) Any plant that qualifies as a pool plant in each of the immediately preceding 3 months pursuant to paragraph (a) of this section or the shipping percentages in paragraph (c) of this section that is unable to meet such performance standards for the current month because of unavoidable circumstances determined by the market administrator to be beyond the control of the handler operating the plant, such as a natural disaster (ice storm, wind storm, flood), fire, breakdown of equipment, or work stoppage, shall be considered to have met the minimum performance standards during the period of such unavoidable circumstances, but such relief shall not be granted for more than 2 consecutive months.

[64 FR 47978, Sept. 1, 1999, as amended at 67 FR 19508, Apr. 22, 2002; 70 FR 31322, June 1, 2005; 71 FR 25499, May 1, 2006; 71 FR 28249, May 16, 2006]

§ 1030.8 Nonpool plant.

See § 1000.8.

§ 1030.9 Handler.

See § 1000.9.

§ 1030.10 Producer-handler.

Producer-handler means a person who:

(a) Operates a dairy farm and a distributing plant from which there is route disposition in the marketing area, and from which total route disposition and packaged sales of fluid milk products to other plants during the month does not exceed 3 million pounds;

(b) Receives fluid milk from own farm production or milk that is fully subject to the pricing and pooling provisions of the order in this part or any other Federal order;

(c) Receives at its plant or acquires for route disposition no more than 150,000 pounds of fluid milk products from handlers fully regulated under any Federal order. This limitation shall not apply if the producer-handler's own farm production is less than 150,000 pounds during the month;

(d) Disposes of no other source milk as Class I milk except by increasing the nonfat milk solids content of the fluid milk products; and

(e) Provides proof satisfactory to the market administrator that the care and management of the dairy animals and other resources necessary to produce all Class I milk handled (excluding receipts from handlers fully regulated under any Federal order) and the processing and packaging operations are the producer-handler's own enterprise and at its own risk.

(f) Any producer-handler with Class I route dispositions and/or transfers of packaged fluid milk products in the marketing area described in § 1131.2 of this chapter shall be subject to payments into the Order 1131 producer settlement fund on such dispositions pursuant to § 1000.76(a) and payments into the Order 1131 administrative fund provided such dispositions are less than three million pounds in the current month and such producer-handler had total Class I route dispositions and/or transfers of packaged fluid milk products from own farm production of three million pounds or more the previous month. If the producer-handler has Class I route dispositions and/or transfers of packaged fluid milk products into the marketing area described in § 1131.2 of this chapter of three million pounds or more during the current month, such producer-handler shall be subject to the provisions described in § 1131.7 of this chapter or § 1000.76(a).

[64 FR 47978, Sept. 1, 1999, as amended at 71 FR 25499, May 1, 2006; 75 FR 21160, Apr. 23, 2010]

§ 1030.11 [Reserved]

§ 1030.12 Producer.

(a) Except as provided in paragraph (b) of this section, *producer* means any person who produces milk approved by a duly constituted regulatory agency for fluid consumption as Grade A milk and whose milk is:

(1) Received at a pool plant directly from the producer or diverted by the plant operator in accordance with § 1030.13; or

(2) Received by a handler described in § 1000.9(c).

(b) Producer shall not include:

(1) A producer-handler as defined in any Federal order;

(2) A dairy farmer whose milk is received at an exempt plant, excluding producer milk diverted to the exempt plant pursuant to § 1030.13(d);

(3) A dairy farmer whose milk is received by diversion at a pool plant from a handler regulated under another Federal order if the other Federal order designates the dairy farmer as a producer under that order and that milk is allocated by request to a utilization other than Class I; and

(4) A dairy farmer whose milk is reported as diverted to a plant fully regulated under another Federal order with respect to that portion of the milk so diverted that is assigned to Class I under the provisions of such other order.

§1030.13 Producer milk.

Except as provided for in paragraph (e) of this section, *Producer milk* means the skim milk (or the skim equivalent of components of skim milk), including nonfat components, and butterfat in milk of a producer that is:

(a) Received by the operator of a pool plant directly from a producer or a handler described in §1000.9(c). All milk received pursuant to this paragraph shall be priced at the location of the plant where it is first physically received;

(b) Received by a handler described in §1000.9(c) in excess of the quantity delivered to pool plants;

(c) Diverted by a pool plant operator to another pool plant. Milk so diverted shall be priced at the location of the plant to which diverted; or

(d) Diverted by the operator of a pool plant or a cooperative association described in §1000.9(c) to a nonpool plant located in the States of Illinois, Iowa, Minnesota, North Dakota, South Dakota, and Wisconsin, and the Upper Peninsula of Michigan, subject to the following conditions:

(1) Milk of a dairy farmer shall not be eligible for diversion unless at least one day's production of such dairy farmer is physically received as producer milk at a pool plant during the first month the dairy farmer is a producer. If a dairy farmer loses producer status under the order in this part (except as a result of a temporary loss of Grade A approval or as a result of the handler of the dairy farmer's milk failing to pool the milk under any order), the dairy farmer's milk shall not be eligible for diversion unless at least one day's production of the dairy farmer has been physically received as producer milk at a pool plant during the first month the dairy farmer is re-associated with the market;

(2) The quantity of milk diverted by a handler described in §1000.9(c) may not exceed 90 percent of the producer milk receipts reported by the handler pursuant to §1030.30(c) provided that not less than 10 percent of such receipts are delivered to plants described in §1030.7(c)(1)(i) through (iii). These percentages are subject to any adjustments that may be made pursuant to §1030.7(g); and

(3) The quantity of milk diverted to nonpool plants by the operator of a pool plant described in §1030.7(a), (b) or (d) may not exceed 90 percent of the Grade A milk received from dairy farmers (except dairy farmers described in §1030.12(b)) including milk diverted pursuant to §1030.13; and

(4) Diverted milk shall be priced at the location of the plant to which diverted.

(e) Producer milk shall not include milk of a producer that is subject to inclusion and participation in a marketwide equalization pool under a milk classification and pricing program imposed under the authority of a State government maintaining marketwide pooling of returns.

(f) The quantity of milk reported by a handler pursuant to either §1030.30(a)(1) or §1030.30(c)(1) for April through February may not exceed 125 percent, and March may not exceed 135 percent of the producer milk receipts pooled by the handler during the prior month. Milk diverted to nonpool plants reported in excess of this limit shall be removed from the pool. Milk in excess of this limit received at pool plants, other than pool distributing plants, shall be classified pursuant to §1000.44(a)(3)(v) and §1000.44(b). The handler must designate, by producer pick-up, which milk is to be removed from the pool. If the handler fails to provide this information, the market administrator will make the determination. The following provisions apply:

(1) Milk shipped to and physically received at pool distributing plants in excess of the previous month's pooled volume shall not be subject to the 125 or 135 percent limitation;

(2) Producer milk qualified pursuant to §____.13 of any other Federal Order and continuously pooled in any Federal Order for the previous six months shall not be included in the computation of the 125 or 135 percent limitation;

(3) The market administrator may waive the 125 or 135 percent limitation:

(i) For a new handler on the order, subject to the provisions of §1030.13(f)(4), or

(ii) For an existing handler with significantly changed milk supply conditions due to unusual circumstances;

113

(4) A bloc of milk may be considered ineligible for pooling if the market administrator determines that handlers altered the reporting of such milk for the purpose of evading the provisions of this paragraph.

[64 FR 47978, Sept. 1, 1999, as amended at 67 FR 19508, Apr. 22, 2002; 70 FR 31322, June 1, 2005; 71 FR 25499, May 1, 2006; 71 FR 63214, Oct. 30, 2006]

§ 1030.14   Other source milk.

See § 1000.14.

§ 1030.15   Fluid milk product.

See § 1000.15.

§ 1030.16   Fluid cream product.

See § 1000.16.

§ 1030.17   [Reserved]

§ 1030.18   Cooperative association.

See § 1000.18.

§ 1030.19   Commercial food processing establishment.

See § 1000.19.

HANDLER REPORTS

§ 1030.30   Reports of receipts and utilization.

Each handler shall report monthly so that the market administrator's office receives the report on or before the 9th day after the end of the month, in the detail and on the prescribed forms, as follows:

(a) Each handler that operates a pool plant shall report for each of its operations the following information:

(1) Product pounds, pounds of butterfat, pounds of protein, pounds of solids-not-fat other than protein (other solids), and the value of the somatic cell adjustment pursuant to § 1000.50(p), contained in or represented by:

(i) Receipts of producer milk, including producer milk diverted by the reporting handler, from sources other than handlers described in § 1000.9(c); and

(ii) Receipts of milk from handlers described in § 1000.9(c);

(2) Product pounds and pounds of butterfat contained in:

(i) Receipts of fluid milk products and bulk fluid cream products from other pool plants;

(ii) Receipts of other source milk; and

(iii) Inventories at the beginning and end of the month of fluid milk products and bulk fluid cream products;

(3) The utilization or disposition of all milk and milk products required to be reported pursuant to this paragraph; and

(4) Such other information with respect to the receipts and utilization of skim milk, butterfat, milk protein, other nonfat solids, and somatic cell information, as the market administrator may prescribe.

(b) Each handler operating a partially regulated distributing plant shall report with respect to such plant in the same manner as prescribed for reports required by paragraph (a) of this section. Receipts of milk that would have been producer milk if the plant had been fully regulated shall be reported in lieu of producer milk. The report shall show also the quantity of any reconstituted skim milk in route disposition in the marketing area.

(c) Each handler described in § 1000.9(c) shall report:

(1) The product pounds, pounds of butterfat, pounds of protein, pounds of solids-not-fat other than protein (other solids), and the value of the somatic cell adjustment pursuant to § 1000.50(p), contained in receipts of milk from producers; and

(2) The utilization or disposition of such receipts.

(d) Each handler not specified in paragraphs (a) through (c) of this section shall report with respect to its receipts and utilization of milk and milk products in such manner as the market administrator may prescribe.

§ 1030.31   Payroll reports.

(a) On or before the 22nd day after the end of each month, each handler that operates a pool plant pursuant to § 1030.7 and each handler described in § 1000.9(c) shall report to the market administrator its producer payroll for the month, in the detail prescribed by the market administrator, showing for each producer the information described in § 1030.73(f).

(b) Each handler operating a partially regulated distributing plant who elects to make payment pursuant to §1000.76(b) shall report for each dairy farmer who would have been a producer if the plant had been fully regulated in the same manner as prescribed for reports required by paragraph (a) of this section.

§1030.32 Other reports.

In addition to the reports required pursuant to §§1030.30 and 1030.31, each handler shall report any information the market administrator deems necessary to verify or establish each handler's obligation under the order.

CLASSIFICATION OF MILK

§1030.40 Classes of utilization.

See §1000.40.

§1030.41 [Reserved]

§1030.42 Classification of transfers and diversions.

See §1000.42.

§1030.43 General classification rules.

See §1000.43.

§1030.44 Classification of producer milk.

See §1000.44.

§1030.45 Market administrator's reports and announcements concerning classification.

See §1000.45.

CLASS PRICES

§1030.50 Class prices, component prices, and advanced pricing factors.

See §1000.50.

§1030.51 Class I differential and price.

The Class I differential shall be the differential established for Cook County, Illinois, which is reported in §1000.52. The Class I price shall be the price computed pursuant to §1000.50(a) for Cook County, Illinois.

§1030.52 Adjusted Class I differentials.

See §1000.52.

§1030.53 Announcement of class prices, component prices, and advanced pricing factors.

See §1000.53.

§1030.54 Equivalent price.

See §1000.54.

§1030.55 Transportation credits and assembly credits.

(a) Each handler operating a pool distributing plant described in §1030.7(a), (b), (d), or (e) that receives bulk milk from another pool plant shall receive a transportation credit for such milk computed as follows:

(1) Determine the hundredweight of milk eligible for the credit by completing the steps in paragraph (c) of this section;

(2) Multiply the hundredweight of milk eligible for the credit by .28 cents times the number of miles, not to exceed 400 miles, between the transferor plant and the transferee plant;

(3) Subtract the effective Class I price at the transferor plant from the effective Class I price at the transferee plant;

(4) Multiply any positive amount resulting from the subtraction in paragraph (a)(3) of this section by the hundredweight of milk eligible for the credit; and

(5) Subtract the amount computed in paragraph (a)(4) of this section from the amount computed in paragraph (a)(2) of this section. If the amount computed in paragraph (a)(4) of this section exceeds the amount computed in paragraph (a)(2) of this section, the transportation credit shall be zero.

(b) Each handler operating a pool distributing plant described in §1030.7(a), (b), (d), or (e) that receives milk from dairy farmers, each handler that transfers or diverts bulk milk from a pool plant to a pool distributing plant, and each handler described in §1000.9(c) that delivers producer milk to a pool distributing plant shall receive an assembly credit on the portion of such milk eligible for the credit pursuant to paragraph (c) of this section. The credit shall be computed by multiplying the hundredweight of milk eligible for the credit by $0.08.

(c) The following procedure shall be used to determine the amount of milk eligible for transportation and assembly credits pursuant to paragraphs (a) and (b) of this section:

(1) At each pool distributing plant, determine the aggregate quantity of Class I milk, excluding beginning inventory of packaged fluid milk products;

(2) Subtract the quantity of packaged fluid milk products received at the pool distributing plant from other pool plants and from nonpool plants if such receipts are assigned to Class I;

(3) Subtract the quantity of bulk milk shipped from the pool distributing plant to other plants to the extent that such milk is classified as Class I milk;

(4) Subtract the quantity of bulk milk received at the pool distributing plant from other order plants and unregulated supply plants that is assigned to Class I pursuant to §§ 1000.43(d) and 1000.44; and

(5) Assign the remaining quantity pro rata to physical receipts during the month from:

(i) Producers;

(ii) Handlers described in § 1000.9(c); and

(iii) Other pool plants.

(d) For purposes of this section, the distances to be computed shall be determined by the market administrator using the shortest available state and/or Federal highway mileage. Mileage determinations are subject to redetermination at all times. In the event a handler requests a redetermination of the mileage pertaining to any plant, the market administrator shall notify the handler of such redetermination within 30 days after the receipt of such request. Any financial obligations resulting from a change in mileage shall not be retroactive for any periods prior to the redetermination by the market administrator.

[64 FR 47978, Sept. 1, 1999, as amended at 70 FR 31322, June 1, 2005; 71 FR 25499, May 1, 2006]

PRODUCER PRICE DIFFERENTIAL

§ 1030.60   Handler's value of milk.

For the purpose of computing a handler's obligation for producer milk, the market administrator shall determine for each month the value of milk of each handler with respect to each of the handler's pool plants and of each handler described in § 1000.9(c) with respect to milk that was not received at a pool plant by adding the amounts computed in paragraphs (a) through (i) of this section and subtracting from that total amount the values computed in paragraphs (j) and (k) of this section. Unless otherwise specified, the skim milk, butterfat, and the combined pounds of skim milk and butterfat referred to in this section shall result from the steps set forth in § 1000.44(a), (b), and (c), respectively, and the nonfat components of producer milk in each class shall be based upon the proportion of such components in producer skim milk. Receipts of nonfluid milk products that are distributed as labeled reconstituted milk for which payments are made to the producer-settlement fund of another Federal order under § 1000.76(a)(4) or (d) shall be excluded from pricing under this section.

(a) Class I value. (1) Multiply the pounds of skim milk in Class I by the Class I skim milk price; and

(2) Add an amount obtained by multiplying the pounds of butterfat in Class I by the Class I butterfat price.

(b) Class II value. (1) Multiply the pounds of nonfat solids in Class II skim milk by the Class II nonfat solids price; and

(2) Add an amount obtained by multiplying the pounds of butterfat in Class II times the Class II butterfat price.

(c) Class III value. (1) Multiply the pounds of protein in Class III skim milk by the protein price;

(2) Add an amount obtained by multiplying the pounds of other solids in Class III skim milk by the other solids price; and

(3) Add an amount obtained by multiplying the pounds of butterfat in Class III by the butterfat price.

(d) Class IV value. (1) Multiply the pounds of nonfat solids in Class IV skim milk by the nonfat solids price; and

(2) Add an amount obtained by multiplying the pounds of butterfat in Class IV by the butterfat price.

(e) Compute an adjustment for the somatic cell content of producer milk

116

by multiplying the values reported pursuant to §1030.30(a)(1) and (c)(1) by the percentage of total producer milk allocated to Class II, Class III, and Class IV pursuant to §1000.44(c);

(f) Multiply the pounds of skim milk and butterfat overage assigned to each class pursuant to §1000.44(a)(11) and the corresponding step of §1000.44(b) by the skim milk prices and butterfat prices applicable to each class.

(g) Multiply the difference between the current month's Class I, II, or III price, as the case may be, and the Class IV price for the preceding month and by the hundredweight of skim milk and butterfat subtracted from Class I, II, or III, respectively, pursuant to §1000.44(a)(7) and the corresponding step of §1000.44(b);

(h) Multiply the difference between the Class I price applicable at the location of the pool plant and the Class IV price by the hundredweight of skim milk and butterfat assigned to Class I pursuant to §1000.43(d) and the hundredweight of skim milk and butterfat subtracted from Class I pursuant to §1000.44(a)(3)(i) through (vi) and the corresponding step of §1000.44(b), excluding receipts of bulk fluid cream products from plants regulated under other Federal orders and bulk concentrated fluid milk products from pool plants, plants regulated under other Federal orders, and unregulated supply plants.

(i) Multiply the difference between the Class I price applicable at the location of the nearest unregulated supply plants from which an equivalent volume was received and the Class III price by the pounds of skim milk and butterfat in receipts of concentrated fluid milk products assigned to Class I pursuant to §1000.43(d) and §1000.44(a)(3)(i) and the corresponding step of §1000.44(b) and the pounds of skim milk and butterfat subtracted from Class I pursuant to §1000.44(a)(8) and the corresponding step of §1000.44(b), excluding such skim milk and butterfat in receipts of fluid milk products from an unregulated supply plant to the extent that an equivalent amount of skim milk or butterfat disposed of to such plant by handlers fully regulated under any Federal milk order is classified and priced as Class I milk

and is not used as an offset for any other payment obligation under any order.

(j) For reconstituted milk made from receipts of nonfluid milk products, multiply $1.00 (but not more than the difference between the Class I price applicable at the location of the pool plant and the Class IV price) by the hundredweight of skim milk and butterfat contained in receipts of nonfluid milk products that are allocated to Class I use pursuant to §1000.43(d).

(k) Compute the amount of credits applicable pursuant to §1030.55.

[64 FR 47978, Sept. 1, 1999, as amended at 65 FR 82836, Dec. 28, 2000; 68 FR 7066, Feb. 12, 2003]

§1030.61 Computation of producer price differential.

For each month the market administrator shall compute a producer price differential per hundredweight. The report of any handler who has not made payments required pursuant to §1030.71 for the preceding month shall not be included in the computation of the producer price differential, and such handler's report shall not be included in the computation for succeeding months until the handler has made full payment of outstanding monthly obligations. Subject to the conditions of this paragraph, the market administrator shall compute the producer price differential in the following manner:

(a) Combine into one total the values computed pursuant to §1030.60 for all handlers required to file reports prescribed in §1030.30;

(b) Subtract the total values obtained by multiplying each handler's total pounds of protein, other solids, and butterfat contained in the milk for which an obligation was computed pursuant to §1030.60 by the protein price, other solids price, and the butterfat price, respectively, and the total value of the somatic cell adjustment pursuant to §1030.30(a)(1) and (c)(1);

(c) Add an amount equal to the minus location adjustments and subtract an amount equal to the plus location adjustments computed pursuant to §1030.75;

(d) Add an amount equal to not less than one-half of the unobligated balance in the producer-settlement fund;

(e) Divide the resulting amount by the sum of the following for all handlers included in these computations:

(1) The total hundredweight of producer milk; and

(2) The total hundredweight for which a value is computed pursuant to § 1030.60(i); and

(f) Subtract not less than 4 cents nor more than 5 cents from the price computed pursuant to paragraph (e) of this section. The result shall be known as the producer price differential for the month.

[68 FR 7066, Feb. 12, 2003]

§ 1030.62 Announcement of producer prices.

On or before the 13th day after the end of each month, the market administrator shall announce publicly the following prices and information:

(a) The producer price differential;

(b) The protein price;

(c) The nonfat solids price;

(d) The other solids price;

(e) The butterfat price;

(f) The somatic cell adjustment rate;

(g) The average butterfat, nonfat solids, protein and other solids content of producer milk; and

(h) The statistical uniform price for milk containing 3.5 percent butterfat, computed by combining the Class III price and the producer price differential.

[64 FR 47978, Sept. 1, 1999, as amended at 65 FR 82837, Dec. 28, 2000; 68 FR 7066, Feb. 12, 2003; 68 FR 13618, Mar. 20, 2003]

PAYMENTS FOR MILK

§ 1030.70 Producer-settlement fund.

See § 1000.70.

§ 1030.71 Payments to the producer-settlement fund.

Each handler shall make payment to the producer-settlement fund in a manner that provides receipt of the funds by the market administrator no later than the 15th day after the end of the month (except as provided in § 1000.90). Payment shall be the amount, if any, by which the amount specified in paragraph (a) of this section exceeds the amount specified in paragraph (b) of this section:

(a) The total value of milk to the handler for the month as determined pursuant to § 1030.60.

(b) The sum of:

(1) An amount obtained by multiplying the total hundredweight of producer milk as determined pursuant to § 1000.44(c) by the producer price differential as adjusted pursuant to § 1030.75;

(2) An amount obtained by multiplying the total pounds of protein, other solids, and butterfat contained in producer milk by the protein, other solids, and butterfat prices respectively;

(3) The total value of the somatic cell adjustment to producer milk; and

(4) An amount obtained by multiplying the pounds of skim milk and butterfat for which a value was computed pursuant to § 1030.60(i) by the producer price differential as adjusted pursuant to § 1030.75 for the location of the plant from which received.

[64 FR 47978, Sept. 1, 1999, as amended at 65 FR 82836, Dec. 28, 2000; 68 FR 7066, Feb. 12, 2003]

§ 1030.72 Payments from the producer-settlement fund.

No later than the 16th day after the end of each month (except as provided in § 1000.90), the market administrator shall pay to each handler the amount, if any, by which the amount computed pursuant to § 1030.71(b) exceeds the amount computed pursuant to § 1030.71(a). If, at such time, the balance in the producer-settlement fund is insufficient to make all payments pursuant to this section, the market administrator shall reduce uniformly such payments and shall complete the payments as soon as the funds are available.

§ 1030.73 Payments to producers and to cooperative associations.

(a) Each handler shall pay each producer for producer milk for which payment is not made to a cooperative association pursuant to paragraph (b) of this section, as follows:

(1) *Partial payment.* For each producer who has not discontinued shipments as of the date of this partial payment, payment shall be made so that it is received by each producer on or before

the 26th day of the month (except as provided in § 1000.90) for milk received during the first 15 days of the month from the producer at not less than the lowest announced class price for the preceding month, less proper deductions authorized in writing by the producer.

(2) *Final payment.* For milk received during the month, payment shall be made so that it is received by each producer no later than the 17th day after the end of the month (except as provided in § 1000.90) in an amount equal to not less than the sum of:

(i) The hundredweight of producer milk received times the producer price differential for the month as adjusted pursuant to § 1030.75;

(ii) The pounds of butterfat received times the butterfat price for the month;

(iii) The pounds of protein received times the protein price for the month;

(iv) The pounds of other solids received times the other solids price for the month;

(v) The hundredweight of milk received times the somatic cell adjustment for the month;

(vi) Less any payment made pursuant to paragraph (a)(1) of this section;

(vii) Less proper deductions authorized in writing by such producer, and plus or minus adjustments for errors in previous payments to such producer subject to approval by the market administrator; and

(viii) Less deductions for marketing services pursuant to § 1000.86.

(b) *Payments for milk received from cooperative association members.* On or before the day prior to the dates specified in paragraphs (a)(1) and (a)(2) of this section (except as provided in § 1000.90), each handler shall pay to a cooperative association for milk from producers who market their milk through the cooperative association and who have authorized the cooperative to collect such payments on their behalf an amount equal to the sum of the individual payments otherwise payable for such producer milk pursuant to paragraphs (a)(1) and (a)(2) of this section.

(c) *Payment for milk received from cooperative association pool plants or from cooperatives as handlers pursuant to § 1000.9(c).* On or before the day prior to

the dates specified in paragraphs (a)(1) and (a)(2) of this section (except as provided in § 1000.90), each handler who receives fluid milk products at its plant from a cooperative association in its capacity as the operator of a pool plant or who receives milk from a cooperative association in its capacity as a handler pursuant to § 1000.9(c), including the milk of producers who are not members of such association and who the market administrator determines have authorized the cooperative association to collect payment for their milk, shall pay the cooperative for such milk as follows:

(1) For bulk fluid milk products and bulk fluid cream products received from a cooperative association in its capacity as the operator of a pool plant and for milk received from a cooperative association in its capacity as a handler pursuant to § 1000.9(c) during the first 15 days of the month, at not less than the lowest announced class prices per hundredweight for the preceding month;

(2) For the total quantity of bulk fluid milk products and bulk fluid cream products received from a cooperative association in its capacity as the operator of a pool plant, at not less than the total value of such products received from the association's pool plants, as determined by multiplying the respective quantities assigned to each class under § 1000.44, as follows:

(i) The hundredweight of Class I skim milk times the Class I skim milk price for the month plus the pounds of Class I butterfat times the Class I butterfat price for the month. The Class I price to be used shall be that price effective at the location of the receiving plant;

(ii) The pounds of nonfat solids in Class II skim milk by the Class II nonfat solids price;

(iii) The pounds of butterfat in Class II times the Class II butterfat price;

(iv) The pounds of nonfat solids in Class IV times the nonfat solids price;

(v) The pounds of butterfat in Class III and Class IV milk times the butterfat price;

(vi) The pounds of protein in Class III milk times the protein price;

(vii) The pounds of other solids in Class III milk times the other solids price;

(viii) The hundredweight of Class II, Class III, and Class IV milk times the somatic cell adjustment; and

(ix) Add together the amounts computed in paragraphs (c)(2)(i) through (viii) of this section and from that sum deduct any payment made pursuant to paragraph (c)(1) of this section; and

(3) For the total quantity of milk received during the month from a cooperative association in its capacity as a handler under § 1000.9(c) as follows:

(i) The hundredweight of producer milk received times the producer price differential as adjusted pursuant to § 1030.75;

(ii) The pounds of butterfat received times the butterfat price for the month;

(iii) The pounds of protein received times the protein price for the month;

(iv) The pounds of other solids received times the other solids price for the month;

(v) The hundredweight of milk received times the somatic cell adjustment for the month; and

(vi) Add together the amounts computed in paragraphs (c)(3)(i) through (v) of this section and from that sum deduct any payment made pursuant to paragraph (c)(1) of this section.

(d) If a handler has not received full payment from the market administrator pursuant to § 1030.72 by the payment date specified in paragraph (a), (b) or (c) of this section, the handler may reduce pro rata its payments to producers or to the cooperative association (with respect to receipts described in paragraph (b) of this section, prorating the underpayment to the volume of milk received from the cooperative association in proportion to the total milk received from producers by the handler), but not by more than the amount of the underpayment. The payments shall be completed on the next scheduled payment date after receipt of the balance due from the market administrator.

(e) If a handler claims that a required payment to a producer cannot be made because the producer is deceased or cannot be located, or because the cooperative association or its lawful successor or assignee is no longer in existence, the payment shall be made to the producer-settlement fund, and in the event that the handler subsequently locates and pays the producer or a lawful claimant, or in the event that the handler no longer exists and a lawful claim is later established, the market administrator shall make the required payment from the producer-settlement fund to the handler or to the lawful claimant, as the case may be.

(f) In making payments to producers pursuant to this section, each handler shall furnish each producer, except a producer whose milk was received from a cooperative association handler described in § 1000.9(a) or (c), a supporting statement in a form that may be retained by the recipient which shall show:

(1) The name, address, Grade A identifier assigned by a duly constituted regulatory agency, and payroll number of the producer;

(2) The daily and total pounds, and the month and dates such milk was received from that producer;

(3) The total pounds of butterfat, protein, and other solids contained in the producer's milk;

(4) The somatic cell count of the producer's milk;

(5) The minimum rate or rates at which payment to the producer is required pursuant to the order in this part;

(6) The rate used in making payment if the rate is other than the applicable minimum rate;

(7) The amount, or rate per hundredweight, or rate per pound of component, and the nature of each deduction claimed by the handler; and

(8) The net amount of payment to the producer or cooperative association.

[64 FR 47978, Sept. 1, 1999, as amended at 65 FR 82836, Dec. 28, 2000; 68 FR 7066, Feb. 12, 2003]

§ 1030.74 [Reserved]

§ 1030.75 Plant location adjustments for producer milk and nonpool milk.

For purposes of making payments for producer milk and nonpool milk, a plant location adjustment shall be determined by subtracting the Class I price specified in § 1030.51 from the Class I price at the plant's location. The difference, plus or minus as the case may be, shall be used to adjust the payments required pursuant to §§ 1030.73 and 1000.76.

§ 1030.76 Payments by a handler operating a partially regulated distributing plant.

See § 1000.76.

§ 1030.77 Adjustment of accounts.

See § 1000.77.

§ 1030.78 Charges on overdue accounts.

See § 1000.78.

ADMINISTRATIVE ASSESSMENT AND MARKETING SERVICE DEDUCTION

§ 1030.85 Assessment for order administration.

On or before the payment receipt date specified under § 1030.71, each handler shall pay to the market administrator its pro rata share of the expense of administration of the order at a rate specified by the market administrator that is no more than 8 cents per hundredweight with respect to:

(a) Receipts of producer milk (including the handler's own production) other than such receipts by a handler described in § 1000.9(c) that were delivered to pool plants of other handlers;

(b) Receipts from a handler described in § 1000.9(c);

(c) Receipts of concentrated fluid milk products from unregulated supply plants and receipts of nonfluid milk products assigned to Class I use pursuant to § 1000.43(d) and other source milk allocated to Class I pursuant to § 1000.44(a)(3) and (8) and the corresponding steps of § 1000.44(b), except other source milk that is excluded from the computations pursuant to § 1030.60(h) and (i); and

(d) Route disposition in the marketing area from a partially regulated distributing plant that exceeds the skim milk and butterfat subtracted pursuant to § 1000.76(a)(1)(i) and (ii).

[71 FR 63215; Oct. 30, 2006]

§ 1030.86 Deduction for marketing services.

See § 1000.86.

PART 1032—MILK IN THE CENTRAL MARKETING AREA

Subpart—Order Regulating Handling

GENERAL PROVISIONS

1032.71 Payments to the producer-settlement fund.
1032.72 Payments from the producer-settlement fund.
1032.73 Payments to producers and to cooperative associations.
1032.74 [Reserved]
1032.75 Plant location adjustments for producer milk and nonpool milk.
1032.76 Payments by a handler operating a partially regulated distributing plant.
1032.77 Adjustment of accounts.
1032.78 Charges on overdue accounts.

ADMINISTRATIVE ASSESSMENT AND MARKETING SERVICE DEDUCTION

1032.85 Assessment for order administration.
1032.86 Deduction for marketing services.

AUTHORITY: 7 U.S.C. 601–674, and 7253.

SOURCE: 64 FR 47985, Sept. 1, 1999, unless otherwise noted.

Subpart—Order Regulating Handling

GENERAL PROVISIONS

§ 1032.1   General provisions.

The terms, definitions, and provisions in part 1000 of this chapter apply to this part 1032. In this part 1032, all references to sections in part 1000 refer to part 1000 of this chapter.

DEFINITIONS

§ 1032.2   Central marketing area.

The marketing area means all territory within the bounds of the following states and political subdivisions, including all piers, docks, and wharves connected therewith and all craft moored thereat, and all territory occupied by government (municipal, State, or Federal) reservations, installations, institutions, or other similar establishments if any part thereof is within any of the listed states or political subdivisions:

COLORADO COUNTIES

Adams, Arapahoe, Baca, Bent, Boulder, Broomfield, Chaffee, Clear Creek, Cheyenne, Crowley, Custer, Delta, Denver, Douglas, Eagle, El Paso, Elbert, Fremont, Garfield, Gilpin, Gunnison, Huerfano, Jefferson, Kiowa, Kit Carson, Lake, Larimer, Las Animas, Lincoln, Logan, Mesa, Montrose, Morgan, Otero, Park, Phillips, Pitkin, Prowers, Pueblo, Sedgwick, Summit, Teller, Washington, Weld, and Yuma.

ILLINOIS COUNTIES

Adams, Alexander, Bond, Brown, Bureau, Calhoun, Cass, Champaign, Christian, Clark, Clay, Clinton, Coles, Crawford, Cumberland, De Witt, Douglas, Edgar, Edwards, Effingham, Fayette, Ford, Franklin, Fulton, Gallatin, Greene, Grundy, Hamilton, Hancock, Hardin, Henderson, Henry, Iroquois, Jackson, Jasper, Jefferson, Jersey, Johnson, Kankakee, Knox, La Salle, Lawrence, Livingston, Logan, McDonough, McLean, Macon, Macoupin, Madison, Marion, Marshall, Mason, Massac, Menard, Mercer, Monroe, Montgomery, Morgan, Moultrie, Peoria, Perry, Piatt, Pike, Pope, Pulaski, Putnam, Randolph, Richland, Rock Island, Saline, Sangamon, Schuyler, Scott, Shelby, St. Clair, Stark, Tazewell, Union, Vermilion, Wabash, Warren, Washington, Wayne, White, Whiteside, Williamson, and Woodford.

IOWA COUNTIES

All Iowa counties except Howard, Kossuth, Mitchell, Winnebago, Winneshiek, and Worth.

KANSAS

All of the State of Kansas.

MINNESOTA COUNTIES

Lincoln, Nobles, Pipestone, and Rock.

MISSOURI COUNTIES AND CITIES

The counties of Andrew, Atchison, Bates, Buchanan, Caldwell, Carroll, Cass, Clay, Clinton, Daviess, De Kalb, Franklin, Gentry, Grundy, Harrison, Henry, Hickory, Holt, Jackson, Jefferson, Johnson, Lafayette, Lincoln, Livingston, Mercer, Nodaway, Pettis, Platte, Putnam, Ray, Saline, Schuyler, St. Charles, St. Clair, Ste. Genevieve, St. Louis, Sullivan, Warren, and Worth; and the city of St. Louis.

NEBRASKA COUNTIES

Adams, Antelope, Boone, Buffalo, Burt, Butler, Cass, Cedar, Chase, Clay, Colfax, Cuming, Custer, Dakota, Dawson, Dixon, Dodge, Douglas, Dundy, Fillmore, Franklin, Frontier, Furnas, Gage, Gosper, Greeley, Hall, Hamilton, Harlan, Hayes, Hitchcock, Howard, Jefferson, Johnson, Kearney, Keith, Knox, Lancaster, Lincoln, Madison, Merrick, Nance, Nemaha, Nuckolls, Otoe, Pawnee, Perkins, Phelps, Pierce, Platte, Polk, Red Willow, Richardson, Saline, Sarpy, Saunders, Seward, Sherman, Stanton, Thayer, Thurston, Valley, Washington, Wayne, Webster, and York.

OKLAHOMA

All of the State of Oklahoma.

SOUTH DAKOTA COUNTIES

Aurora, Beadle, Bon Homme, Brookings, Clark, Clay, Codington, Davison, Deuel, Douglas, Hamlin, Hanson, Hutchinson, Jerauld, Kingsbury, Lake, Lincoln, McCook, Miner, Minnehaha, Moody, Sanborn, Spink, Turner, Union, and Yankton.

WISCONSIN COUNTIES

Crawford and Grant.

[64 FR 47985, Sept. 1, 1999, as amended at 68 FR 48771, Aug. 15, 2003]

§ 1032.3 Route disposition.

See § 1000.3.

§ 1032.4 Plant.

See § 1000.4.

§ 1032.5 Distributing plant.

See § 1000.5.

§ 1032.6 Supply plant.

See § 1000.6.

§ 1032.7 Pool plant.

Pool plant means a plant, unit of plants, or system of plants as specified in paragraphs (a) through (f) of this section, or a plant specified in paragraph (i) of this section, but excluding a plant specified in paragraph (h) of this section. The pooling standards described in paragraphs (c) and (d) and (f) of this section are subject to modification pursuant to paragraph (g) of this section:

(a) A distributing plant, other than a plant qualified as a pool plant pursuant to paragraph (b) of this section or _____.7(b) of any other Federal milk order, from which during the month 25 percent or more of the total quantity of fluid milk products physically received at the plant (excluding concentrated milk received from another plant by agreement for other than Class I use) are disposed of as route disposition or are transferred in the form of packaged fluid milk products to other distributing plants. At least 25 percent of such route disposition and transfers must be to outlets in the marketing area.

(b) Any distributing plant located in the marketing area which during the month processed at least 25 percent of the total quantity of fluid milk products physically received at the plant (excluding concentrated milk received from another plant by agreement for other than Class I use) into ultra-pasteurized or aseptically-processed fluid milk products.

(c) A supply plant from which the quantity of bulk fluid milk products shipped to (and physically unloaded into) plants described in paragraph (c)(1) of this section is not less than 25 percent during the months of August through February and 20 percent in all other months of the Grade A milk received from dairy farmers (except dairy farmers described in § 1032.12(b)) and from handlers described in § 1000.9(c), including milk diverted pursuant to § 1032.13, subject to the following conditions:

(1) Qualifying shipments may be made to plants described in paragraphs (a) or (b) of this section;

(2) The operator of a pool plant located in the marketing area may include as qualifying shipments milk delivered directly from producer's farms pursuant to § 1000.9(c) or § 1032.13(c). Handlers may not use shipments pursuant to § 1000.9(c) or § 1032.13(c) to qualify plants located outside the marketing area;

(3) Concentrated milk transferred from the supply plant to a distributing plant for an agreed-upon use other than Class I shall be excluded from the supply plant's shipments in computing the supply plant's shipping percentage;

(4) No plant may qualify as a pool plant due to a reduction in the shipping percentage pursuant to paragraph (g) of this section unless it has been a pool supply plant during each of the immediately preceding 3 months.

(5) Shipments used in determining qualifying percentages shall be milk transferred or diverted to and physically received by pool distributing plants, less any transfers or diversions of bulk fluid milk products from such pool distributing plants.

(d) A plant located in the marketing area and operated by a cooperative association if, during the month or the immediately preceding 12-month period, 35 percent or more of the producer milk of members of the association (and any producer milk of nonmembers and members of another cooperative association which may be marketed by the cooperative association) is physically received in the form of bulk fluid milk products (excluding concentrated milk transferred to a distributing plant for an agreed-upon use other than Class I) at plants specified in paragraph (a) or (b) of this section either directly from farms or by transfer from supply plants operated by the cooperative association and from plants of the cooperative association for which pool plant status has been requested under this paragraph subject to the following conditions:

(1) The plant does not qualify as a pool plant under paragraph (a), (b) or (c) of this section or under comparable provisions of another Federal order; and

(2) The plant is approved by a duly constituted regulatory agency for the handling of milk approved for fluid consumption in the marketing area.

(e) Two or more plants operated by the same handler and located in the marketing area may qualify for pool status as a unit by meeting the total and in-area route disposition requirements of a pool distributing plant specified in paragraph (a) of this section subject to the following additional requirements:

(1) At least one of the plants in the unit must qualify as a pool plant pursuant to paragraph (a) of this section;

(2) Other plants in the unit must process Class I or Class II products, using 50 percent or more of the total Grade A fluid milk products received in bulk form at such plant or diverted therefrom by the plant operator in Class I or Class II products, and must be located in a pricing zone providing the same or a lower Class I price than the price applicable at the distributing plant included in the unit pursuant to paragraph (e)(1) of this section; and

(3) The operator of the unit has filed a written request with the market administrator prior to the first day of the month for which such status is desired to be effective. The unit shall continue from month to month thereafter without further notification. The handler shall notify the market administrator in writing prior to the first day of any month for which termination or any change of the unit is desired.

(f) A system of supply plants may qualify for pooling if 2 or more plants operated by one or more handlers meet the applicable percentage requirements of paragraph (c) of this section in the same manner as a single plant, subject to the following additional requirements:

(1) Each plant in the system is located within the marketing area;

(2) The handler(s) establishing the system submits a written request to the market administrator on or before September 1 requesting that such plants qualify as a system for the period of September through August of the following year. Such request will contain a list of the plants participating in the system;

(3) Each plant included within a pool supply plant system shall continue each month as a plant in the system through the following August unless the handler(s) establishing the system submits a written request to the market administrator that the plant be deleted from the system or that the system be discontinued. Any plant that has been so deleted from a system, or that has failed to qualify in any month, will not be part of any system for the remaining months through August. No plant may be added in any subsequent month through the following August to a system that qualifies in September; and

(4) If a system fails to qualify under the requirements of this paragraph, the handler responsible for qualifying the system shall notify the market administrator which plant or plants will be deleted from the system so that the remaining plants may be pooled as a system. If the handler fails to do so, the market administrator shall exclude one or more plants, beginning at the bottom of the list of plants in the system and continuing up the list as necessary until the deliveries are sufficient to qualify the remaining plants in the system.

(g) The applicable shipping percentages of paragraphs (c), (d), and (f) of this section may be increased or decreased, for all or part of the marketing area, by the market administrator if the market administrator finds that such adjustment is necessary to encourage needed shipments or to prevent uneconomic shipments. Before making such a finding, the market administrator shall investigate the need for adjustment either on the market administrator's own initiative or at the request of interested parties if the request is made in writing at least 15 days prior to the month for which the requested revision is desired effective. If the investigation shows that an adjustment of the shipping percentages might be appropriate, the market administrator shall issue a notice stating that an adjustment is being considered and invite data, views and arguments. Any decision to revise an applicable shipping percentage must be issued in writing at least one day before the effective date.

(h) The term pool plant shall not apply to the following plants:

(1) A producer-handler as defined under any Federal order;

(2) An exempt plant as defined in §1000.8(e);

(3) A plant located within the marketing area and qualified pursuant to paragraph (a) of this section which meets the pooling requirements of another Federal order, and from which more than 50 percent of its route disposition has been in the other Federal order marketing area for 3 consecutive months. On the basis of a written application made by the plant operator at least 15 days prior to the date for which a determination of the market administrator is to be effective, the market administrator may determine that the route disposition in the respective marketing areas to be used for purposes of this paragraph shall exclude (for a specified period of time) route disposition made under limited term contracts to governmental bases and institutions;

(4) A plant located outside any Federal order marketing area and qualified pursuant to paragraph (a) of this section that meets the pooling requirements of another Federal order and has had greater route disposition in such other Federal order's marketing area for 3 consecutive months;

(5) A plant located in another Federal order marketing area and qualified pursuant to paragraph (a) of this section that meets the pooling requirements of such other Federal order and

does not have a majority of its route distribution in this marketing area for 3 consecutive months or if the plant is required to be regulated under such other Federal order without regard to its route disposition in any other Federal order marketing area;

(6) A plant qualified pursuant to paragraph (c) of this section which also meets the pooling requirements of another Federal order and from which greater qualifying shipments are made to plants regulated under the other Federal order than are made to plants regulated under the order in this part, or the plant has automatic pooling status under the other Federal order; and

(7) That portion of a regulated plant designated as a nonpool plant that is physically separate and operated separately from the pool portion of such plant. The designation of a portion of a plant must be requested in advance and in writing by the handler and must be approved by the market administrator. Such nonpool status shall be effective on the first day of the month following approval of the request by the market administrator and thereafter for the longer of twelve (12) consecutive months or until notification of the desire to requalify as a pool plant, in writing, is received by the market administrator. Requalification will require deliveries to a pool distributing plant(s) as provided for in §1032.7(c). For requalification, handlers may not use milk delivered directly from producer's farms pursuant to §1000.9(c) or §1032.13(c) for the first month.

(i) Any distributing plant, located within the marketing area as described on May 1, 2006, in §1032.2;

(1) From which there is route disposition and/or transfers of packaged fluid milk products in any non-federally regulated marketing area(s) located within one or more States that require handlers to pay minimum prices for raw milk provided that 25 percent or more of the total quantity of fluid milk products physically received at such plant (excluding concentrated milk received from another plant by agreement for other than Class I use) is disposed of as route disposition and/or is transferred in the form of packaged fluid milk products to other plants. At

least 25 percent of such route disposition and/or transfers, in aggregate, are in any non-federally regulated marketing area(s) located within one or more States that require handlers to pay minimum prices for raw milk. Subject to the following exclusions:

(i) The plant is described in § 1032.7(a), (b), or (e);

(ii) The plant is subject to the pricing provisions of a State-operated milk pricing plan which provides for the payment of minimum class prices for raw milk;

(iii) The plant is described in § 1000.8(a) or (e); or

(iv) A producer-handler described in § 1032.10 with less than three million pounds during the month of route disposition and/or transfers of packaged fluid milk products to other plants.

(2) [Reserved]

[64 FR 47985, Sept. 1, 1999, as amended at 68 FR 7072, Feb. 12, 2003; 71 FR 25500, May 1, 2006; 71 FR 28249, May 16, 2006; 71 FR 63216, Oct. 30, 2006]

§ 1032.8 Nonpool plant.

See § 1000.8.

§ 1032.9 Handler.

See § 1000.9.

§ 1032.10 Producer-handler.

Producer-handler means a person who:

(a) Operates a dairy farm and a distributing plant from which there is route disposition in the marketing area, and from which total route disposition and packaged sales of fluid milk products to other plants during the month does not exceed 3 million pounds;

(b) Receives fluid milk from own farm production or milk that is fully subject to the pricing and pooling provisions of the order in this part or any other Federal order;

(c) Receives at its plant or acquires for route disposition no more than 150,000 pounds of fluid milk products from handlers fully regulated under any Federal order. This limitation shall not apply if the producer-handler's own farm production is less than 150,000 pounds during the month;

(d) Disposes of no other source milk as Class I milk except by increasing

the nonfat milk solids content of the fluid milk products; and

(e) Provides proof satisfactory to the market administrator that the care and management of the dairy animals and other resources necessary to produce all Class I milk handled (excluding receipts from handlers fully regulated under any Federal order) and the processing and packaging operations are the producer-handler's own enterprise and at its own risk.

(f) Any producer-handler with Class I route dispositions and/or transfers of packaged fluid milk products in the marketing area described in § 1131.2 of this chapter shall be subject to payments into the Order 1131 producer settlement fund on such dispositions pursuant to § 1000.76(a) and payments into the Order 1131 administrative fund provided such dispositions are less than three million pounds in the current month and such producer-handler had total Class I route dispositions and/or transfers of packaged fluid milk products from own farm production of three million pounds or more the previous month. If the producer-handler has Class I route dispositions and/or transfers of packaged fluid milk products into the marketing area described in § 1131.2 of this chapter of three million pounds or more during the current month, such producer-handler shall be subject to the provisions described in § 1131.7 of this chapter or 1000.76(a).

[64 FR 47985, Sept. 1, 1999, as amended at 71 FR 25500, May 1, 2006; 75 FR 21160, Apr. 23, 2010]

§ 1032.11 [Reserved]

§ 1032.12 Producer.

(a) Except as provided in paragraph (b) of this section, *producer* means any person who produces milk approved by a duly constituted regulatory agency for fluid consumption as Grade A milk and whose milk (or components of milk) is:

(1) Received at a pool plant directly from the producer or diverted by the plant operator in accordance with § 1032.13; or (2) Received by a handler described in § 1000.9(c).

(b) Producer shall not include:

(1) A producer-handler as defined in any Federal order;

(2) A dairy farmer whose milk is received at an exempt plant, excluding producer milk diverted to the exempt plant pursuant to § 1032.13(d);

(3) A dairy farmer whose milk is received by diversion at a pool plant from a handler regulated under another Federal order if the other Federal order designates the dairy farmer as a producer under that order and that milk is allocated by request to a utilization other than Class I; and

(4) A dairy farmer whose milk is reported as diverted to a plant fully regulated under another Federal order with respect to that portion of the milk so diverted that is assigned to Class I under the provisions of such other order.

§ 1032.13 Producer milk.

Producer milk means the skim milk (or the skim equivalent of components of skim milk), including nonfat components, and butterfat in milk of a producer that is:

(a) Received by the operator of a pool plant directly from a producer or a handler described in § 1000.9(c). All milk received pursuant to this paragraph shall be priced at the location of the plant where it is first physically received;

(b) Received by a handler described in § 1000.9(c) in excess of the quantity delivered to pool plants;

(c) Diverted by a pool plant operator to another pool plant. Milk so diverted shall be priced at the location of the plant to which diverted; or

(d) Diverted by the operator of a pool plant or a cooperative association described in § 1000.9(c) to a nonpool plant, subject to the following conditions:

(1) Milk of a dairy farmer shall not be eligible for diversion until milk of such dairy farmer has been physically received as producer milk at a pool plant and the dairy farmer has continuously retained producer status since that time. If a dairy farmer loses producer status under the order in this part (except as a result of a temporary loss of Grade A approval), the dairy farmer's milk shall not be eligible for diversion until milk of the dairy farmer has been physically received as producer milk at a pool plant;

(2) The equivalent of at least one day's milk production is caused by the handler to be physically received at a pool plant in each of the months of January and February, and August through November;

(3) The equivalent of at least one day's milk production is caused by the handler to be physically received at a pool plant in each of the months of March through July and December if the requirement of paragraph (d)(2) of this section (§ 1032.13) in each of the prior months of August through November and January through February are not met, except in the case of a dairy farmer who marketed no Grade A milk during each of the prior months of August through November or January through February.

(4) Of the quantity of producer milk received during the month (including diversions, but excluding the quantity of producer milk received from a handler described in § 1000.9(c)) the handler diverts to nonpool plants not more than 75 percent during the months of August through February, and not more than 80 percent during the months of March through July, provided that not less than 25 percent of such receipts in the months of August through February and 20 percent of the remaining months' receipts are delivered to plants described in § 1032.7(a), (b), or (i).;

(5) Receipts used in determining qualifying percentages shall be milk transferred to or diverted to or physically received by a plant described in § 1032.7(a), (b) or (i) less any transfer or diversion of bulk fluid milk products from such plants;

(6) Diverted milk shall be priced at the location of the plant to which diverted;

(7) Any milk diverted in excess of the limits prescribed in paragraph (d)(2) of this section shall not be producer milk. If the diverting handler or cooperative association fails to designate the dairy farmers' deliveries that are not to be producer milk, no milk diverted by the handler or cooperative association during the month to a nonpool plant shall be producer milk; and

(8) The applicable diversion limits in paragraph (d)(2) of this section may be increased or decreased by the market

administrator if the market administrator finds that such revision is necessary to assure orderly marketing and efficient handling of milk in the marketing area. Before making such a finding, the market administrator shall investigate the need for the revision either on the market administrator's own initiative or at the request of interested persons if the request is made in writing at least 15 days prior to the month for which the requested revision is desired effective. If the investigation shows that a revision might be appropriate, the market administrator shall issue a notice stating that the revision is being considered and inviting written data, views, and arguments. Any decision to revise an applicable percentage must be issued in writing at least one day before the effective date.

(e) Producer milk shall not include milk of a producer that is subject to inclusion and participation in a marketwide equalization pool under a milk classification and pricing program imposed under the authority of a State government maintaining marketwide pooling of returns.

(f) The quantity of milk reported by a handler pursuant to § 1032.30(a)(1) and/or § 1032.30(c)(1) for the current month may not exceed 125 percent of the producer milk receipts pooled by the handler during the prior month. Milk diverted to nonpool plants reported in excess of this limit shall be removed from the pool. Milk received at pool plants in excess of the 125 percent limit, other than pool distributing plants, shall be classified pursuant to § 1000.44(a)(3)(v). The handler must designate, by producer pick-up, which milk is to be removed from the pool. If the handler fails to provide this information the provisions of paragraph (d)(5) of this section shall apply. The following provisions apply:

(1) Milk shipped to and physically received at pool distributing plants shall not be subject to the 125 percent limitation;

(2) Producer milk qualified pursuant to § ___ .13 of any other Federal Order in the previous month shall not be included in the computation of the 125 percent limitation; provided that the producers comprising the milk supply

have been continuously pooled on any Federal Order for the entirety of the most recent three consecutive months.

(3) The market administrator may waive the 125 percent limitation:

(i) For a new handler on the order, subject to the provisions of paragraph (f)(3) of this section, or

(ii) For an existing handler with significantly changed milk supply conditions due to unusual circumstances;

(4) A bloc of milk may be considered ineligible for pooling if the market administrator determines that handlers altered the reporting of such milk for the purpose of evading the provisions of this paragraph.

[64 FR 47985, Sept. 1, 1999, as amended at 68 FR 7072, Feb. 12, 2003; 71 FR 25500, May 1, 2006; 71 FR 63217, Oct. 30, 2006]

§ 1032.14   Other source milk.

See § 1000.14.

§ 1032.15   Fluid milk product.

See § 1000.15.

§ 1032.16   Fluid cream product.

See § 1000.16.

§ 1032.17   [Reserved]

§ 1032.18   Cooperative association.

See § 1000.18.

§ 1032.19   Commercial food processing establishment.

See § 1000.19.

HANDLER REPORTS

§ 1032.30   Reports of receipts and utilization.

Each handler shall report monthly so that the market administrator's office receives the report on or before the 7th day after the end of the month, in the detail and on the prescribed forms, as follows:

(a) Each handler that operates a pool plant pursuant to § 1032.7 shall report for each of its operations the following information:

(1) Product pounds, pounds of butterfat, pounds of protein, pounds of solids-not-fat other than protein (other solids), and the value of the somatic cell adjustment pursuant to § 1000.50(p), contained in or represented by:

(i) Receipts of producer milk, including producer milk diverted by the reporting handler, from sources other than handlers described in § 1000.9(c); and

(ii) Receipts of milk from handlers described in § 1000.9(c);

(2) Product pounds and pounds of butterfat contained in:

(i) Receipts of fluid milk products and bulk fluid cream products from other pool plants;

(ii) Receipts of other source milk; and

(iii) Inventories at the beginning and end of the month of fluid milk products and bulk fluid cream products;

(3) The utilization or disposition of all milk and milk products required to be reported pursuant to this paragraph; and

(4) Such other information with respect to the receipts and utilization of skim milk, butterfat, milk protein, other nonfat solids, and somatic cell information, as the market administrator may prescribe.

(b) Each handler operating a partially regulated distributing plant shall report with respect to such plant in the same manner as prescribed for reports required by paragraph (a) of this section. Receipts of milk that would have been producer milk if the plant had been fully regulated shall be reported in lieu of producer milk. The report shall show also the quantity of any reconstituted skim milk in route disposition in the marketing area.

(c) Each handler described in § 1000.9(c) shall report:

(1) The product pounds, pounds of butterfat, pounds of protein, pounds of solids-not-fat other than protein (other solids), and the value of the somatic cell adjustment pursuant to § 1000.50(p), contained in receipts of milk from producers; and

(2) The utilization or disposition of such receipts.

(d) Each handler not specified in paragraphs (a) through (c) of this section shall report with respect to its receipts and utilization of milk and milk products in such manner as the market administrator may prescribe.

§ 1032.31 Payroll reports.

(a) On or before the 20th day after the end of each month, each handler that operates a pool plant pursuant to § 1032.7 and each handler described in § 1000.9(c) shall report to the market administrator its producer payroll for the month, in the detail prescribed by the market administrator, showing for each producer the information described in § 1032.73(f).

(b) Each handler operating a partially regulated distributing plant who elects to make payment pursuant to § 1000.76(b) shall report for each dairy farmer who would have been a producer if the plant had been fully regulated in the same manner as prescribed for reports required by paragraph (a) of this section.

§ 1032.32 Other reports.

In addition to the reports required pursuant to §§ 1032.30 and 1032.31, each handler shall report any information the market administrator deems necessary to verify or establish each handler's obligation under the order.

CLASSIFICATION OF MILK

§ 1032.40 Classes of utilization.

See § 1000.40.

§ 1032.41 [Reserved]

§ 1032.42 Classification of transfers and diversions.

See § 1000.42.

§ 1032.43 General classification rules.

See § 1000.43.

§ 1032.44 Classification of producer milk.

See § 1000.44.

§ 1032.45 Market administrator's reports and announcements concerning classification.

See § 1000.45.

CLASS PRICES

§ 1032.50 Class prices, component prices, and advanced pricing factors.

See § 1000.50.

§ 1032.51　Class I differential and price.

The Class I differential shall be the differential established for Jackson County, Missouri, which is reported in § 1000.52. The Class I price shall be the price computed pursuant to § 1000.50(a) for Jackson County, Missouri.

§ 1032.52　Adjusted Class I differentials.

See § 1000.52.

§ 1032.53　Announcement of class prices, component prices, and advanced pricing factors.

See § 1000.53.

§ 1032.54　Equivalent price.

See § 1000.54.

PRODUCER PRICE DIFFERENTIAL

§ 1032.60　Handler's value of milk.

For the purpose of computing a handler's obligation for producer milk, the market administrator shall determine for each month the value of milk of each handler with respect to each of the handler's pool plants and of each handler described in § 1000.9(c) with respect to milk that was not received at a pool plant by adding the amounts computed in paragraphs (a) through (i) of this section and subtracting from that total amount the value computed in paragraph (j) of this section. Unless otherwise specified, the skim milk, butterfat, and the combined pounds of skim milk and butterfat referred to in this section shall result from the steps set forth in § 1000.44(a), (b), and (c), respectively, and the nonfat components of producer milk in each class shall be based upon the proportion of such components in producer skim milk. Receipts of nonfluid milk products that are distributed as labeled reconstituted milk for which payments are made to the producer-settlement fund of another Federal order under § 1000.76(a)(4) or (d) shall be excluded from pricing under this section.

(a) *Class I value.* (1) Multiply the pounds of skim milk in Class I by the Class I skim milk price; and

(2) Add an amount obtained by multiplying the pounds of butterfat in Class I by the Class I butterfat price.

(b) *Class II value.* (1) Multiply the pounds of nonfat solids in Class II skim milk by the Class II nonfat solids price; and

(2) Add an amount obtained by multiplying the pounds of butterfat in Class II times the Class II butterfat price.

(c) *Class III value.* (1) Multiply the pounds of protein in Class III skim milk by the protein price;

(2) Add an amount obtained by multiplying the pounds of other solids in Class III skim milk by the other solids price; and

(3) Add an amount obtained by multiplying the pounds of butterfat in Class III by the butterfat price.

(d) *Class IV value.* (1) Multiply the pounds of nonfat solids in Class IV skim milk by the nonfat solids price; and

(2) Add an amount obtained by multiplying the pounds of butterfat in Class IV by the butterfat price.

(e) Compute an adjustment for the somatic cell content of producer milk by multiplying the values reported pursuant to § 1032.30(a)(1) and (c)(1) by the percentage of total producer milk allocated to Class II, Class III, and Class IV pursuant to § 1000.44(c);

(f) Multiply the pounds of skim milk and butterfat overage assigned to each class pursuant to § 1000.44(a)(11) and the corresponding step of § 1000.44(b) by the skim milk prices and butterfat prices applicable to each class.

(g) Multiply the difference between the current month's Class I, II, or III price, as the case may be, and the Class IV price for the preceding month by the hundredweight of skim milk and butterfat subtracted from Class I, II, or III, respectively, pursuant to § 1000.44(a)(7) and the corresponding step of § 1000.44(b);

(h) Multiply the difference between the Class I price applicable at the location of the pool plant and the Class IV price by the hundredweight of skim milk and butterfat assigned to Class I pursuant to § 1000.43(d) and the hundredweight of skim milk and butterfat subtracted from Class I pursuant to § 1000.44(a)(3)(i) through (vi) and the corresponding step of § 1000.44(b), excluding receipts of bulk fluid cream products from a plant regulated under other Federal orders and bulk concentrated fluid milk products from pool plants, plants regulated under other Federal orders, and unregulated supply plants.

(i) Multiply the difference between the Class I price applicable at the location of the nearest unregulated supply plants from which an equivalent volume was received and the Class III price by the pounds of skim milk and butterfat in receipts of concentrated fluid milk products assigned to Class I pursuant to §1000.43(d) and §1000.44(a)(3)(i) and the corresponding step of §1000.44(b) and the pounds of skim milk and butterfat subtracted from Class I pursuant to §1000.44(a)(8) and the corresponding step of §1000.44(b), excluding such skim milk and butterfat in receipts of fluid milk products from an unregulated supply plant to the extent that an equivalent amount of skim milk or butterfat disposed of to such plant by handlers fully regulated under any Federal milk order is classified and priced as Class I milk and is not used as an offset for any other payment obligation under any order.

(j) For reconstituted milk made from receipts of nonfluid milk products, multiply $1.00 (but not more than the difference between the Class I price applicable at the location of the pool plant and the Class IV price) by the hundredweight of skim milk and butterfat contained in receipts of nonfluid milk products that are allocated to Class I use pursuant to §1000.43(d).

[64 FR 47985, Sept. 1, 1999, as amended at 65 FR 82837, Dec. 28, 2000; 68 FR 7067, Feb. 12, 2003]

§1032.61 Computation of producer price differential.

For each month the market administrator shall compute a producer price differential per hundredweight. The report of any handler who has not made payments required pursuant to §1032.71 for the preceding month shall not be included in the computation of the producer price differential, and such handler's report shall not be included in the computation for succeeding months until the handler has made full payment of outstanding monthly obligations. Subject to the conditions of this paragraph, the market administrator shall compute the producer price differential in the following manner:

(a) Combine into one total the values computed pursuant to §1032.60 for all handlers required to file reports prescribed in §1032.30;

(b) Subtract the total values obtained by multiplying each handler's total pounds of protein, other solids, and butterfat contained in the milk for which an obligation was computed pursuant to §1032.60 by the protein price, the other solids price, and the butterfat price, respectively, and the total value of the somatic cell adjustment pursuant to §1032.30(a)(1) and (c)(1);

(c) Add an amount equal to the minus location adjustments and subtract an amount equal to the plus location adjustments computed pursuant to §1032.75;

(d) Add an amount equal to not less than one-half of the unobligated balance in the producer-settlement fund;

(e) Divide the resulting amount by the sum of the following for all handlers included in these computations:

(1) The total hundredweight of producer milk; and

(2) The total hundredweight for which a value is computed pursuant to §1032.60(i); and

(f) Subtract not less than 4 cents nor more than 5 cents from the price computed pursuant to paragraph (e) of this section. The result shall be known as the producer price differential for the month.

[68 FR 7067, Feb. 12, 2003]

§1032.62 Announcement of producer prices.

On or before the 11th day after the end of each month, the market administrator shall announce publicly the following prices and information:

(a) The producer price differential;

(b) The protein price;

(c) The nonfat solids price;

(d) The other solids price;

(e) The butterfat price;

(f) The somatic cell adjustment rate;

(g) The average butterfat, protein, nonfat solids, and other solids content of producer milk; and

(h) The statistical uniform price for milk containing 3.5 percent butterfat, computed by combining the Class III price and the producer price differential.

[64 FR 47985, Sept. 1, 1999, as amended at 65 FR 82838, Dec. 28, 2000; 68 FR 7067, Feb. 12, 2003]

PAYMENTS FOR MILK

§ 1032.70   Producer-settlement fund.

See § 1000.70.

§ 1032.71   Payments to the producer-settlement fund.

Each handler shall make payment to the producer-settlement fund in a manner that provides receipt of the funds by the market administrator no later than the 14th day after the end of the month (except as provided in § 1000.90). Payment shall be the amount, if any, by which the amount specified in paragraph (a) of this section exceeds the amount specified in paragraph (b) of this section:

(a) The total value of milk to the handler for the month as determined pursuant to § 1032.60.

(b) The sum of:

(1) An amount obtained by multiplying the total hundredweight of producer milk as determined pursuant to § 1000.44(c) by the producer price differential as adjusted pursuant to § 1032.75;

(2) An amount obtained by multiplying the total pounds of protein, other solids, and butterfat contained in producer milk by the protein, other solids, and butterfat prices respectively;

(3) The total value of the somatic cell adjustment to producer milk; and

(4) An amount obtained by multiplying the pounds of skim milk and butterfat for which a value was computed pursuant to § 1032.60(i) by the producer price differential as adjusted pursuant to § 1032.75 for the location of the plant from which received.

[64 FR 47985, Sept. 1, 1999, as amended at 65 FR 82838, Dec. 28, 2000; 68 FR 7067, Feb. 12, 2003]

§ 1032.72   Payments from the producer-settlement fund.

No later than the 15th day after the end of each month (except as provided in § 1000.90), the market administrator shall pay to each handler the amount, if any, by which the amount computed pursuant to § 1032.71(b) exceeds the amount computed pursuant to § 1032.71(a). If, at such time, the balance in the producer-settlement fund is insufficient to make all payments pursuant to this section, the market administrator shall reduce uniformly such payments and shall complete the payments as soon as the funds are available.

§ 1032.73   Payments to producers and to cooperative associations.

(a) Each handler shall pay each producer for producer milk for which payment is not made to a cooperative association pursuant to paragraph (b) of this section, as follows:

(1) *Partial payment.* For each producer who has not discontinued shipments as of the date of this partial payment, payment shall be made so that it is received by each producer on or before the 26th day of the month (except as provided in § 1000.90) for milk received during the first 15 days of the month from the producer at not less than the lowest announced class price for the preceding month, less proper deductions authorized in writing by the producer.

(2) *Final payment.* For milk received during the month, payment shall be made so that it is received by each producer no later than the 17th day after the end of the month (except as provided in § 1000.90) in an amount equal to not less than the sum of:

(i) The hundredweight of producer milk received times the producer price differential for the month as adjusted pursuant to § 1032.75;

(ii) The pounds of butterfat received times the butterfat price for the month;

(iii) The pounds of protein received times the protein price for the month;

(iv) The pounds of other solids received times the other solids price for the month;

(v) The hundredweight of milk received times the somatic cell adjustment for the month;

(vi) Less any payment made pursuant to paragraph (a)(1) of this section;

(vii) Less proper deductions authorized in writing by such producer and plus or minus adjustments for errors in

132

previous payments to such producer; and

(viii) Less deductions for marketing services pursuant to §1000.86.

(b) *Payments for milk received from cooperative association members.* On or before the day prior to the dates specified in paragraphs (a)(1) and (a)(2) of this section (except as provided in §1000.90), each handler shall pay to a cooperative association for milk from producers who market their milk through the cooperative association and who have authorized the cooperative to collect such payments on their behalf an amount equal to the sum of the individual payments otherwise payable for such producer milk pursuant to paragraphs (a)(1) and (a)(2) of this section.

(c) *Payment for milk received from cooperative association pool plants or from cooperatives as handlers pursuant to §1000.9(c).* On or before the day prior to the dates specified in paragraphs (a)(1) and (a)(2) of this section (except as provided in §1000.90), each handler who receives fluid milk products at its plant from a cooperative association in its capacity as the operator of a pool plant or who receives milk from a cooperative association in its capacity as a handler pursuant to §1000.9(c), including the milk of producers who are not members of such association and who the market administrator determines have authorized the cooperative association to collect payment for their milk, shall pay the cooperative for such milk as follows:

(1) For bulk fluid milk products and bulk fluid cream products received from a cooperative association in its capacity as the operator of a pool plant and for milk received from a cooperative association in its capacity as a handler pursuant to §1000.9(c) during the first 15 days of the month, at not less than the lowest announced class prices per hundredweight for the preceding month;

(2) For the total quantity of bulk fluid milk products and bulk fluid cream products received from a cooperative association in its capacity as the operator of a pool plant, at not less than the total value of such products received from the association's pool plants, as determined by multiplying

the respective quantities assigned to each class under §1000.44 as follows:

(i) The hundredweight of Class I skim milk times the Class I skim milk price for the month plus the pounds of Class I butterfat times the Class I butterfat price for the month. The Class I prices to be used shall be the prices effective at the location of the receiving plant;

(ii) The pounds of nonfat solids in Class II skim milk by the Class II nonfat solids price;

(iii) The pounds of butterfat in Class II times the Class II butterfat price;

(iv) The pounds of nonfat solids in Class IV times the nonfat solids price;

(v) The pounds of butterfat in Class III and Class IV milk times the butterfat price;

(vi) The pounds of protein in Class III milk times the protein price;

(vii) The pounds of other solids in Class III milk times the other solids price;

(viii) The hundredweight of Class II, Class III, and Class IV milk times the somatic cell adjustment; and

(ix) Add together the amounts computed in paragraphs (c)(2)(i) through (viii) of this section and from that sum deduct any payment made pursuant to paragraph (c)(1) of this section; and

(3) For the total quantity of milk received during the month from a cooperative association in its capacity as a handler under §1000.9(c) as follows:

(i) The hundredweight of producer milk received times the producer price differential as adjusted pursuant to §1032.75;

(ii) The pounds of butterfat received times the butterfat price for the month;

(iii) The pounds of protein received times the protein price for the month;

(iv) The pounds of other solids received times the other solids price for the month;

(v) The hundredweight of milk received times the somatic cell adjustment for the month; and

(vi) Add together the amounts computed in paragraphs (c)(3)(i) through (v) of this section and from that sum deduct any payment made pursuant to paragraph (c)(1) of this section.

(d) If a handler has not received full payment from the market administrator pursuant to § 1032.72 by the payment date specified in paragraph (a), (b) or (c) of this section, the handler may reduce pro rata its payments to producers or to the cooperative association (with respect to receipts described in paragraph (b) of this section, prorating the underpayment to the volume of milk received from the cooperative association in proportion to the total milk received from producers by the handler), but not by more than the amount of the underpayment. The payments shall be completed on the next scheduled payment date after receipt of the balance due from the market administrator.

(e) If a handler claims that a required payment to a producer cannot be made because the producer is deceased or cannot be located, or because the cooperative association or its lawful successor or assignee is no longer in existence, the payment shall be made to the producer-settlement fund, and in the event that the handler subsequently locates and pays the producer or a lawful claimant, or in the event that the handler no longer exists and a lawful claim is later established, the market administrator shall make the required payment from the producer-settlement fund to the handler or to the lawful claimant, as the case may be.

(f) In making payments to producers pursuant to this section, each handler shall furnish each producer, except a producer whose milk was received from a cooperative association handler described in § 1000.9(a) or (c), a supporting statement in a form that may be retained by the recipient which shall show:

(1) The name, address, Grade A identifier assigned by a duly constituted regulatory agency, and payroll number of the producer;

(2) The daily and total pounds, and the month and dates such milk was received from that producer;

(3) The total pounds of butterfat, protein, and other solids contained in the producer's milk;

(4) The somatic cell count of the producer's milk;

(5) The minimum rate or rates at which payment to the producer is required pursuant to the order in this part;

(6) The rate used in making payment if the rate is other than the applicable minimum rate;

(7) The amount, or rate per hundredweight, or rate per pound of component, and the nature of each deduction claimed by the handler; and

(8) The net amount of payment to the producer or cooperative association.

[64 FR 47985, Sept. 1, 1999, as amended at 65 FR 82838, Dec. 28, 2000; 68 FR 7067, Feb. 12, 2003]

§ 1032.74 [Reserved]

§ 1032.75 Plant location adjustments for producer milk and nonpool milk.

For purposes of making payments for producer milk and nonpool milk, a plant location adjustment shall be determined by subtracting the Class I price specified in § 1032.51 from the Class I price at the plant's location. The difference, plus or minus as the case may be, shall be used to adjust the payments required pursuant to §§ 1032.73 and 1000.76.

§ 1032.76 Payments by a handler operating a partially regulated distributing plant.

See § 1000.76.

§ 1032.77 Adjustment of accounts.

See § 1000.77.

§ 1032.78 Charges on overdue accounts.

See § 1000.78.

ADMINISTRATIVE ASSESSMENT AND MARKETING SERVICE DEDUCTION

§ 1032.85 Assessment for order administration.

See § 1000.85.

§ 1032.86 Deduction for marketing services.

See § 1000.86.

AUTHORITY: 7 U.S.C. 601–674, and 7253.

SOURCE: 64 FR 47991, Sept. 1, 1999, unless otherwise noted.

Subpart—Order Regulating Handling

GENERAL PROVISIONS

§ 1033.1 General provisions.

The terms, definitions, and provisions in part 1000 of this chapter apply to this part 1033. In this part 1033, all references to sections in part 1000 refer to part 1000 of this chapter.

DEFINITIONS

§ 1033.2 Mideast marketing area.

The marketing area means all territory within the bounds of the following states and political subdivisions, including all piers, docks, and wharves connected therewith and all craft moored thereat, and all territory occupied by government (municipal, State, or Federal) reservations, installations, institutions, or other similar establishments if any part thereof is within any of the listed states or political subdivisions:

INDIANA COUNTIES

Adams, Allen, Bartholomew, Benton, Blackford, Boone, Brown, Carroll, Cass, Clay, Clinton, Dearborn, Decatur, De Kalb, Delaware, Elkhart, Fayette, Fountain, Franklin, Fulton, Grant, Hamilton, Hancock, Hendricks, Henry, Howard, Huntington, Jackson, Jasper, Jay, Jefferson, Jennings, Johnson, Kosciusko, Lagrange, Lake, La Porte, Lawrence, Madison, Marion, Marshall,

Miami, Monroe, Montgomery, Morgan, Newton, Noble, Ohio, Owen, Parke, Porter, Pulaski, Putnam, Randolph, Ripley, Rush, Shelby, St. Joseph, Starke, Steuben, Switzerland, Tippecanoe, Tipton, Union, Vermillion, Vigo, Wabash, Warren, Wayne, Wells, White, and Whitley.

KENTUCKY COUNTIES

Boone, Boyd, Bracken, Campbell, Floyd, Grant, Greenup, Harrison, Johnson, Kenton, Lawrence, Lewis, Magoffin, Martin, Mason, Pendleton, Pike, and Robertson.

MICHIGAN COUNTIES

All counties except Delta, Dickinson, Gogebic, Iron, Menominee, and Ontonagon.

OHIO

The townships of Woodville and Madison in Sandusky County and all other counties in Ohio except Erie, Huron, and Ottawa.

PENNSYLVANIA COUNTIES

Allegheny, Armstrong, Beaver, Butler, Crawford, Erie, Fayette, Greene, Lawrence, Mercer, Venango, and Washington.

In Clarion County only the townships of Ashland, Beaver, Licking, Madison, Perry, Piney, Richland, Salem, and Toby.

All of Westmoreland County except the townships of Cook, Donegal, Fairfield, Ligonier, and St. Clair, and the boroughs of Bolivar, Donegal, Ligonier, New Florence, and Seward.

WEST VIRGINIA COUNTIES

Barbour, Boone, Brooke, Cabell, Calhoun, Doddridge, Fayette, Gilmer, Hancock, Harrison, Jackson, Kanawha, Lewis, Lincoln, Logan, Marion, Marshall, Mason, Mingo, Monongalia, Ohio, Pleasants, Preston, Putnam, Raleigh, Randolph, Ritchie, Roane, Taylor, Tucker, Tyler, Upshur, Wayne, Wetzel, Wirt, Wood, and Wyoming.

§ 1033.3 Route disposition.

See § 1000.3.

§ 1033.4 Plant.

See § 1000.4.

§ 1033.5 Distributing plant.

See § 1000.5.

§ 1033.6 Supply plant.

See § 1000.6.

§ 1033.7 Pool plant.

Pool plant means a plant, unit of plants, or system of plants as specified in paragraphs (a) through (f) of this section, or a plant specified in paragraph (j) of this section, but excluding a plant specified in paragraph (h) of this section. The pooling standards described in paragraphs (c) through (f) of this section are subject to modification pursuant to paragraph (g) of this section:

(a) A distributing plant, other than a plant qualified as a pool plant pursuant to paragraph (b) of this section or § ___.7(b) of any other Federal milk order, from which during the month 30 percent or more of the total quantity of fluid milk products physically received at the plant (excluding concentrated milk received from another plant by agreement for other than class I use) are disposed of as route disposition or are transferred in the form of packaged fluid milk products to other distributing plants. At least 25 percent of such route disposition and transfers must be to outlets in the marketing area. Plants located within the marketing area that meet the 30 percent route disposition standard contained above, and have combined route disposition and transfers of at least 50 percent into Federal order marketing areas will be regulated as a distributing plant in this order.

(b) Any distributing plant located in the marketing area which during the month processed at least 30 percent of the total quantity of fluid milk products physically received at the plant (excluding concentrated milk received from another plant by agreement for other than Class I use) into ultra-pasteurized or aseptically-processed fluid milk products.

(c) A supply plant from which the quantity of bulk fluid milk products shipped to, received at, and physically unloaded into plants described in paragraph (a) or (b) of this section as a percent of the Grade A milk received at the plant from dairy farmers (except dairy farmers described in § 1033.12(b)) and handlers described in § 1000.9(c), as reported in § 1033.30(a), is not less than 40 percent of the milk received from dairy farmers, including milk diverted pursuant to § 1033.13, subject to the following conditions:

(1) Qualifying shipments pursuant to this paragraph may be made to the following plants, except whenever the authority provided in paragraph (g) of

this section is applied to increase the shipping requirements specified in this section, only shipments to pool plants described in §1033.7(a) and (b), shall count as qualifying shipments for the purpose of meeting the increased shipments:

(i) Pool plants described in §1033.7(a) and (b);

(ii) Plants of producer-handlers;

(iii) Partially regulated distributing plants, except that credit for such shipments shall be limited to the amount of such milk classified as Class I at the transferee plant.

(2) The operator of a supply plant located within the marketing area may include deliveries to pool distributing plants directly from farms of producers pursuant to §1033.13(c) as up to 90 percent of the supply plant's qualifying shipments. Handlers may not use shipments pursuant to §1033.13(c) to qualify plants located outside the marketing area.

(3) Concentrated milk transferred from the supply plant to a distributing plant for an agreed-upon use other than Class I shall be excluded from the supply plant's shipments in computing the supply plant's shipping percentage.

(4) Shipments used in determining qualifying percentages shall be milk transferred or diverted and physically received by pool distributing plants, less any transfers or diversions of bulk fluid milk products from such pool distributing plants.

(5) A supply plant that does not meet the minimum delivery requirements specified in this paragraph to qualify for pool status in the current month because a distributing plant to which the supply plant delivered its fluid milk products during such month failed to qualify as a pool plant pursuant to paragraph (a) or (b) of this section shall continue to be a pool plant for the current month if such supply plant qualified as a pool plant in the 3 immediately preceding months.

(d) A plant located in the marketing area and operated by a cooperative association if, during the months of December through July 30 percent, during the month of August 35 percent and during the months of September through November 40 percent or more of the producer milk of members of the association is delivered to a distributing pool plant(s) or to a nonpool plant(s) and classified as Class I. Deliveries for qualification purposes may be made directly from the farm or by transfer from such association's plant, subject to the following conditions:

(1) The cooperative requests pool status for such plant;

(2) The 30 percent delivery requirement for the months of December through July may be met for the current month or it may be met on the basis of deliveries during the preceding 12-month period ending with the current month.

(3) The plant is approved by a duly constituted regulatory authority to handle milk for fluid consumption; and

(4) The plant does not qualify as a pool plant under paragraph (a), (b), or (c) of this section or under the similar provisions of another Federal order applicable to a distributing plant or supply plant.

(e) A plant located inside the marketing area which has been a pool plant under this order for twelve consecutive months, but is not otherwise qualified under this paragraph, if it has a marketing agreement with a cooperative association and it fulfills the following conditions:

(1) The aggregate monthly quantity supplied by all parties to such an agreement as a percentage of the producer milk receipts included in the unit during the months of August through November is not less than 45 percent and during the months of December through July is not less than 35 percent;

(2) Shipments for qualification purposes shall include both transfers from supply plants to plants described in paragraph (c)(1) of this section, and deliveries made direct from the farm to plants qualified under paragraph (a) of this section.

(f) A system of supply plants may qualify for pooling if 2 or more plants operated by one or more handlers meet the applicable percentage requirements of paragraph (c) of this section in the same manner as a single plant subject to the following additional requirements:

(1) Each plant in the system is located within the marketing area, or

was a pool supply plant for each of the 3 months immediately preceding the effective date of this paragraph so long as it continues to maintain pool status. Cooperative associations may not use shipments pursuant to §1033.9(c) to qualify plants located outside the marketing area;

(2) A written notification to the market administrator listing the plants to be included in the system and the handler that is responsible for meeting the performance requirements of this paragraph under a marketing agreement certified to the market administrator by the designated handler and any others included in the system, and the period during which such consideration shall apply. Such notice, and notice of any change in designation, shall be furnished on or before the 5th working day following the month to which the notice applies. The listed plants included in the system shall also be in the sequence in which they shall qualify for pool plant status based on the minimum deliveries required. If the deliveries made are insufficient to qualify the entire system for pooling, the last listed plant shall be excluded from the system, followed by the plant next-to-last on the list, and continuing in this sequence until remaining listed plants have met the minimum shipping requirements; and

(3) Each plant that qualifies as a pool plant within a system shall continue each month as a plant in the system unless the plant subsequently fails to qualify for pooling, or the responsible handler submits a written notification to the market administrator prior to the first day of the month that the plant is to be deleted from the system, or that the system is to be discontinued. In any month of March through August, a system shall not contain any plant which was not qualified under this paragraph, either individually or as a member of a system, during the previous September through February.

(g) The applicable shipping percentages of paragraphs (c) through (f) of this section may be increased or decreased by the market administrator if the market administrator finds that such adjustment is necessary to encourage needed shipments or to prevent uneconomic shipments. Before making such a finding, the market administrator shall investigate the need for adjustment either on the market administrator's own initiative or at the request of interested parties if the request is made in writing at least 15 days prior to the month for which the requested revision is desired effective. If the investigation shows that an adjustment of the shipping percentages might be appropriate, the market administrator shall issue a notice stating that an adjustment is being considered and invite data, views and arguments. Any decision to revise an applicable shipping percentage must be issued in writing at least one day before the effective date.

(h) The term pool plant shall not apply to the following plants:

(1) A producer-handler as defined under any Federal order;

(2) An exempt plant as defined in §1000.8(e);

(3) A plant located within the marketing area and qualified pursuant to paragraph (a) of this section that meets the pooling requirements of another Federal order, and from which more than 50 percent of its route disposition has been in the other Federal order marketing area for 3 consecutive months;

(4) A plant located outside any Federal order marketing area and qualified pursuant to paragraph (a) of this section that meets the pooling requirements of another Federal order and has had greater route disposition in such other Federal order's marketing area for 3 consecutive months;

(5) A plant located in another Federal order marketing area and qualified pursuant to paragraph (a) of this section that meets the pooling requirements of such other Federal order and does not have a majority of its route distribution in this marketing area for 3 consecutive months or if the plant is required to be regulated under such other Federal order without regard to its route disposition in any other Federal order marketing area;

(6) A plant qualified pursuant to paragraph (c) of this section that also meets the pooling requirements of another Federal order and from which greater qualifying shipments are made to plants regulated under the other

Federal order than are made to plants regulated under the order in this part, or the plant has automatic pooling status under the other Federal order.

(i) Any plant that qualifies as a pool plant in each of the immediately preceding 3 months pursuant to paragraph (a) of this section or the shipping percentages in paragraph (c) of this section that is unable to meet such performance standards for the current month because of unavoidable circumstances determined by the market administrator to be beyond the control of the handler operating the plant, such as a natural disaster (ice storm, wind storm, flood), fire, breakdown of equipment, or work stoppage, shall be considered to have met the minimum performance standards during the period of such unavoidable circumstances, but such relief shall not be granted for more than 2 consecutive months.

(j) Any distributing plant, located within the marketing area as described on May 1, 2006, in § 1033.2;

(1) From which there is route disposition and/or transfers of packaged fluid milk products in any non-federally regulated marketing area(s) located within one or more States that require handlers to pay minimum prices for raw milk provided that 25 percent or more of the total quantity of fluid milk products physically received at such plant (excluding concentrated milk received from another plant by agreement for other than Class I use) is disposed of as route disposition and/or is transferred in the form of packaged fluid milk products to other plants. At least 25 percent of such route disposition and/or transfers, in aggregate, are in any non-federally regulated marketing area(s) located within one or more States that require handlers to pay minimum prices for raw milk. Subject to the following exclusions:

(i) The plant is described in § 1033.7(a) or (b);

(ii) The plant is subject to the pricing provisions of a State-operated milk pricing plan which provides for the payment of minimum class prices for raw milk;

(iii) The plant is described in § 1000.8(a) or (e); or

(iv) A producer-handler described in § 1033.10 with less than three million pounds during the month of route disposition and/or transfers of packaged fluid milk products to other plants.

(2) [Reserved]

[64 FR 47991, Sept. 1, 1999, as amended at 67 FR 48744, July 26, 2002; 69 FR 34555, June 22, 2004; 70 FR 56112, Sept. 26, 2005; 71 FR 25500, May 1, 2006; 71 FR 28249, May 16, 2006; 77 FR 51695, Aug. 27, 2012]

§ 1033.8 Nonpool plant.

See § 1000.8.

§ 1033.9 Handler.

See § 1000.9.

§ 1033.10 Producer-handler.

Producer-handler means a person who:

(a) Operates a dairy farm and a distributing plant from which there is route disposition in the marketing area, and from which total route disposition and packaged sales of fluid milk products to other plants during the month does not exceed 3 million pounds;

(b) Receives fluid milk from own farm production or that is fully subject to the pricing and pooling provisions of the order in this part or any other Federal order;

(c) Receives at its plant or acquires for route disposition no more than 150,000 pounds of fluid milk products from handlers fully regulated under any Federal order. This limitation shall not apply if the producer-handler's own farm production is less than 150,000 pounds during the month;

(d) Disposes of no other source milk as Class I milk except by increasing the nonfat milk solids content of the fluid milk products; and

(e) Provides proof satisfactory to the market administrator that the care and management of the dairy animals and other resources necessary to produce all Class I milk handled (excluding receipts from handlers fully regulated under any Federal order) and the processing and packaging operations are the producer-handler's own enterprise and at its own risk.

(f) Any producer-handler with Class I route dispositions and/or transfers of packaged fluid milk products in the marketing area described in § 1131.2 of this chapter shall be subject to payments into the Order 1131 producer settlement fund on such dispositions pursuant to § 1000.76(a) and payments into the Order 1131 administrative fund provided·such dispositions are less than three million pounds in the current month and such producer-handler had total Class I route dispositions and/or transfer of packaged fluid milk products from own farm production of three million pounds or more the previous month. If the producer-handler has Class I route dispositions and/or transfers of packaged fluid milk products into the marketing area described in § 1131.2 of this chapter of three million pounds or more during the current month, such producer-handler shall be subject to the provisions described in § 1131.7 of this chapter or § 1000.76(a).

[64 FR 47991, Sept. 1, 1999, as amended at 71 FR 25500, May 1, 2006; 75 FR 21160, Apr. 23, 2010]

§ 1033.11　[Reserved]

§ 1033.12　Producer.

(a) Except as provided in paragraph (b) of this section, *producer* means any person who produces milk approved by a duly constituted regulatory agency for fluid consumption as Grade A milk and whose milk is:

(1) Received at a pool plant directly from the producer or diverted by the plant operator in accordance with § 1033.13; or

(2) Received by a handler described in § 1033.9(c).

(b) Producer shall not include:

(1) A producer-handler as defined in any Federal order;

(2) A dairy farmer whose milk is received at an exempt plant, excluding producer milk diverted to the exempt plant pursuant to § 1033.13(d);

(3) A dairy farmer whose milk is received by diversion at a pool plant from a handler regulated under another Federal order if the other Federal order designates the dairy farmer as a producer under that order and that milk is allocated by request to a utilization other than Class I; and

(4) A dairy farmer whose milk is reported as diverted to a plant fully regulated under another Federal order with respect to that portion of the milk so diverted that is assigned to Class I under the provisions of such other order.

§ 1033.13　Producer milk.

Producer milk means the skim milk (or the skim equivalent of components of skim milk), including nonfat components, and butterfat in milk of a producer that is:

(a) Received by the operator of a pool plant directly from a producer or a handler described in § 1000.9(c). All milk received pursuant to this paragraph shall be priced at the location of the plant where it is first physically received;

(b) Received by a handler described in § 1000.9(c) in excess of the quantity delivered to pool plants;

(c) Diverted by a pool plant operator to another pool plant. Milk so diverted shall be priced at the location of the plant to which diverted; or

(d) Diverted by the operator of a pool plant or by a cooperative association described in § 1000.9(c) to a nonpool plant, subject to the following conditions:

(1) Milk of a dairy farmer shall not be eligible for diversion until milk of such dairy farmer has been physically received as producer milk at a pool plant and the dairy farmer has continuously retained producer status since that time. If a dairy farmer loses producer status under the order in this part (except as a result of a temporary loss of Grade A approval), the dairy farmer's milk shall not be eligible for diversion until milk of the dairy farmer has been physically received as producer milk at a pool plant;

(2) The equivalent of at least two days' milk production is caused by the handler to be physically received at a pool plant in each of the months of August through November;

(3) The equivalent of at least two days' milk production is caused by the handler to be physically received at a pool plant in each of the months of December through July if the requirement of paragraph (d)(2) of this section (§ 1033.13) in each of the prior months of August through November are not met, except in the case of a dairy farmer who marketed no Grade A milk during

each of the prior months of August through November.

(4) Of the total quantity of producer milk received during the month (including diversions but excluding the quantity of producer milk received from a handler described in § 1000.9(c) or which is diverted to another pool plant), the handler diverted to nonpool plants not more than 50 percent in each of the months of August through February and 60 percent in each of the months of March through July.

(5) Diverted milk shall be priced at the location of the plant to which diverted;

(6) Any milk diverted in excess of the limits set forth in paragraph (d)(3) of this section shall not be producer milk. The diverting handler shall designate the dairy farmer deliveries that shall not be producer milk. If the handler fails to designate the dairy farmer deliveries which are ineligible, producer milk status shall be forfeited with respect to all milk diverted to nonpool plants by such handler; and

(7) The delivery day requirement in paragraphs (d)(2) and (d)(3) of this section and the diversion percentages in paragraph (d)(4) of this section may be increased or decreased by the market administrator if the market administrator finds that suhc revision is necessary to assure orderly marketing and efficient handling of milk in the marketing area. Before making such a finding, the market administrator shall investigate the need for the revision either on the market administrator's own initiative or at the request of interested persons if the request is made in writing at least 15 days prior to the month for which the requested revision is desired effective. If the investigation shows that a revision might be appropriate, the market administrator shall issue a notice stating that the revision is being considered and inviting written data, views, and arguments. Any decision to revise an applicable percentage must be issued in writing at least one day before the effective date.

(e) Producer milk shall not include milk of a producer that is subject to inclusion and participation in a marketwide equalization pool under a milk classification and pricing plan imposed under the authority of another government entity.

(f) Producer milk of a handler shall not exceed the limits as established in § 1033.13(f)(1) through § 1033.13(f)(3).

(1) Producer milk for the months of April through February may not exceed 115 percent of the producer milk receipts of the prior month. Producer milk for March may not exceed 120 percent of producer receipts of the prior month; plus

(2) Milk shipped to and physically received at pool distributing plants and allocated to Class I use in excess of the volume allocated to Class I in the prior month; plus

(3) If a producer did not have any milk delivered to any plant as other than producer milk as defined under the order in this part or any other Federal milk order for the preceding three months; and the producer had milk qualified as producer milk on any other Federal order in the previous month, add the lesser of the following:

(i) Any positive difference of the volume of milk qualified as producer milk on any other Federal order in the previous month, less the volume of milk qualified as producer milk on any other Federal order in the current month, or

(ii) Any positive difference of the volume of milk qualified as producer milk under the order in this part in the current month, less the volume of milk qualified as producer milk under the order in this part in the previous month.

(4) Milk received at pool plants in excess of these limits shall be classified pursuant to § 1000.44(a)(3)(v) and § 1000.44(b). Milk diverted to nonpool plants reported in excess of this limit shall not be producer milk. The handler must designate, by producer pickup, which milk shall not be producer milk. If the handler fails to provide this information the provisions of § 1033.13(d)(6) shall apply.

(5) The market administrator may waive these limitations:

(i) For a new handler on the order, subject to the provisions of § 1033.13(f)(6), or

(ii) For an existing handler with significantly changed milk supply conditions due to unusual circumstances;

(6) Milk may not be considered producer milk if the market administrator determines that handlers altered the reporting of such milk for the purpose of evading the provisions of this paragraph.

[64 FR 47991, Sept. 1, 1999, as amended at 67 FR 48744, July 26, 2002; 69 FR 34555, June 22, 2004; 70 FR 56112, Sept. 26, 2005; 71 FR 63219, Oct. 30, 2006]

§ 1033.14 Other source milk.

See § 1000.14.

§ 1033.15 Fluid milk products.

See § 1000.15.

§ 1033.16 Fluid cream product.

See § 1000.16.

§ 1033.17 [Reserved]

§ 1033.18 Cooperative association.

See § 1000.18.

§ 1033.19 Commercial food processing establishment.

See § 1000.19.

HANDLER REPORTS

§ 1033.30 Reports of receipts and utilization.

Each handler shall report monthly so that the market administrator's office receives the report on or before the 7th day after the end of the month, in the detail and on the prescribed forms, as follows:

(a) Each handler that operates a pool plant pursuant to § 1033.7 shall report for each of its operations the following information:

(1) Product pounds, pounds of butterfat, pounds of protein, pounds of solids-not-fat other than protein (other solids), and the value of the somatic cell adjustment pursuant to § 1000.50(p), contained in or represented by:

(i) Receipts of producer milk, including producer milk diverted by the reporting handler, from sources other than handlers described in § 1000.9(c); and

(ii) Receipts of milk from handlers described in § 1000.9(c);

(2) Product pounds and pounds of butterfat contained in:

(i) Receipts of fluid milk products and bulk fluid cream products from other pool plants;

(ii) Receipts of other source milk; and

(iii) Inventories at the beginning and end of the month of fluid milk products and bulk fluid cream products;

(3) The utilization or disposition of all milk and milk products required to be reported pursuant to this paragraph; and

(4) Such other information with respect to the receipts and utilization of skim milk, butterfat, milk protein, other nonfat solids, and somatic cell information as the market administrator may prescribe.

(b) Each handler operating a partially regulated distributing plant shall report with respect to such plant in the same manner as prescribed for reports required by paragraph (a) of this section. Receipts of milk that would have been producer milk if the plant had been fully regulated shall be reported in lieu of producer milk. The report shall show also the quantity of any reconstituted skim milk in route disposition in the marketing area.

(c) Each handler described in § 1000.9(c) shall report:

(1) The product pounds, pounds of butterfat, pounds of protein, pounds of solids-not-fat other than protein (other solids), and the value of the somatic cell adjustment pursuant to § 1000.50(p), contained in receipts of milk from producers; and

(2) The utilization or disposition of such receipts.

(d) Each handler not specified in paragraphs (a) through (c) of this section shall report with respect to its receipts and utilization of milk and milk products in such manner as the market administrator may prescribe.

§ 1033.31 Payroll reports.

(a) On or before the 22nd day after the end of each month, each handler that operates a pool plant pursuant to § 1033.7 and each handler described in § 1000.9(c) shall report to the market administrator its producer payroll for the month, in the detail prescribed by the market administrator, showing for each producer the information described in § 1033.73(e).

(b) Each handler operating a partially regulated distributing plant who elects to make payment pursuant to §1000.76(b) shall report for each dairy farmer who would have been a producer if the plant had been fully regulated in the same manner as prescribed for reports required by paragraph (a) of this section.

§1033.32 Other reports.

In addition to the reports required pursuant to §§1033.30 and 1033.31, each handler shall report any information the market administrator deems necessary to verify or establish each handler's obligation under the order.

CLASSIFICATION OF MILK

§1033.40 Classes of utilization.

See §1000.40.

§1033.41 [Reserved]

§1033.42 Classification of transfers and diversions.

See §1000.42.

§1033.43 General classification rules.

See §1000.43.

§1033.44 Classification of producer milk.

See §1000.44.

§1033.45 Market administrator's reports and announcements concerning classification.

See §1000.45.

CLASS PRICES

§1033.50 Class prices, component prices, and advanced pricing factors.

See §1000.50.

§1033.51 Class I differential and price.

· The Class I differential shall be the differential established for Cuyahoga County, Ohio which is reported in §1000.52. The Class I price shall be the price computed pursuant to §1000.50(a) for Cuyahoga County, Ohio.

§1033.52 Adjusted Class I differentials.

See §1000.52.

§1033.53 Announcement of class prices, component prices, and advanced pricing factors.

See §1000.53.

§1033.54 Equivalent price.

See §1000.54.

PRODUCER PRICE DIFFERENTIAL

§1033.60 Handler's value of milk.

For the purpose of computing a handler's obligation for producer milk, the market administrator shall determine for each month the value of milk of each handler with respect to each of the handler's pool plants and of each handler described in §1000.9(c) with respect to milk that was not received at a pool plant by adding the amounts computed in paragraphs (a) through (i) of this section and subtracting from that total amount the value computed in paragraph (j) of this section. Unless otherwise specified, the skim milk, butterfat, and the combined pounds of skim milk and butterfat referred to in this section shall result from the steps set forth in §1000.44(a), (b), and (c), respectively, and the nonfat components of producer milk in each class shall be based upon the proportion of such components in producer skim milk. Receipts of nonfluid milk products that are distributed as labeled reconstituted milk for which payments are made to the producer-settlement fund of another Federal order under §1000.76(a)(4) or (d) shall be excluded from pricing under this section.

(a) Class I value.

(1) Multiply the pounds of skim milk in Class I by the Class I skim milk price; and

(2) Add an amount obtained by multiplying the pounds of butterfat in Class I by the Class I butterfat price.

(b) Class II value.

(1) Multiply the pounds of nonfat solids in Class II skim milk by the Class II nonfat solids price; and

(2) Add an amount obtained by multiplying the pounds of butterfat in Class II times the Class II butterfat price.

(c) Class III value.

(1) Multiply the pounds of protein in Class III skim milk by the protein price;

(2) Add an amount obtained by multiplying the pounds of other solids in Class III skim milk by the other solids price; and

(3) Add an amount obtained by multiplying the pounds of butterfat in Class III by the butterfat price.

(d) Class IV value.

(1) Multiply the pounds of nonfat solids in Class IV skim milk by the nonfat solids price; and

(2) Add an amount obtained by multiplying the pounds of butterfat in Class IV by the butterfat price.

(e) Compute an adjustment for the somatic cell content of producer milk by multiplying the values reported pursuant to § 033.30(a)(1) and (c)(1) by the percentage of total producer milk allocated to Class II, Class III, and Class IV pursuant to § 1000.44(c);

(f) Multiply the pounds of skim milk and butterfat overage assigned to each class pursuant to § 1000.44(a)(11) and the corresponding step of § 1000.44(b) by the skim milk prices and butterfat prices applicable to each class.

(g) Multiply the difference between the current month's Class I, II, or III price, as the case may be, and the Class IV price for the preceding month by the hundredweight of skim milk and butterfat subtracted from Class I, II, or III, respectively, pursuant to § 1000.44(a)(7) and the corresponding step of § 1000.44(b);

(h) Multiply the difference between the Class I price applicable at the location of the pool plant and the Class IV price by the hundredweight of skim milk and butterfat assigned to Class I pursuant to § 1000.43(d) and the hundredweight of skim milk and butterfat subtracted from Class I pursuant to § 1000.44(a)(3)(i) through (vi) and the corresponding step of § 1000.44(b), excluding receipts of bulk fluid cream products from a plant regulated under other Federal orders and bulk concentrated fluid milk products from pool plants, plants regulated under other Federal orders, and unregulated supply plants.

(i) Multiply the difference between the Class I price applicable at the location of the nearest unregulated supply plants from which an equivalent volume was received and the Class III price by the pounds of skim milk and butterfat in receipts of concentrated fluid milk products assigned to Class I pursuant to § 1000.43(d) and § 1000.44(a)(3)(i) and the corresponding step of § 1000.44(b) and the pounds of skim milk and butterfat subtracted from Class I pursuant to § 1000.44(a)(8) and the corresponding step of § 1000.44(b), excluding such skim milk and butterfat in receipts of fluid milk products from an unregulated supply plant to the extent that an equivalent amount of skim milk or butterfat disposed of to such plant by handlers fully regulated under any Federal milk order is classified and priced as Class I milk and is not used as an offset for any other payment obligation under any order.

(j) For reconstituted milk made from receipts of nonfluid milk products, multiply $1.00 (but not more than the difference between the Class I price applicable at the location of the pool plant and the Class IV price) by the hundredweight of skim milk and butterfat contained in receipts of nonfluid milk products that are allocated to Class I use pursuant to § 1000.43(d).

[64 FR 47991, Sept. 1, 1999, as amended at 65 FR 82838, Dec. 28, 2000; 68 FR 7067, Feb. 12, 2003]

§ 1033.61 Computation of producer price differential.

For each month the market administrator shall compute a producer price differential per hundredweight. The report of any handler who has not made payments required pursuant to § 1033.71 for the preceding month shall not be included in the computation of the producer price differential, and such handler's report shall not be included in the computation for succeeding months until the handler has made full payment of outstanding monthly obligations. Subject to the conditions of this paragraph, the market administrator shall compute the producer price differential in the following manner:

(a) Combine into one total the values computed pursuant to § 1033.60 for all handlers required to file reports prescribed in § 1033.30;

(b) Subtract the total values obtained by multiplying each handler's total pounds of protein, other solids, and butterfat contained in the milk for which an obligation was computed pursuant to § 1033.60 by the protein price, the other solids price, and the butterfat price, respectively, and the total value of the somatic cell adjustment pursuant to § 1033.30(a)(1) and (c)(1);

(c) Add an amount equal to the minus location adjustments and subtract an amount equal to the plus location adjustments computed pursuant to § 1033.75;

(d) Add an amount equal to not less than one-half of the unobligated balance in the producer-settlement fund;

(e) Divide the resulting amount by the sum of the following for all handlers included in these computations:

(1) The total hundredweight of producer milk; and

(2) The total hundredweight for which a value is computed pursuant to § 1033.60(i); and

(f) Subtract not less than 4 cents nor more than 5 cents from the price computed pursuant to paragraph (e) of this section. The result shall be known as the producer price differential for the month.

[68 FR 7067, Feb. 12, 2003]

§ 1033.62 Announcement of producer prices.

On or before the 13th day after the end of each month, the market administrator shall announce publicly the following prices and information:

(a) The producer price differential;

(b) The protein price;

(c) The nonfat solids price;

(d) The other solids price;

(e) The butterfat price;

(f) The somatic cell adjustment rate;

(g) The average butterfat, protein, nonfat solids, and other solids content of producer milk; and

(h) The statistical uniform price for milk containing 3.5 percent butterfat, computed by combining the Class III price and the producer price differential.

[64 FR 47991, Sept. 1, 1999, as amended at 65 FR 82839, Dec. 28, 2000; 68 FR 7068, Feb. 12, 2003]

PAYMENTS FOR MILK

§ 1033.70 Producer-settlement fund.

See § 1000.70.

§ 1033.71 Payments to the producer-settlement fund.

Each handler shall make payment to the producer-settlement fund in a manner that provides receipt of the funds by the market administrator no later than the 15th day after the end of the month (except as provided in § 1000.90). Payment shall be the amount, if any, by which the amount specified in paragraph (a) of this section exceeds the amount specified in paragraph (b) of this section:

(a) The total value of milk to the handler for the month as determined pursuant to § 1033.60.

(b) The sum of:

(1) An amount obtained by multiplying the total hundredweight of producer milk as determined pursuant to § 1000.44(c) by the producer price differential as adjusted pursuant to § 1033.75;

(2) An amount obtained by multiplying the total pounds of protein, other solids, and butterfat contained in producer milk by the protein, other solids, and butterfat prices, respectively;

(3) The total value of the somatic cell adjustment to producer milk; and

(4) An amount obtained by multiplying the pounds of skim milk and butterfat for which a value was computed pursuant to § 1033.60(i) by the producer price differential as adjusted pursuant to § 1033.75 for the location of the plant from which received.

[64 FR 47991, Sept. 1, 1999, as amended at 65 FR 82839, Dec. 28, 2000; 68 FR 7068, Feb. 12, 2003]

§ 1033.72 Payments from the producer-settlement fund.

No later than the 16th day after the end of each month (except as provided in § 1000.90), the market administrator shall pay to each handler the amount, if any, by which the amount computed pursuant to § 1033.71(b) exceeds the amount computed pursuant to § 1033.71(a). If, at such time, the balance in the producer-settlement fund is insufficient to make all payments pursuant to this section, the market administrator shall reduce uniformly such payments and shall complete the payments as soon as the funds are available.

§ 1033.73 Payments to producers and to cooperative associations.

(a) Each handler shall pay each producer for producer milk for which payment is not made to a cooperative association pursuant to paragraph (b) of this section, as follows:

(1) *Partial payment.* For each producer who has not discontinued shipments as of the date of this partial payment, payment shall be made so that it is received by each producer on or before the 26th day of the month (except as provided in § 1000.90) for milk received during the first 15 days of the month from the producer at not less than the lowest announced class price for the preceding month, less proper deductions authorized in writing by the producer.

(2) *Final payment.* For milk received during the month, payment shall be made so that it is received by each producer no later than the 17th day after the end of the month (except as provided in § 1000.90) in an amount equal to not less than the sum of:

(i) The hundredweight of producer milk received times the producer price differential for the month as adjusted pursuant to § 1033.75;

(ii) The pounds of butterfat received times the butterfat price for the month;

(iii) The pounds of protein received times the protein price for the month;

(iv) The pounds of other solids received times the other solids price for the month;

(v) The hundredweight of milk received times the somatic cell adjustment for the month;

(vi) Less any payment made pursuant to paragraph (a)(1) of this section;

(vii) Less proper deductions authorized in writing by such producer and plus or minus adjustments for errors in previous payments to such producer; and

(viii) Less deductions for marketing services pursuant to § 1000.86.

(b) *Payments for milk received from cooperative associations.* On or before the day prior to the dates specified in paragraphs (a)(1) and (a)(2) of this section (except as provided in § 1000.90), each handler shall pay to a cooperative association for milk received as follows:

(1) *Partial payment to a cooperative association.* For bulk fluid milk/skimmed milk received during the first 15 days of the month from a cooperative association in any capacity, except as the operator of a pool plant, the partial payment shall be equal to the hundredweight of milk received multiplied by the lowest announced class price for the preceding month.

(2) *Partial payment to a cooperative association for milk transferred from its pool plant.* For bulk fluid milk/skimmed milk products received during the first 15 days of the month from a cooperative association in its capacity as the operator of a pool plant, the partial payment shall be at the pool plant operator's estimated use value of the milk using the most recent class prices available at the receiving plant's location.

(3) *Final payment to a cooperative association for milk transferred from its pool plant.* Following the classification of bulk fluid milk products and bulk fluid cream products received during the month from a cooperative association in its capacity as the operator of a pool plant, the final payment for such receipts shall be determined as follows:

(i) The hundredweight of Class I skim milk times the Class I skim milk price for the month plus the pounds of Class I butterfat times the Class I butterfat price for the month. The Class I prices to be used shall be the prices effective at the location of the receiving plant;

(ii) The pounds of nonfat solids in Class II skim milk by the Class II nonfat solids price;

(iii) The pounds of butterfat in Class II times the Class II butterfat price;

(iv) The pounds of nonfat solids in Class IV times the nonfat solids price;

(v) The pounds of butterfat in Class III and Class IV milk times the butterfat price;

(vi) The pounds of protein in Class III milk times the protein price;

(vii) The pounds of other solids in Class III milk times the other solids price;

(viii) The hundredweight of Class II, Class III, and Class IV milk times the somatic cell adjustment; and

(ix) Add together the amounts computed in paragraphs (b)(3)(i) through (viii) of this section and from that sum deduct any payment made pursuant to paragraph (b)(2) of this section; and

(4) *Final payment to a cooperative association for bulk milk received directly from producers' farms.* For bulk milk received from a cooperative association during the month, including the milk of producers who are not members of such association and who the market administrator determines have authorized the cooperative association to collect payment for their milk, the final payment for such milk shall be an amount equal to the sum of the individual payments otherwise payable for such milk pursuant to paragraph (a)(2) of this section.

(c) If a handler has not received full payment from the market administrator pursuant to §1033.72 by the payment date specified in paragraph (a) or (b) of this section, the handler may reduce payments pursuant to paragraphs (a) and (b) of this section, but not by more than the amount of the underpayment. The payments shall be completed on the next scheduled payment date after receipt of the balance due from the market administrator.

(d) If a handler claims that a required payment to a producer cannot be made because the producer is deceased or cannot be located, or because the cooperative association or its lawful successor or assignee is no longer in existence, the payment shall be made to the producer-settlement fund, and in the event that the handler subsequently locates and pays the producer or a lawful claimant, or in the event that the handler no longer exists and a lawful claim is later established, the market administrator shall make the required payment from the producer-settlement fund to the handler or to the lawful claimant, as the case may be.

(e) In making payments to producers pursuant to this section, each handler shall furnish each producer, except a producer whose milk was received from a cooperative association handler described in §1000.9(a) or (c), a supporting statement in a form that may be retained by the recipient which shall show:

(1) The name, address, Grade A identifier assigned by a duly constituted regulatory agency, and payroll number of the producer;

(2) The daily and total pounds, and the month and dates such milk was received from that producer;

(3) The total pounds of butterfat, protein, and other solids contained in the producer's milk;

(4) The somatic cell count of the producer's milk;

(5) The minimum rate or rates at which payment to the producer is required pursuant to the order in this part;

(6) The rate used in making payment if the rate is other than the applicable minimum rate;

(7) The amount, or rate per hundredweight, or rate per pound of component, and the nature of each deduction claimed by the handler; and

(8) The net amount of payment to the producer or cooperative association.

[64 FR 47991, Sept. 1, 1999, as amended at 65 FR 82839, Dec. 28, 2000; 68 FR 7068, Feb. 12, 2003]

§1033.74 [Reserved]

§1033.75 Plant location adjustments for producer milk and nonpool milk.

For purposes of making payments for producer milk and nonpool milk, a plant location adjustment shall be determined by subtracting the Class I price specified in §1033.51 from the Class I price at the plant's location. The difference, plus or minus as the case may be, shall be used to adjust the payments required pursuant to §§1033.73 and 1000.76.

§1033.76 Payments by a handler operating a partially regulated distributing plant.

See §1000.76.

§1033.77 Adjustment of accounts.

See §1000.77.

§1033.78 Charges on overdue accounts.

See §1000.78.

ADMINISTRATIVE ASSESSMENT AND MARKETING SERVICE DEDUCTION

§1033.85 Assessment for order administration.

See §1000.85.

§ 1033.86 Deduction for marketing services.

See § 1000.86.

PARTS 1036–1050 [RESERVED]

PART 1051—MILK IN THE CALI-FORNIA MILK MARKETING AREA

Subpart A—Order Regulating Handling

AUTHORITY: 7 U.S.C. 601–674, and 7253.

SOURCE: 83 FR 26548, June 8, 2018, unless otherwise noted.

Subpart A—Order Regulating Handling

GENERAL PROVISIONS

§ 1051.1 General provisions.

The terms, definitions, and provisions in part 1000 of this chapter apply to this part unless otherwise specified. In this part, all references to sections in part 1000 refer to part 1000 of this chapter.

DEFINITIONS

§1051.2 California marketing area.

The marketing area means all territory within the bounds of the following states and political subdivisions, including all piers, docks, and wharves connected therewith and all craft moored thereat, and all territory occupied by government (municipal, State, or Federal) reservations, installations, institutions, or other similar establishments if any part thereof is within any of the listed states or political subdivisions:

California

All of the State of California.

§1051.3 Route disposition.

See §1000.3.

§1051.4 Plant.

See §1000.4.

§1051.5 Distributing plant.

See §1000.5.

§1051.6 Supply plant.

See §1000.6.

§1051.7 Pool plant.

Pool plant means a plant, unit of plants, or system of plants as specified in paragraphs (a) through (f) of this section, but excluding a plant specified in paragraph (h) of this section. The pooling standards described in paragraphs (c) and (f) of this section are subject to modification pursuant to paragraph (g) of this section:

(a) A distributing plant, other than a plant qualified as a pool plant pursuant to paragraph (b) of this section or §_____.7(b) of any other Federal milk order, from which during the month 25 percent or more of the total quantity of fluid milk products physically received at the plant (excluding concentrated milk received from another plant by agreement for other than Class I use) are disposed of as route disposition or are transferred in the form of packaged fluid milk products to other distributing plants. At least 25 percent of such route disposition and transfers must be to outlets in the marketing area.

(b) Any distributing plant located in the marketing area which during the month processed at least 25 percent of the total quantity of fluid milk products physically received at the plant (excluding concentrated milk received from another plant by agreement for other than Class I use) into ultra-pasteurized or aseptically-processed fluid milk products.

(c) A supply plant from which the quantity of bulk fluid milk products shipped to (and physically unloaded into) plants described in paragraph (c)(1) of this section is not less than 10 percent of the Grade A milk received from dairy farmers (except dairy farmers described in §1051.12(b)) and handlers described in §1000.9(c), including milk diverted pursuant to §1051.13, subject to the following conditions:

(1) Qualifying shipments may be made to plants described in paragraphs (c)(1)(i) through (iv) of this section, except that whenever shipping requirements are increased pursuant to paragraph (g) of this section, only shipments to pool plants described in paragraphs (a), (b), and (d) of this section shall count as qualifying shipments for the purpose of meeting the increased shipments:

(i) Pool plants described in paragraphs (a), (b), and (d) of this section;

(ii) Plants of producer-handlers;

(iii) Partially regulated distributing plants, except that credit for such shipments shall be limited to the amount of such milk classified as Class I at the transferee plant; and

(iv) Distributing plants fully regulated under other Federal orders, except that credit for shipments to such plants shall be limited to the quantity shipped to (and physically unloaded into) pool distributing plants during the month and credits for shipments to other order plants shall not include any such shipments made on the basis of agreed-upon Class II, Class III, or Class IV utilization.

(2) Concentrated milk transferred from the supply plant to a distributing plant for an agreed-upon use other than Class I shall be excluded from the supply plant's shipments in computing the supply plant's shipping percentage.

(d) Two or more plants operated by the same handler and located in the marketing area may qualify for pool status as a unit by meeting the total and in-area route disposition requirements of a pool distributing plant specified in paragraph (a) of this section and subject to the following additional requirements:

(1) At least one of the plants in the unit must qualify as a pool plant pursuant to paragraph (a) of this section;

(2) Other plants in the unit must process Class I or Class II products, using 50 percent or more of the total Grade A fluid milk products received in bulk form at such plant or diverted therefrom by the plant operator in Class I or Class II products; and

(3) The operator of the unit has filed a written request with the market administrator prior to the first day of the month for which such status is desired to be effective. The unit shall continue from month-to-month thereafter without further notification. The handler shall notify the market administrator in writing prior to the first day of any month for which termination or any change of the unit is desired.

(e) A system of two or more supply plants operated by one or more handlers may qualify for pooling by meeting the shipping requirements of paragraph (c) of this section in the same manner as a single plant subject to the following additional requirements:

(1) Each plant in the system is located within the marketing area. Cooperative associations or other handlers may not use shipments pursuant to § 1000.9(c) to qualify supply plants located outside the marketing area;

(2) The handler(s) establishing the system submits a written request to the market administrator on or before July 15 requesting that such plants qualify as a system for the period of August through July of the following year. Such request will contain a list of the plants participating in the system in the order, beginning with the last plant, in which the plants will be dropped from the system if the system fails to qualify. Each plant that qualifies as a pool plant within a system shall continue each month as a plant in the system through the following July unless the handler(s) establishing the system submits a written request to the market administrator that the

plant be deleted from the system or that the system be discontinued. Any plant that has been so deleted from a system, or that has failed to qualify in any month, will not be part of any system for the remaining months through July. The handler(s) that have established a system may add a plant operated by such handler(s) to a system if such plant has been a pool plant each of the 6 prior months and would otherwise be eligible to be in a system, upon written request to the market administrator no later than the 15th day of the prior month. In the event of an ownership change or the business failure of a handler who is a participant in a system, the system may be reorganized to reflect such changes if a written request to file a new marketing agreement is submitted to the market administrator; and

(3) If a system fails to qualify under the requirements of this paragraph (e), the handler responsible for qualifying the system shall notify the market administrator which plant or plants will be deleted from the system so that the remaining plants may be pooled as a system. If the handler fails to do so, the market administrator shall exclude one or more plants, beginning at the bottom of the list of plants in the system and continuing up the list as necessary until the deliveries are sufficient to qualify the remaining plants in the system.

(f) Any distributing plant, located within the marketing area as described in § 1051.2:

(1) From which there is route disposition and/or transfers of packaged fluid milk products in any non-federally regulated marketing area(s) located within one or more States that require handlers to pay minimum prices for raw milk, provided that 25 percent or more of the total quantity of fluid milk products physically received at such plant (excluding concentrated milk received from another plant by agreement for other than Class I use) is disposed of as route disposition and/or is transferred in the form of packaged fluid milk products to other plants. At least 25 percent of such route disposition and/or transfers, in aggregate, are in any non-federally regulated marketing area(s) located within one or

more States that require handlers to pay minimum prices for raw milk. Subject to the following exclusions:

(i) The plant is described in paragraph (a), (b), or (e) of this section;

(ii) The plant is subject to the pricing provisions of a State-operated milk pricing plan which provides for the payment of minimum class prices for raw milk;

(iii) The plant is described in § 1000.8(a) or (e); or

(iv) A producer-handler described in § 1051.10 with less than three million pounds during the month of route disposition and/or transfers of packaged fluid milk products to other plants.

(2) [Reserved]

(g) The applicable shipping percentages of paragraphs (c) and (e) of this section and § 1051.13(d)(2) and (3) may be increased or decreased, for all or part of the marketing area, by the market administrator if the market administrator finds that such adjustment is necessary to encourage needed shipments or to prevent uneconomic shipments. Before making such a finding, the market administrator shall investigate the need for adjustment either on the market administrator's own initiative or at the request of interested parties if the request is made in writing at least 15 days prior to the month for which the requested revision is desired effective. If the investigation shows that an adjustment of the shipping percentages might be appropriate, the market administrator shall issue a notice stating that an adjustment is being considered and invite data, views, and arguments. Any decision to revise an applicable shipping or diversion percentage must be issued in writing at least one day before the effective date.

(h) The term pool plant shall not apply to the following plants:

(1) A producer-handler as defined under any Federal order;

(2) An exempt plant as defined in § 1000.8(e);

(3) A plant located within the marketing area and qualified pursuant to paragraph (a) of this section which meets the pooling requirements of another Federal order, and from which more than 50 percent of its route disposition has been in the other Federal

order marketing area for 3 consecutive months;

(4) A plant located outside any Federal order marketing area and qualified pursuant to paragraph (a) of this section that meets the pooling requirements of another Federal order and has had greater route disposition in such other Federal order's marketing area for 3 consecutive months;

(5) A plant located in another Federal order marketing area and qualified pursuant to paragraph (a) of this section that meets the pooling requirements of such other Federal order and does not have a majority of its route disposition in this marketing area for 3 consecutive months, or if the plant is required to be regulated under such other Federal order without regard to its route disposition in any other Federal order marketing area;

(6) A plant qualified pursuant to paragraph (c) of this section which also meets the pooling requirements of another Federal order and from which greater qualifying shipments are made to plants regulated under the other Federal order than are made to plants regulated under the order in this part, or the plant has automatic pooling status under the other Federal order; and

(7) That portion of a regulated plant designated as a nonpool plant that is physically separate and operated separately from the pool portion of such plant. The designation of a portion of a regulated plant as a nonpool plant must be requested in advance and in writing by the handler and must be approved by the market administrator.

(i) Any plant that qualifies as a pool plant in each of the immediately preceding 3 months pursuant to paragraph (a) of this section or the shipping percentages in paragraph (c) of this section that is unable to meet such performance standards for the current month because of unavoidable circumstances determined by the market administrator to be beyond the control of the handler operating the plant, such as a natural disaster (ice storm, wind storm, flood, fire, earthquake, breakdown of equipment, or work stoppage, shall be considered to have met the minimum performance standards during the period of such unavoidable circumstances, but such relief shall not be granted for more than 2 consecutive months.

§ 1051.8 Nonpool plant.

See § 1000.8.

§ 1051.9 Handler.

See § 1000.9.

§ 1051.10 Producer-handler.

Producer-handler means a person who operates a dairy farm and a distributing plant from which there is route disposition in the marketing area, from which total route disposition and packaged sales of fluid milk products to other plants during the month does not exceed 3 million pounds, and who the market administrator has designated a producer-handler after determining that all of the requirements of this section have been met.

(a) *Requirements for designation.* Designation of any person as a producer-handler by the market administrator shall be contingent upon meeting the conditions set forth in paragraphs (a)(1) through (5) of this section. Following the cancellation of a previous producer-handler designation, a person seeking to have their producer-handler designation reinstated must demonstrate that these conditions have been met for the preceding month:

(1) The care and management of the dairy animals and the other resources and facilities designated in paragraph (b)(1) of this section necessary to produce all Class I milk handled (excluding receipts from handlers fully regulated under any Federal order) are under the complete and exclusive control, ownership, and management of the producer-handler and are operated as the producer-handler's own enterprise and at its sole risk.

(2) The plant operation designated in paragraph (b)(2) of this section at which the producer-handler processes and packages, and from which it distributes, its own milk production is under the complete and exclusive control, ownership, and management of the producer-handler and is operated as the producer-handler's own enterprise and at its sole risk.

(3) The producer-handler neither receives at its designated milk production resources and facilities nor receives, handles, processes, or distributes at or through any of its designated milk handling, processing, or distributing resources and facilities other source milk products for reconstitution into fluid milk products or fluid milk products derived from any source other than:

(i) Its designated milk production resources and facilities (own farm production);

(ii) Pool handlers and plants regulated under any Federal order within the limitation specified in paragraph (c)(2) of this section; or

(iii) Nonfat milk solids which are used to fortify fluid milk products.

(4) The producer-handler is neither directly nor indirectly associated with the business control or management of, nor has a financial interest in, another handler's operation; nor is any other handler so associated with the producer-handler's operation.

(5) No milk produced by the herd(s) or on the farm(s) that supplies milk to the producer-handler's plant operation is:

(i) Subject to inclusion and participation in a marketwide equalization pool under a milk classification and pricing program under the authority of a State government maintaining marketwide pooling of returns; or

(ii) Marketed in any part as Class I milk to the non-pool distributing plant of any other handler.

(b) *Designation of resources and facilities.* Designation of a person as a producer-handler shall include the determination of what shall constitute milk production, handling, processing, and distribution resources and facilities, all of which shall be considered an integrated operation, under the sole and exclusive ownership of the producer-handler.

(1) Milk production resources and facilities shall include all resources and facilities (milking herd(s), buildings housing such herd(s), and the land on which such buildings are located) used for the production of milk which are solely owned, operated, and which the producer-handler has designated as a source of milk supply for the producer-handler's plant operation. However, for purposes of this paragraph (b)(1), any

such milk production resources and facilities which do not constitute an actual or potential source of milk supply for the producer-handler's operation shall not be considered a part of the producer-handler's milk production resources and facilities.

(2) Milk handling, processing, and distribution resources and facilities shall include all resources and facilities (including store outlets) used for handling, processing, and distributing fluid milk products which are solely owned by, and directly operated or controlled by the producer-handler or in which the producer-handler in any way has an interest, including any contractual arrangement, or over which the producer-handler directly or indirectly exercises any degree of management control.

(3) All designations shall remain in effect until canceled pursuant to paragraph (c) of this section.

(c) *Cancellation.* The designation as a producer-handler shall be canceled upon determination by the market administrator that any of the requirements of paragraphs (a)(1) through (5) of this section are not continuing to be met, or under any of the conditions described in paragraph (c)(1), (2), or (3) of this section. Cancellation of a producer-handler's status pursuant to this paragraph (c) shall be effective on the first day of the month following the month in which the requirements were not met or the conditions for cancellation occurred.

(1) Milk from the milk production resources and facilities of the producer-handler, designated in paragraph (b)(1) of this section, is delivered in the name of another person as producer milk to another handler.

(2) The producer-handler handles fluid milk products derived from sources other than the milk production facilities and resources designated in paragraph (b)(1) of this section, except that it may receive at its plant, or acquire for route disposition, fluid milk products from fully regulated plants and handlers under any Federal order if such receipts do not exceed 150,000 pounds monthly. This limitation shall not apply if the producer-handler's own-farm production is less than 150,000 pounds during the month.

(3) Milk from the milk production resources and facilities of the producer-handler is subject to inclusion and participation in a marketwide equalization pool under a milk classification and pricing plan operating under the authority of a State government.

(d) *Public announcement.* The market administrator shall publicly announce:

(1) The name, plant location(s), and farm location(s) of persons designated as producer-handlers;

(2) The names of those persons whose designations have been cancelled; and

(3) The effective dates of producer-handler status or loss of producer-handler status for each. Such announcements shall be controlling with respect to the accounting at plants of other handlers for fluid milk products received from any producer-handler.

(e) *Burden of establishing and maintaining producer-handler status.* The burden rests upon the handler who is designated as a producer-handler to establish through records required pursuant to §1000.27 that the requirements set forth in paragraph (a) of this section have been and are continuing to be met, and that the conditions set forth in paragraph (c) of this section for cancellation of the designation do not exist.

(f) *Payments subject to Order 1131.* Any producer-handler with Class I route dispositions and/or transfers of packaged fluid milk products in the marketing area described in §1131.2 of this chapter shall be subject to payments into the Order 1131 producer settlement fund on such dispositions pursuant to §1000.76(a) and payments into the Order 1131 administrative fund, provided such dispositions are less than three million pounds in the current month and such producer-handler had total Class I route dispositions and/or transfers of packaged fluid milk products from own farm production of three million pounds or more the previous month. If the producer-handler has Class I route dispositions and/or transfers of packaged fluid milk products into the marketing area described in §1131.2 of this chapter of three million pounds or more during the current month, such producer-handler shall be subject to the provisions described in §1131.7 of this chapter or §1000.76(a).

§ 1051.11   California quota program.

California Quota Program means the applicable provisions of the California Food and Agriculture Code, and related provisions of the pooling plan administered by the California Department of Food and Agriculture (CDFA).

§ 1051.12   Producer.

(a) Except as provided in paragraph (b) of this section, *producer* means any person who produces milk approved by a duly constituted regulatory agency for fluid consumption as Grade A milk and whose milk is:

(1) Received at a pool plant directly from the producer or diverted by the plant operator in accordance with § 1051.13; or

(2) Received by a handler described in § 1000.9(c).

(b) Producer shall not include:

(1) A producer-handler as defined in any Federal order;

(2) A dairy farmer whose milk is received at an exempt plant, excluding producer milk diverted to the exempt plant pursuant to § 1051.13(d);

(3) A dairy farmer whose milk is received by diversion at a pool plant from a handler regulated under another Federal order if the other Federal order designates the dairy farmer as a producer under that order and that milk is allocated by request to a utilization other than Class I; and

(4) A dairy farmer whose milk is reported as diverted to a plant fully regulated under another Federal order with respect to that portion of the milk so diverted that is assigned to Class I under the provisions of such other order.

§ 1051.13   Producer milk.

Except as provided for in paragraph (e) of this section, *producer milk* means the skim milk (or the skim equivalent of components of skim milk), including nonfat components, and butterfat in milk of a producer that is:

(a) Received by the operator of a pool plant directly from a producer or a handler described in § 1000.9(c). All milk received pursuant to this paragraph (a) shall be priced at the location of the plant where it is first physically received;

(b) Received by a handler described in § 1000.9(c) in excess of the quantity delivered to pool plants;

(c) Diverted by a pool plant operator to another pool plant. Milk so diverted shall be priced at the location of the plant to which diverted; or

(d) Diverted by the operator of a pool plant or a cooperative association described in § 1000.9(c) to a nonpool plant located in the States of California, Arizona, Nevada, or Oregon, subject to the following conditions:

(1) Milk of a dairy farmer shall not be eligible for diversion unless at least one day's production of such dairy farmer is physically received as producer milk at a pool plant during the first month the dairy farmer is a producer. If a dairy farmer loses producer status under the order in this part (except as a result of a temporary loss of Grade A approval or as a result of the handler of the dairy farmer's milk failing to pool the milk under any order), the dairy farmer's milk shall not be eligible for diversion unless at least one day's production of the dairy farmer has been physically received as producer milk at a pool plant during the first month the dairy farmer is re-associated with the market;

(2) The quantity of milk diverted by a handler described in § 1000.9(c) may not exceed 90 percent of the producer milk receipts reported by the handler pursuant to § 1051.30(c) provided that not less than 10 percent of such receipts are delivered to plants described in § 1051.7(c)(1)(i) through (iii). These percentages are subject to any adjustments that may be made pursuant to § 1051.7(g); and

(3) The quantity of milk diverted to nonpool plants by the operator of a pool plant described in § 1051.7(a), (b) or (d) may not exceed 90 percent of the Grade A milk received from dairy farmers (except dairy farmers described in § 1051.12(b)) including milk diverted pursuant to this section. These percentages are subject to any adjustments that may be made pursuant to § 1051.7(g).

(4) Diverted milk shall be priced at the location of the plant to which diverted.

(e) Producer milk shall not include milk of a producer that is subject to inclusion and participation in a marketwide equalization pool under a milk classification and pricing program imposed under the authority of a State government maintaining marketwide pooling of returns.

(f) The quantity of milk reported by a handler pursuant to either § 1051.30(a)(1) or (c)(1) for April through February may not exceed 125 percent, and for March may not exceed 135 percent, of the producer milk receipts pooled by the handler during the prior month. Milk diverted to nonpool plants reported in excess of this limit shall be removed from the pool. Milk in excess of this limit received at pool plants, other than pool distributing plants, shall be classified pursuant to § 1000.44(a)(3)(v) and (b). The handler must designate, by producer pick-up, which milk is to be removed from the pool. If the handler fails to provide this information, the market administrator will make the determination. The following provisions apply:

(1) Milk shipped to and physically received at pool distributing plants in excess of the previous month's pooled volume shall not be subject to the 125 or 135 percent limitation;

(2) Producer milk qualified pursuant to § _____.13 of any other Federal Order and continuously pooled in any Federal Order for the previous six months shall not be included in the computation of the 125 or 135 percent limitation;

(3) The market administrator may waive the 125 or 135 percent limitation:

(i) For a new handler on the order, subject to the provisions of paragraph (f)(4) of this section; or

(ii) For an existing handler with significantly changed milk supply conditions due to unusual circumstances; and

(4) A bloc of milk may be considered ineligible for pooling if the market administrator determines that handlers altered the reporting of such milk for the purpose of evading the provisions of this paragraph (f).

§ 1051.14 Other source milk.

See § 1000.14.

§ 1051.15 Fluid milk product.

See § 1000.15.

§ 1051.16 Fluid cream product.

See § 1000.16.

§ 1051.17 [Reserved]

§ 1051.18 Cooperative association.

See § 1000.18.

§ 1051.19 Commercial food processing establishment.

See § 1000.19.

MARKET ADMINISTRATOR, CONTINUING OBLIGATIONS, AND HANDLER RESPONSIBILITIES

§ 1051.25 Market administrator.

See § 1000.25.

§ 1051.26 Continuity and separability of provisions.

See § 1000.26.

§ 1051.27 Handler responsibility for records and facilities.

See § 1000.27.

§ 1051.28 Termination of obligations.

See § 1000.28.

HANDLER REPORTS

§ 1051.30 Reports of receipts and utilization.

Each handler shall report monthly so that the market administrator's office receives the report on or before the 9th day after the end of the month, in the detail and on the prescribed forms, as follows:

(a) Each handler that operates a pool plant shall report for each of its operations the following information:

(1) Product pounds, pounds of butterfat, pounds of protein, and pounds of solids-not-fat other than protein (other solids) contained in or represented by:

(i) Receipts of producer milk, including producer milk diverted by the reporting handler, from sources other than handlers described in § 1000.9(c); and

(ii) Receipts of milk from handlers described in § 1000.9(c);

(2) Product pounds and pounds of butterfat contained in:

(i) Receipts of fluid milk products and bulk fluid cream products from other pool plants;

(ii) Receipts of other source milk; and

(iii) Inventories at the beginning and end of the month of fluid milk products and bulk fluid cream products;

(3) The utilization or disposition of all milk and milk products required to be reported pursuant to this paragraph (a); and

(4) Such other information with respect to the receipts and utilization of skim milk, butterfat, milk protein, and other nonfat solids as the market administrator may prescribe.

(b) Each handler operating a partially regulated distributing plant shall report with respect to such plant in the same manner as prescribed for reports required by paragraph (a) of this section. Receipts of milk that would have been producer milk if the plant had been fully regulated shall be reported in lieu of producer milk. The report shall show also the quantity of any reconstituted skim milk in route disposition in the marketing area.

(c) Each handler described in § 1000.9(c) shall report:

(1) The product pounds, pounds of butterfat, pounds of protein, pounds of solids-not-fat other than protein (other solids) contained in receipts of milk from producers; and

(2) The utilization or disposition of such receipts.

(d) Each handler not specified in paragraphs (a) through (c) of this section shall report with respect to its receipts and utilization of milk and milk products in such manner as the market administrator may prescribe.

§ 1051.31 Payroll reports.

(a) On or before the 20th day after the end of each month, each handler that operates a pool plant pursuant to § 1051.7 and each handler described in § 1000.9(c) shall report to the market administrator its producer payroll for the month, in the detail prescribed by the market administrator, showing for each producer the information described in § 1051.73(f).

(b) Each handler operating a partially regulated distributing plant who elects to make payment pursuant to

§ 1000.76(b) shall report for each dairy farmer who would have been a producer if the plant had been fully regulated in the same manner as prescribed for reports required by paragraph (a) of this section.

§ 1051.32 Other reports.

In addition to the reports required pursuant to §§ 1051.30 and 1051.31, each handler shall report any information the market administrator deems necessary to verify or establish each handler's obligation under the order.

Subpart B—Milk Pricing

CLASSIFICATION OF MILK

§ 1051.40 Classes of utilization.

See § 1000.40.

§ 1051.41 [Reserved]

§ 1051.42 Classification of transfers and diversions.

See § 1000.42.

§ 1051.43 General classification rules.

See § 1000.43.

§ 1051.44 Classification of producer milk.

See § 1000.44.

§ 1051.45 Market administrator's reports and announcements concerning classification.

See § 1000.45.

CLASS PRICES

§ 1051.50 Class prices, component prices, and advanced pricing factors.

See § 1000.50.

§ 1051.51 Class I differential and price.

The Class I differential shall be the differential established for Los Angeles County, California, which is reported in § 1000.52. The Class I price shall be the price computed pursuant to § 1000.50(a) for Los Angeles County, California.

§ 1051.52 Adjusted Class I differentials.

See § 1000.52.

§ 1051.53 Announcement of class prices, component prices, and advanced pricing factors.

See § 1000.53.

§ 1051.54 Equivalent price.

See § 1000.54.

PRODUCER PRICE DIFFERENTIAL

§ 1051.60 Handler's value of milk.

For the purpose of computing a handler's obligation for producer milk, the market administrator shall determine for each month the value of milk of each handler with respect to each of the handler's pool plants and of each handler described in § 1000.9(c) with respect to milk that was not received at a pool plant by adding the amounts computed in paragraphs (a) through (h) of this section and subtracting from that total amount the values computed in paragraphs (i) and (j) of this section. Unless otherwise specified, the skim milk, butterfat, and the combined pounds of skim milk and butterfat referred to in this section shall result from the steps set forth in § 1000.44(a), (b), and (c), respectively, and the nonfat components of producer milk in each class shall be based upon the proportion of such components in producer skim milk. Receipts of nonfluid milk products that are distributed as labeled reconstituted milk for which payments are made to the producer-settlement fund of another Federal order under § 1000.76(a)(4) or (d) shall be excluded from pricing under this section.

(a) *Class I value.* (1) Multiply the hundredweight of skim milk in Class I by the Class I skim milk price; and

(2) Add an amount obtained by multiplying the pounds of butterfat in Class I by the Class I butterfat price; and

(b) *Class II value.* (1) Multiply the pounds of nonfat solids in Class II skim milk by the Class II nonfat solids price; and

(2) Add an amount obtained by multiplying the pounds of butterfat in Class II times the Class II butterfat price.

(c) *Class III value.* (1) Multiply the pounds of protein in Class III skim milk by the protein price;

(2) Add an amount obtained by multiplying the pounds of other solids in Class III skim milk by the other solids price; and

(3) Add an amount obtained by multiplying the pounds of butterfat in Class III by the butterfat price.

(d) *Class IV value.* (1) Multiply the pounds of nonfat solids in Class IV skim milk by the nonfat solids price; and

(2) Add an amount obtained by multiplying the pounds of butterfat in Class IV by the butterfat price.

(e) *Classification of overage.* Multiply the pounds of skim milk and butterfat overage assigned to each class pursuant to § 1000.44(a)(11) and the corresponding step of § 1000.44(b) by the skim milk prices and butterfat prices applicable to each class.

(f) *Reclassification of inventory.* Multiply the difference between the current month's Class I, II, or III price, as the case may be, and the Class IV price for the preceding month and by the hundredweight of skim milk and butterfat subtracted from Class I, II, or III, respectively, pursuant to § 1000.44(a)(7) and the corresponding step of § 1000.44(b).

(g) *Class I calculation applicable to unregulated milk.* Multiply the difference between the Class I price applicable at the location of the pool plant and the Class IV price by the hundredweight of skim milk and butterfat assigned to Class I pursuant to § 1000.43(d) and the hundredweight of skim milk and butterfat subtracted from Class I pursuant to § 1000.44(a)(3)(i) through (vi) and the corresponding step of § 1000.44(b), excluding receipts of bulk fluid cream products from plants regulated under other Federal orders and bulk concentrated fluid milk products from pool plants, plants regulated under other Federal orders, and unregulated supply plants.

(h) *Class I calculation applicable to unregulated supply plant milk.* Multiply the difference between the Class I price applicable at the location of the nearest unregulated supply plants from which an equivalent volume was received and the Class III price by the pounds of skim milk and butterfat in receipts of concentrated fluid milk products assigned to Class I pursuant to §§ 1000.43(d) and 1000.44(a)(3)(i) and the corresponding step of § 1000.44(b) and the pounds of skim milk and butterfat subtracted from Class I pursuant to §1000.44(a)(8) and the corresponding step of §1000.44(b), excluding such skim milk and butterfat in receipts of fluid milk products from an unregulated supply plant to the extent that an equivalent amount of skim milk or butterfat disposed of to such plant by handlers fully regulated under any Federal milk order is classified and priced as Class I milk and is not used as an offset for any other payment obligation under any order.

(i) *Calculation of nonfluid milk receipts for reconstitution.* For reconstituted milk made from receipts of nonfluid milk products, multiply $1.00 (but not more than the difference between the Class I price applicable at the location of the pool plant and the Class IV price) by the hundredweight of skim milk and butterfat contained in receipts of nonfluid milk products that are allocated to Class I use pursuant to §1000.43(d).

§1051.61 Computation of producer price differential.

For each month the market administrator shall compute a producer price differential per hundredweight. The report of any handler who has not made payments required pursuant to §1051.71 for the preceding month shall not be included in the computation of the producer price differential, and such handler's report shall not be included in the computation for succeeding months until the handler has made full payment of outstanding monthly obligations. Subject to the conditions of this introductory text, the market administrator shall compute the producer price differential in the following manner:

(a) Combine into one total the values computed pursuant to §1051.60 for all handlers required to file reports prescribed in §1051.30;

(b) Subtract the total values obtained by multiplying each handler's total pounds of protein, other solids, and butterfat contained in the milk for which an obligation was computed pursuant to §1051.60 by the protein price, other solids price, and the butterfat price, respectively;

(c) Add an amount equal to the minus location adjustments and subtract an amount equal to the plus location adjustments computed pursuant to §1051.75;

(d) Add an amount equal to not less than one-half of the unobligated balance in the producer-settlement fund;

(e) Divide the resulting amount by the sum of the following for all handlers included in these computations:

(1) The total hundredweight of producer milk; and

(2) The total hundredweight for which a value is computed pursuant to §1051.60(i); and

(f) Subtract not less than 4 cents nor more than 5 cents from the price computed pursuant to paragraph (e) of this section. The result shall be known as the producer price differential for the month.

§1051.62 Announcement of producer prices.

On or before the 14th day after the end of each month, the market administrator shall announce publicly the following prices and information:

(a) The producer price differential;

(b) The protein price;

(c) The nonfat solids price;

(d) The other solids price;

(e) The butterfat price;

(f) The average butterfat, nonfat solids, protein and other solids content of producer milk; and

(g) The statistical uniform price for milk containing 3.5 percent butterfat, computed by combining the Class III price and the producer price differential.

Subpart C—Payments for Milk

PRODUCER PAYMENTS

§1051.70 Producer-settlement fund.

See §1000.70.

§1051.71 Payments to the producer-settlement fund.

Each handler shall make payment to the producer-settlement fund in a manner that provides receipt of the funds by the market administrator no later than the 16th day after the end of the month (except as provided in §1000.90). Payment shall be the amount, if any, by which the amount specified in paragraph (a) of this section exceeds the

4 4

amount specified in paragraph (b) of this section:

(a) The total value of milk to the handler for the month as determined pursuant to §1051.60.

(b) The sum of:

(1) An amount obtained by multiplying the total hundredweight of producer milk as determined pursuant to §1000.44(c) by the producer price differential as adjusted pursuant to §1051.75;

(2) An amount obtained by multiplying the total pounds of protein, other solids, and butterfat contained in producer milk by the protein, other solids, and butterfat prices respectively; and

(3) An amount obtained by multiplying the pounds of skim milk and butterfat for which a value was computed pursuant to §1051.60(i) by the producer price differential as adjusted pursuant to §1051.75 for the location of the plant from which received.

§1051.72 Payments from the producer-settlement fund.

No later than the 18th day after the end of each month (except as provided in §1000.90), the market administrator shall pay to each handler the amount, if any, by which the amount computed pursuant to §1051.71(b) exceeds the amount computed pursuant to §1051.71(a). If, at such time, the balance in the producer-settlement fund is insufficient to make all payments pursuant to this section, the market administrator shall reduce uniformly such payments and shall complete the payments as soon as the funds are available.

§1051.73 Payments to producers and to cooperative associations.

(a) *Handler payment responsibility.* Each handler shall pay each producer for producer milk for which payment is not made to a cooperative association pursuant to paragraph (b) of this section, as follows:

(1) *Partial payment.* For each producer who has not discontinued shipments as of the date of this partial payment, payment shall be made so that it is received by each producer on or before the last day of the month (except as provided in §1000.90) for milk received during the first 15 days of the month from the producer at not less than the lowest announced class price for the preceding month, less proper deductions authorized in writing by the producer.

(2) *Final payment.* For milk received during the month, payment shall be made so that it is received by each producer no later than the 19th day after the end of the month (except as provided in §1000.90) in an amount not less than the sum of:

(i) The hundredweight of producer milk received times the producer price differential for the month as adjusted pursuant to §1051.75;

(ii) The pounds of butterfat received times the butterfat price for the month;

(iii) The pounds of protein received times the protein price for the month;

(iv) The pounds of other solids received times the other solids price for the month;

(v) Less any payment made pursuant to paragraph (a)(1) of this section;

(vi) Less proper deductions authorized in writing by such producer, and plus or minus adjustments for errors in previous payments to such producer subject to approval by the market administrator;

(vii) Less deductions for marketing services pursuant to §1000.86; and

(viii) Less deductions authorized by CDFA for the California Quota Program pursuant to §1051.11.

(b) *Payments for milk received from cooperative association members.* On or before the day prior to the dates specified in paragraphs (a)(1) and (2) of this section (except as provided in §1000.90), each handler shall pay to a cooperative association for milk from producers who market their milk through the cooperative association and who have authorized the cooperative to collect such payments on their behalf an amount equal to the sum of the individual payments otherwise payable for such producer milk pursuant to paragraphs (a)(1) and (2) of this section.

(c) *Payment for milk received from cooperative association pool plants or from cooperatives as handlers pursuant to §1000.9(c).* On or before the day prior to the dates specified in paragraphs (a)(1)

159

and (2) of this section (except as provided in § 1000.90), each handler who receives fluid milk products at its plant from a cooperative association in its capacity as the operator of a pool plant or who receives milk from a cooperative association in its capacity as a handler pursuant to § 1000.9(c), including the milk of producers who are not members of such association and who the market administrator determines have authorized the cooperative association to collect payment for their milk, shall pay the cooperative for such milk as follows:

(1) For bulk fluid milk products and bulk fluid cream products received from a cooperative association in its capacity as the operator of a pool plant and for milk received from a cooperative association in its capacity as a handler pursuant to § 1000.9(c) during the first 15 days of the month, at not less than the lowest announced class prices per hundredweight for the preceding month;

(2) For the total quantity of bulk fluid milk products and bulk fluid cream products received from a cooperative association in its capacity as the operator of a pool plant, at not less than the total value of such products received from the association's pool plants, as determined by multiplying the respective quantities assigned to each class under § 1000.44, as follows:

(i) The hundredweight of Class I skim milk times the Class I skim milk price for the month plus the pounds of Class I butterfat times the Class I butterfat price for the month. The Class I price to be used shall be that price effective at the location of the receiving plant;

(ii) The pounds of nonfat solids in Class II skim milk by the Class II nonfat solids price;

(iii) The pounds of butterfat in Class II times the Class II butterfat price;

(iv) The pounds of nonfat solids in Class IV times the nonfat solids price;

(v) The pounds of butterfat in Class III and Class IV milk times the butterfat price;

(vi) The pounds of protein in Class III milk times the protein price;

(vii) The pounds of other solids in Class III milk times the other solids price; and

(viii) Add together the amounts computed in paragraphs (c)(2)(i) through (vii) of this section and from that sum deduct any payment made pursuant to paragraph (c)(1) of this section; and

(3) For the total quantity of milk received during the month from a cooperative association in its capacity as a handler under § 1000.9(c) as follows:

(i) The hundredweight of producer milk received times the producer price differential as adjusted pursuant to § 1051.75;

(ii) The pounds of butterfat received times the butterfat price for the month;

(iii) The pounds of protein received times the protein price for the month;

(iv) The pounds of other solids received times the other solids price for the month; and

(v) Add together the amounts computed in paragraphs (c)(3)(i) through (v) of this section and from that sum deduct any payment made pursuant to paragraph (c)(1) of this section.

(d) *Handler underpayment proration.* If a handler has not received full payment from the market administrator pursuant to § 1051.72 by the payment date specified in paragraph (a), (b), or (c) of this section, the handler may reduce pro rata its payments to producers or to the cooperative association (with respect to receipts described in paragraph (b) of this section, prorating the underpayment to the volume of milk received from the cooperative association in proportion to the total milk received from producers by the handler), but not by more than the amount of the underpayment. The payments shall be completed on the next scheduled payment date after receipt of the balance due from the market administrator.

(e) *Payments to missing or deceased producers.* If a handler claims that a required payment to a producer cannot be made because the producer is deceased or cannot be located, or because the cooperative association or its lawful successor or assignee is no longer in existence, the payment shall be made to the producer-settlement fund, and in the event that the handler subsequently locates and pays the producer or a lawful claimant, or in the event that the handler no longer exists and a lawful claim is later established, the market administrator shall make the required payment from the producer-settlement fund to the handler or to the lawful claimant, as the case may be.

(f) *Producer payment record.* In making payments to producers pursuant to this section, each handler shall furnish each producer, except a producer whose milk was received from a cooperative association handler described in § 1000.9(a) or (c), a supporting statement in a form that may be retained by the recipient which shall show:

(1) The name, address, Grade A identifier assigned by a duly constituted regulatory agency, and payroll number of the producer;

(2) The daily and total pounds, and the month and dates such milk was received from that producer;

(3) The total pounds of butterfat, protein, and other solids contained in the producer's milk;

(4) The minimum rate or rates at which payment to the producer is required pursuant to the order in this part;

(5) The rate used in making payment if the rate is other than the applicable minimum rate;

(6) The amount, or rate per hundredweight, or rate per pound of component, and the nature of each deduction claimed by the handler; and

(7) The net amount of payment to the producer or cooperative association.

§ 1051.74 [Reserved]

§ 1051.75 Plant location adjustments for producer milk and nonpool milk.

For purposes of making payments for producer milk and nonpool milk, a plant location adjustment shall be determined by subtracting the Class I price specified in § 1051.51 from the Class I price at the plant's location. The difference, plus or minus as the case may be, shall be used to adjust the payments required pursuant to §§ 1051.73 and 1000.76.

§ 1051.76 Payments by a handler operating a partially regulated distributing plant.

See § 1000.76.

§ 1051.77 Adjustment of accounts.

See § 1000.77.

§ 1051.78 Charges on overdue accounts.

See § 1000.78.

ADMINISTRATIVE ASSESSMENT AND
MARKETING SERVICE DEDUCTION

§ 1051.85 Assessment for order administration.

On or before the payment receipt date specified under § 1051.71, each handler shall pay to the market administrator its pro rata share of the expense of administration of the order at a rate specified by the market administrator that is no more than 8 cents per hundredweight with respect to:

(a) Receipts of producer milk (including the handler's own production) other than such receipts by a handler described in § 1000.9(c) that were delivered to pool plants of other handlers;

(b) Receipts from a handler described in § 1000.9(c);

(c) Receipts of concentrated fluid milk products from unregulated supply plants and receipts of nonfluid milk products assigned to Class I use pursuant to § 1000.43(d) and other source milk allocated to Class I, pursuant to § 1000.44(a)(3) and (8) and the corresponding steps of § 1000.44(b), except other source milk that is excluded from the computations pursuant to § 1051.60(h) and (i); and

(d) Route disposition in the marketing area from a partially regulated distributing plant that exceeds the skim milk and butterfat subtracted pursuant to § 1000.76(a)(1)(i) and (ii).

§ 1051.86 Deduction for marketing services.

See § 1000.86.

Subpart D—Miscellaneous Provisions

§ 1051.90 Dates.

See § 1000.90.

PARTS 1052–1120 [RESERVED]

PART 1124—MILK IN THE PACIFIC NORTHWEST MARKETING AREA

Subpart—Order Regulating Handling

AUTHORITY: 7 U.S.C. 601–674, and 7253.

SOURCE: 64 FR 47998, Sept. 1, 1999, unless otherwise noted.

Subpart—Order Regulating Handling

GENERAL PROVISIONS

§ 1124.1 General provisions.

The terms, definitions, and provisions in part 1000 of this chapter apply to this part 1124. In this part 1124, all references to sections in part 1000 refer to part 1000 of this chapter.

DEFINITIONS

§ 1124.2 Pacific Northwest marketing area.

The marketing area means all territory within the bounds of the following states and political subdivisions, including all piers, docks, and wharves connected therewith and all craft moored thereat, and all territory occupied by government (municipal, State, or Federal) reservations, installations, institutions, or other similar establishments if any part thereof is within any of the listed states or political subdivisions:

IDAHO COUNTIES

Benewah, Bonner, Boundary, Kootenai, Latah, and Shoshone.

OREGON COUNTIES

Benton, Clackamas, Clatsop, Columbia, Coos, Crook, Curry, Deschutes, Douglas, Gilliam, Hood River, Jackson, Jefferson, Josephine, Klamath, Lake, Lane, Lincoln,

Linn, Marion, Morrow, Multnomah, Polk, Sherman, Tillamook, Umatilla, Wasco, Washington, Wheeler, and Yamhill.

WASHINGTON

All of the State of Washington.

§ 1124.3 Route disposition.

See § 1000.3.

§ 1124.4 Plant.

See § 1000.4.

§ 1124.5 Distributing plant.

See § 1000.5.

§ 1124.6 Supply plant.

See § 1000.6.

§ 1124.7 Pool plant.

Pool plant means a plant, unit of plants, or a system of plants as specified in paragraphs (a) through (f) of this section, but excluding a plant specified in paragraph (h) of this section. The pooling standards described in paragraph (c) of this section are subject to modification pursuant to paragraph (g) of this section:

(a) A distributing plant, other than a plant qualified as a pool plant pursuant to paragraph (b) of this section or §_____.7(b) of any other Federal milk order, from which during the month 25 percent or more of the total quantity of fluid milk products physically received at the plant (excluding concentrated milk received from another plant by agreement for other than Class I use) are disposed of as route disposition or are transferred in the form of packaged fluid milk products to other distributing plants. At least 25 percent of such route disposition and transfers must be to outlets in the marketing area.

(b) Any distributing plant located in the marketing area which during the month processed at least 25 percent of the total quantity of fluid milk products physically received at the plant (excluding concentrated milk received from another plant by agreement for other than Class I use) into ultra-pasteurized or aseptically-processed fluid milk products.

(c) A supply plant from which during any month not less than 20 percent of the total quantity of milk that is phys-ically received at such plant from dairy farmers eligible to be producers pursuant to § 1124.12 (excluding milk received at such plant as diverted milk from another plant, which milk is classified other than Class I under the order in this part and is subject to the pricing and pooling provisions of this or another order issued pursuant to the Act) or diverted as producer milk to another plant pursuant to § 1124.13, is shipped in the form of a fluid milk product (excluding concentrated milk transferred by agreement for other than Class I use) to a pool distributing plant or is a route disposition in the marketing area of fluid milk products processed and packaged at such plant;

(1) A supply plant that has qualified as a pool plant during each of the immediately preceding months of September through February shall continue to so qualify in each of the following months of March through August, unless the plant operator files a written request with the market administrator that such plant not be a pool plant, such nonpool status to be effective the first month following such request and thereafter until the plant qualifies as a pool plant on the basis of milk shipments;

(2) No plant may qualify as a pool plant due to a reduction in the shipping percentage pursuant to paragraph (g) of this section unless it has been a pool supply plant during each of the immediately preceding 3 months.

(d) A manufacturing plant located within the marketing area and operated by a cooperative association, or its wholly owned subsidiary, if, during the month, or the immediately preceding 12-month period ending with the current month, 20 percent or more of the producer milk of members of the association (and any producer milk of nonmembers and members of another cooperative association which may be marketed by the cooperative association) is physically received in the form of bulk fluid milk products (excluding concentrated milk transferred to a distributing plant for an agreed-upon use other that Class I) at plants specified in paragraph (a), (b), or (e) of this section either directly from farms or by transfer from supply plants operated by the cooperative association, or its

163

wholly owned subsidiary, and from plants of the cooperative association, or its wholly owned subsidiary, for which pool plant status has been requested under this paragraph subject to the following conditions:

(1) The plant does not qualify as a pool plant under paragraph (a), (b), (c), or (e) of this section or under comparable provisions of another Federal order; and

(2) The plant is approved by a duly constituted regulatory agency for the handling of milk approved for fluid consumption in the marketing area.

(3) A request is filed in writing with the market administrator before the first day of the month for which it is to be effective. The request will remain in effect until a cancellation request is filed in writing with the market administrator before the first day of the month for which the cancellation is to be effective.

(e) Any distributing plant, located within the marketing area as described on May 1, 2006, in § 1124.2;

(1) From which there is route disposition and/or transfers of packaged fluid milk products in any non-federally regulated marketing area(s) located within one or more States that require handlers to pay minimum prices for raw milk provided that 25 percent or more of the total quantity of fluid milk products physically received at such plant (excluding concentrated milk received from another plant by agreement for other than Class I use) is disposed of as route disposition and/or is transferred in the form of packaged fluid milk products to other plants. At least 25 percent of such route disposition and/or transfers, in aggregate, are in any non-federally regulated marketing area(s) located within one or more States that require handlers to pay minimum prices for raw milk. Subject to the following exclusions:

(i) The plant is described in § 1124.7(a) or (b);

(ii) The plant is subject to the pricing provisions of a State-operated milk pricing plan which provides for the payment of minimum class prices for raw milk;

(iii) The plant is described in § 1000.8(a) or (e); or

(iv) A producer-handler described in § 1124.10 with less than three million pounds during the month of route dispositions and/or transfers of packaged fluid milk products to other plants.

(2) [Reserved]

(f) A system of two or more plants identified in § 1124.7(d) operated by one or more cooperative handlers may qualify for pooling by meeting the above shipping requirements subject to the following additional requirements:

(1) The cooperative handler(s) establishing the system submits a written request to the market administrator on or before the first day of the month for which the system is to be effective requesting that such plants qualify as a system. Such request will contain a list of the plants participating in the system in the order, beginning with the last plant, in which the plants will be dropped from the system if the system fails to qualify. Each plant that qualifies as a pool plant within a system shall continue each month as a plant in the system until the handler(s) establishing the system submits a written request before the first day of the month to the market administrator that the plant be deleted from the system or that the system be discontinued. Any plant that has been so deleted from a system, or that has failed to qualify in any month, will not be part of any system. In the event of an ownership change or the business failure of a handler that is a participant in the system, the system may be reorganized to reflect such a change if a written request to file a new marketing agreement is submitted to the market administrator; and

(2) If a system fails to qualify under the requirement of this paragraph, the handler responsible for qualifying the system shall notify the market administrator of which plant or plants will be deleted from the system so that the remaining plants may be pooled as a system. If the handler fails to do so, the market administrator shall exclude one or more plants, beginning at the bottom of the list of plants in the system and continue up the list as necessary until the deliveries are sufficient to qualify the remaining plants in the system.

(g) The applicable shipping percentage of paragraphs (c) and (d) of this section may be increased or decreased by the market administrator if the market administrator finds that such adjustment is necessary to encourage needed shipments or to prevent uneconomic shipments. Before making such a finding, the market administrator shall investigate the need for adjustment either on the market administrator's own initiative or at the request of interested parties if the request is made in writing at least 15 days prior to the month for which the requested revision is desired to be effective. If the investigation shows that an adjustment of the shipping percentages might be appropriate, the market administrator shall issue a notice stating that an adjustment is being considered and invite data, views and arguments. Any decision to revise an applicable shipping percentage must be issued in writing at least one day before the effective date.

(h) The term pool plant shall not apply to the following plants:

(1) A producer-handler as defined under any Federal order;

(2) An exempt plant as defined in §1000.8(e);

(3) A plant located within the marketing area and qualified pursuant to paragraph (a) of this section which meets the pooling requirements of another Federal order, and from which more than 50 percent of its route disposition has been in the other Federal order marketing area for 3 consecutive months;

(4) A plant located outside any Federal order marketing area and qualified pursuant to paragraph (a) of this section that meets the pooling requirements of another Federal order and has had greater route disposition in such other Federal order's marketing area for 3 consecutive months;

(5) A plant located in another Federal order marketing area and qualified pursuant to paragraph (a) of this section that meets the pooling requirements of such other Federal order and does not have a majority of its route distribution in this marketing area for 3 consecutive months or if the plant is required to be regulated under such other Federal order without regard to

its route disposition in any other Federal order marketing area; and

(6) A plant qualified pursuant to paragraph (c) of this section which also meets the pooling requirements of another Federal order and from which greater qualifying shipments are made to plants regulated under the other Federal order than are made to plants regulated under the order in this part, or the plant has automatic pooling status under the other Federal order.

[64 FR 47998, Sept. 1, 1999, as amended at 67 FR 69669, Nov. 19, 2002; 71 FR 25501, May 1, 2006; 71 FR 28249, May 16, 2006]

§1124.8 Nonpool plant.

See §1000.8.

§1124.9 Handler.

See §1000.9.

§1124.10 Producer-handler.

Producer-handler means a person who operates a dairy farm and a distributing plant from which there is route disposition in the marketing area, from which total route disposition and packaged sales of fluid milk products to other plants during the month does not exceed 3 million pounds, and who the market administrator has designated a producer-handler after determining that all of the requirements of this section have been met.

(a) *Requirements for designation.* Designation of any person as a producer-handler by the market administrator shall be contingent upon meeting the conditions set forth in paragraphs (a)(1) through (5) of this section. Following the cancellation of a previous producer-handler designation, a person seeking to have their producer-handler designation reinstated must demonstrate that these conditions have been met for the preceding month.

(1) The care and management of the dairy animals and the other resources and facilities designated in paragraph (b)(1) of this section necessary to produce all Class I milk handled (excluding receipts from handlers fully regulated under any Federal order) are under the complete and exclusive control, ownership and management of the producer-handler and are operated as the producer-handler's own enterprise and its own risk.

(2) The plant operation designated in paragraph (b)(2) of this section at which the producer-handler processes and packages, and from which it distributes, its own milk production is under the complete and exclusive control, ownership and management of the producer-handler and is operated as the producer-handler's own enterprise and at its sole risk.

(3) The producer-handler neither receives at its designated milk production resources and facilities nor receives, handles, processes, or distributes at or through any of its designated milk handling, processing, or distributing resources and facilities other source milk products for reconstitution into fluid milk products or fluid milk products derived from any source other than:

(i) Its designated milk production resources and facilities (own farm production);

(ii) Pool handlers and plants regulated under any Federal order within the limitation specified in paragraph (c)(2) of this section; or

(iii) Nonfat milk solids which are used to fortify fluid milk products.

(4) The producer-handler is neither directly nor indirectly associated with the business control or management of, nor has a financial interest in, another handler's operation; nor is any other handler so associated with the producer-handler's operation.

(5) No milk produced by the herd(s) or on the farm(s) that supply milk to the producer-handler's plant operation is:

(i) Subject to inclusion and participation in a marketwide equalization pool under a milk classification and pricing program under the authority of a State government maintaining marketwide pooling of returns, or

(ii) Marketed in any part as Class I milk to the non-pool distributing plant of any other handler.

(b) *Designation of resources and facilities.* Designation of a person as a producer-handler shall include the determination of what shall constitute milk production, handling, processing, and distribution resources and facilities, all of which shall be considered an integrated operation, under the sole and exclusive ownership of the producer-handler.

(1) Milk production resources and facilities shall include all resources and facilities (milking herd(s), buildings housing such herd(s), and the land on which such buildings are located) used for the production of milk which are solely owned, operated, and which the producer-handler has designated as a source of milk supply for the producer-handler's plant operation. However, for purposes of this paragraph, any such milk production resources and facilities which do not constitute an actual or potential source of milk supply for the producer-handler's operation shall not be considered a part of the producer-handler's milk production resources and facilities.

(2) Milk handling, processing, and distribution resources and facilities shall include all resources and facilities (including store outlets) used for handling, processing, and distributing fluid milk products which are solely owned by, and directly operated or controlled by the producer-handler or in which the producer-handler in any way has an interest, including any contractual arrangement, or over which the producer-handler directly or indirectly exercises any degree of management control.

(3) All designations shall remain in effect until canceled, pursuant to paragraph (c) of this section.

(c) *Cancellation.* The designation as a producer-handler shall be canceled upon determination by the market administrator that any of the requirements of paragraph (a)(1) through (5) of this section are not continuing to be met, or under any of the conditions described in paragraphs (c)(1), (2) or (3) of this section. Cancellation of a producer-handler's status pursuant to this paragraph shall be effective on the first day of the month following the month in which the requirements were not met or the conditions for cancellation occurred.

(1) Milk from the milk production resources and facilities of the producer-handler, designated in paragraph (b)(1) of this section, is delivered in the name of another person as producer milk to another handler.

(2) The producer-handler handles fluid milk products derived from sources other than the milk production facilities and resources designated in paragraph (b)(1) of this section, except that it may receive at its plant, or acquire for route disposition, fluid milk products from fully regulated plants and handlers under any Federal order if such receipts do not exceed 150,000 pounds monthly. This limitation shall not apply if the producer-handler's own-farm production is less than 150,000 pounds during the month.

(3) Milk from the milk production resources and facilities of the producer-handler is subject to inclusion and participation in a marketwide equalization pool under a milk classification and pricing plan operating under the authority of a State government.

(d) *Public announcement.* The market administrator shall publicly announce:

(1) The name, plant location(s), and farm location(s) of persons designated as producer-handlers;

(2) The names of those persons whose designations have been cancelled; and

(3) The effective dates of producer-handler status or loss of producer-handler status for each. Such announcements shall be controlling with respect to the accounting at plants of other handlers for fluid milk products received from any producer-handler.

(e) *Burden of establishing and maintaining producer-handler status.* The burden rests upon the handler who is designated as a producer-handler to establish through records required pursuant to §1000.27 that the requirements set forth in paragraph (a) of this section have been and are continuing to be met, and that the conditions set forth in paragraph (c) of this section for cancellation of the designation do not exist.

(f) Any producer-handler with Class I route dispositions and/or transfers of packaged fluid milk products in the marketing area described in §1131.2 of this chapter shall be subject to payments into the Order 1131 producer settlement fund on such dispositions pursuant to §1000.76(a) and payments into the Order 1131 administrative fund provided such dispositions are less than three million pounds in the current month and such producer-handler had total Class I route dispositions and/or transfers of packaged fluid milk products from own farm production of three million pounds or more the previous month. If the producer-handler has Class I route dispositions and/or transfers of packaged fluid milk products into the marketing area described in §1131.2 of this chapter of three million pounds or more during the current month, such producer-handler shall be subject to the provisions described in §1131.7 of this chapter or §1000.76(a).

[71 FR 9432, Feb. 24, 2006, as amended at 71 FR 25501, May 1, 2006; 75 FR 21160, Apr. 23, 2010]

§1124.11 Cooperative reserve supply unit.

Cooperative reserve supply unit means any cooperative association or its agent that is a handler pursuant to §1000.9(c) that does not own or operate a plant, if such cooperative has been qualified to receive payments pursuant to §1124.73 and has been a handler of producer milk under the order in this part or its predecessor order during each of the 12 previous months, and if a majority of the cooperative's member producers are located within 125 miles of a plant described in §1124.7(a). A cooperative reserve supply unit shall be subject to the following conditions:

(a) The cooperative shall file a request with the market administrator for cooperative reserve supply unit status at least 15 days prior to the first day of the month in which such status is desired to be effective. Once qualified as a cooperative reserve supply unit pursuant to this paragraph, such status shall continue to be effective unless the cooperative requests termination prior to the first day of the month that change of status is requested, or the cooperative fails to meet all of the conditions of this section.

(b) The cooperative reserve supply unit supplies fluid milk products to pool distributing plants located within 125 miles of a majority of the cooperative's member producers in compliance with any announcement by the market administrator requesting a minimum level of shipments as follows:

(1) The market administrator may require such supplies of bulk fluid milk

from cooperative reserve supply units whenever the market administrator finds that milk supplies for Class I use are needed for plants defined in §1124.7(a) or (b). Before making such a finding, the market administrator shall investigate the need for such shipments either on the market administrator's own initiative or at the request of interested persons if the request is made in writing at least 15 days prior to the month for which the requested revision is desired effective. If the market administrator's investigation shows that such shipments might be appropriate, the market administrator shall issue a notice stating that a shipping announcement is being considered and inviting data, views and arguments with respect to the proposed shipping announcement. Any decision on the required shipment of bulk fluid milk from cooperative reserve supply units must be made in writing at least one day before the effective date.

(2) Failure of a cooperative reserve supply unit to comply with any announced shipping requirements, including making any significant change in the unit's marketing operation that the market administrator determines has the impact of evading or forcing such an announcement, shall result in immediate loss of cooperative reserve supply unit status until such time as the unit has been a handler pursuant to §1000.9(c) for at least 12 consecutive months.

§1124.12 Producer.

(a) Except as provided in paragraph (b) of this section, *producer* means any person who produces milk approved by a duly constituted regulatory agency for fluid consumption as Grade A milk and whose milk (or components of milk) is:

(1) Received at a pool plant directly from the producer or diverted by the plant operator in accordance with §1124.13; or

(2) Received by a handler described in §1000.9(c).

(b) Producer shall not include:

(1) A producer-handler as defined in any Federal order;

(2) A dairy farmer whose milk is received at an exempt plant, excluding producer milk diverted to the exempt plant pursuant to §1124.13(e);

(3) A dairy farmer whose milk is received by diversion at a pool plant from a handler regulated under another Federal order if the other Federal order designates the dairy farmer as a producer under that order and that milk is allocated by request to a utilization other than Class I;

(4) A dairy farmer whose milk is reported as diverted to a plant fully regulated under another Federal order with respect to that portion of the milk so diverted that is assigned to Class I under the provisions of such other order; and

(5) A dairy farmer whose milk was received at a nonpool plant during the month from the same farm as other than producer milk under the order in this part or any other Federal order. Such a dairy farmer shall be known as a *dairy farmer for other markets.*

§1124.13 Producer milk.

Except as provided for in paragraph (f) of this section, *Producer milk* means the skim milk (or skim milk equivalent of components of skim milk), including nonfat components, and butterfat in milk of a producer that is:

(a) Received by the operator of a pool plant directly from a producer or a handler described in §1000.9(c). All milk received pursuant to this paragraph shall be priced at the location of the plant where it is first physically received;

(b) Received by a cooperative reserve supply unit described in §1124.11. All milk received pursuant to this paragraph shall be priced at the location of the plant where it is first physically received and shall not be subject to the conditions specified in paragraph (e) of this section;

(c) Received by a handler described in §1000.9(c) in excess of the quantity delivered to pool plants;

(d) Diverted by a pool plant operator to another pool plant. Milk so diverted shall be priced at the location of the plant to which diverted; or

(e) Diverted by the operator of a pool plant or a cooperative association described in §1000.9(c), excluding a cooperative reserve supply unit described in

§1124.11, to a nonpool plant, subject to the following conditions:

(1) Milk of a dairy farmer shall not be eligible for diversion unless at least 3 days' production of such dairy farmer's production is physically received at a pool plant during the month.

(2) Of the quantity of producer milk received during the month (including diversions, but excluding the quantity of producer milk received from a handler described in §1000.9(c)) the handler diverts to nonpool plants not more than 80 percent.

(3) Two or more handlers described in §1000.9(c) may have their allowable diversions computed on the basis of their combined deliveries of producer milk which they caused to be delivered to pool plants or diverted during the month if each has filed a request in writing with the market administrator before the first day of the month the agreement is to be effective. The request shall specify the basis for assigning overdiverted milk to the producer deliveries of each according to a method approved by the market administrator.

(4) Diverted milk shall be priced at the location of the plant to which diverted;

(5) Any milk diverted in excess of the limits prescribed in paragraph (e)(2) of this section shall not be producer milk. If the diverting handler or cooperative association fails to designate the dairy farmers' deliveries that are not to be producer milk, no milk diverted by the handler or cooperative association during the month to a nonpool plant shall be producer milk. In the event some of the milk of any producer is determined not to be producer milk pursuant to this paragraph, other milk delivered by such producer as producer milk during the month will not be subject to §1124.12(b)(5).

(6) The delivery day requirement in paragraph (e)(1) of this section and the diversion percentage in paragraph (e)(2) of this section may be increased or decreased by the market administrator if the market administrator finds that such revision is necessary to assure the orderly marketing and efficient handling of milk in the marketing area. Before making such finding, the market administrator shall in-vestigate the need for the revision either on the market administrator's own initiative or at the request of interested persons if the request is made in writing at least 15 days prior to the month for which the requested revision is desired to be effective. If the investigation shows that a revision might be appropriate, the market administrator shall issue a notice stating that the revision is being considered and inviting written data, views, and arguments. Any decision to revise the delivery day requirement or the diversion percentage must be issued in writing at least one day before the effective date.

(f) Producer milk shall not include milk of a producer that is subject to inclusion and participation in a marketwide equalization pool under a milk classification and pricing program imposed under the authority of a State government maintaining marketwide pooling of returns.

[64 FR 47998, Sept. 1, 1999, as amended at 67 FR 69669, Nov. 19, 2002; 69 FR 1655, Jan. 12, 2004]

§1124.14   Other source milk.

See §1000.14.

§1124.15   Fluid milk product.

See §1000.15.

§1124.16   Fluid cream product.

See §1000.16.

§1124.17   [Reserved]

§1124.18   Cooperative association.

See §1000.18.

§1124.19   Commercial food processing establishment.

See §1000.19.

HANDLER REPORTS

§1124.30   Reports of receipts and utilization.

Each handler shall report monthly so that the market administrator's office receives the report on or before the 9th day after the end of the month, in the detail and on the prescribed forms, as follows:

(a) Each handler that operates a pool plant pursuant to §1124.7 shall report for each of its operations the following information:

(1) Product pounds, pounds of butterfat, pounds of protein, and pounds of solids-not-fat other than protein (other solids) contained in or represented by:

(i) Receipts of producer milk, including producer milk diverted by the reporting handler, from sources other than handlers described in § 1000.9(c); and

(ii) Receipts of milk from handlers described in § 1000.9(c);

(2) Product pounds and pounds of butterfat contained in:

(i) Receipts of fluid milk products and bulk fluid cream products from other pool plants;

(ii) Receipts of other source milk; and

(iii) Inventories at the beginning and end of the month of fluid milk products and bulk fluid cream products;

(3) The utilization or disposition of all milk and milk products required to be reported pursuant to this paragraph; and

(4) Such other information with respect to the receipts and utilization of skim milk, butterfat, milk protein, and other nonfat solids, as the market administrator may prescribe.

(b) Each handler operating a partially regulated distributing plant shall report with respect to such plant in the same manner as prescribed for reports required by paragraph (a) of this section. Receipts of milk that would have been producer milk if the plant had been fully regulated shall be reported in lieu of producer milk. The report shall show also the quantity of any reconstituted skim milk in route disposition in the marketing area.

(c) Each handler described in § 1000.9(c) shall report:

(1) The product pounds, pounds of butterfat, pounds of protein, and the pounds of solids-not-fat other than protein (other solids) contained in receipts of milk from producers; and

(2) The utilization or disposition of such receipts.

(d) Each handler not specified in paragraphs (a) through (c) of this section shall report with respect to its receipts and utilization of milk and milk products in such manner as the market administrator may prescribe.

§ 1124.31 Payroll reports.

(a) On or before the 20th day after the end of each month, each handler that operates a pool plant pursuant to § 1124.7 and each handler described in § 1000.9(c) shall report to the market administrator its producer payroll for the month, in the detail prescribed by the market administrator, showing for each producer the information described in § 1124.73(f).

(b) Each handler operating a partially regulated distributing plant who elects to make payment pursuant to § 1000.76(b) shall report for each dairy farmer who would have been a producer if the plant had been fully regulated in the same manner as prescribed for reports required by paragraph (a) of this section.

§ 1124.32 Other reports.

In addition to the reports required pursuant to §§ 1124.30 and 1124.31, each handler shall report any information the market administrator deems necessary to verify or establish each handler's obligation under the order.

CLASSIFICATION OF MILK

§ 1124.40 Classes of utilization.

See § 1000.40.

§ 1124.41 [Reserved]

§ 1124.42 Classification of transfers and diversions.

See § 1000.42.

§ 1124.43 General classification rules.

See § 1000.43.

§ 1124.44 Classification of producer milk.

In addition to the provisions provided in § 1000.44, for purposes of this part 1124, § 1000.44(a)(3)(iv) applies to fluid milk products and bulk fluid cream products received or acquired for distribution from a producer-handler.

§ 1124.45 Market administrator's reports and announcements concerning classification.

See § 1000.45.

CLASS PRICES

§1124.50 Class prices, component prices, and advanced pricing factors.

See §1000.50.

§1124.51 Class I differential and price.

The Class I differential shall be the differential established for King County, Washington, which is reported in §1000.52. The Class I price shall be the price computed pursuant to §1000.50(a) for King County, Washington.

§1124.52 Adjusted Class I differentials.

See §1000.52.

§1124.53 Announcement of class prices, component prices, and advanced pricing factors.

See §1000.53.

§1124.54 Equivalent price.

See §1000.54.

PRODUCER PRICE DIFFERENTIAL

§1124.60 Handler's value of milk.

For the purpose of computing a handler's obligation for producer milk, the market administrator shall determine for each month the value of milk of each handler with respect to each of the handler's pool plants and of each handler described in §1000.9(c) with respect to milk that was not received at a pool plant by adding the amounts computed in paragraphs (a) through (h) of this section and subtracting from that total amount the value computed in paragraph (i) of this section. Unless otherwise specified, the skim milk, butterfat, and the combined pounds of skim milk and butterfat referred to in this section shall result from the steps set forth in §1000.44 (a), (b), and (c), respectively, and the nonfat components of producer milk in each class shall be based upon the proportion of such components in producer skim milk. Receipts of nonfluid milk products that are distributed as labeled reconstituted milk for which payments are made to the producer-settlement fund of another Federal order under §1000.76 (a)(4) or (d) shall be excluded from pricing under this section.

(a) Class I value.

(1) Multiply the hundredweight of skim milk in Class I by the Class I skim milk price; and

(2) Add an amount obtained by multiplying the pounds of butterfat in Class I by the Class I butterfat price.

(b) Class II value.

(1) Multiply the pounds of nonfat solids in Class II skim milk by the Class II nonfat solids price; and

(2) Add an amount obtained by multiplying the pounds of butterfat in Class II times the Class II butterfat price.

(c) Class III value.

(1) Multiply the pounds of protein in Class III skim milk by the protein price;

(2) Add an amount obtained by multiplying the pounds of other solids in Class III skim milk by the other solids price; and

(3) Add an amount obtained by multiplying the pounds of butterfat in Class III by the butterfat price.

(d) Class IV value.

(1) Multiply the pounds of nonfat solids in Class IV skim milk by the nonfat solids price; and

(2) Add an amount obtained by multiplying the pounds of butterfat in Class IV by the butterfat price.

(e) Multiply the pounds of skim milk and butterfat overage assigned to each class pursuant to §1000.44(a)(11) and the corresponding steps of §1000.44(b) by the skim milk prices and butterfat prices applicable to each class.

(f) Multiply the difference between the current month's Class I, II, or III price, as the case may be, and the Class IV price for the preceding month by the hundredweight of skim milk and butterfat subtracted from Class I, II, or III, respectively, pursuant to §1000.44(a)(7) and the corresponding step of §1000.44(b);

(g) Multiply the difference between the Class I price applicable at the location of the pool plant and the Class IV price by the hundredweight of skim milk and butterfat assigned to Class I pursuant to §1000.43(d) and the hundredweight of skim milk and butterfat subtracted from Class I pursuant to §1000.44(a)(3) (i) through (vi) and the corresponding step of §1000.44(b), excluding receipts of bulk fluid cream products from plants regulated under

171

other Federal orders and bulk concentrated fluid milk products from pool plants, plants regulated under other Federal orders, and unregulated supply plants.

(h) Multiply the difference between the Class I price applicable at the location of the nearest unregulated supply plants from which an equivalent volume was received and the Class III price by the pounds of skim milk and butterfat in receipts of concentrated fluid milk products assigned to Class I pursuant to § 1000.43(d) and § 1000.44(a)(3)(i) and the corresponding step of § 1000.44(b) and the pounds of skim milk and butterfat subtracted from Class I pursuant to § 1000.44(a)(8) and the corresponding step of § 1000.44(b), excluding such skim milk and butterfat in receipts of fluid milk products from an unregulated supply plant to the extent that an equivalent amount of skim milk or butterfat disposed of to such plant by handlers fully regulated under any Federal milk order is classified and priced as Class I milk and is not used as an offset for any other payment obligation under any order.

(i) For reconstituted milk made from receipts of nonfluid milk products, multiply $1.00 (but not more than the difference between the Class I price applicable at the location of the pool plant and the Class IV price) by the hundredweight of skim milk and butterfat contained in receipts of nonfluid milk products that are allocated to Class I use pursuant to § 1000.43(d).

[64 FR 47998, Sept. 1, 1999, as amended at 65 FR 82839, Dec. 28, 2000; 68 FR 7068, Feb. 12, 2003]

§ 1124.61 Computation of producer price differential.

For each month the market administrator shall compute a producer price differential per hundredweight. The report of any handler who has not made payments required pursuant to § 1124.71 for the preceding month shall not be included in the computation of the producer price differential, and such handler's report shall not be included in the computation for succeeding months until the handler has made full payment of outstanding monthly obligations. Subject to the conditions of this

paragraph, the market administrator shall compute the producer price differential in the following manner:

(a) Combine into one total the values computed pursuant to § 1124.60 for all handlers required to file reports prescribed in § 1124.30;

(b) Subtract the total values obtained by multiplying each handler's total pounds of protein, other solids, and butterfat contained in the milk for which an obligation was computed pursuant to § 1124.60 by the protein price, the other solids price, and the butterfat price, respectively;

(c) Add an amount equal to the minus location adjustments and subtract an amount equal to the plus location adjustments computed pursuant to § 1124.75;

(d) Add an amount equal to not less than one-half of the unobligated balance in the producer-settlement fund;

(e) Divide the resulting amount by the sum of the following for all handlers included in these computations:

(1) The total hundredweight of producer milk; and

(2) The total hundredweight for which a value is computed pursuant to § 1124.60(h); and

(f) Subtract not less than 4 cents nor more than 5 cents from the price computed pursuant to paragraph (e) of this section. The result shall be known as the producer price differential for the month.

[68 FR 7068, Feb. 12, 2003]

§ 1124.62 Announcement of producer prices.

On or before the 14th day after the end of each month, the market administrator shall announce publicly the following prices and information:

(a) The producer price differential;

(b) The protein price;

(c) The nonfat solids price;

(d) The other solids price;

(e) The butterfat price;

(f) The average butterfat, protein, nonfat solids, and other solids content of producer milk; and

(g) The statistical uniform price for milk containing 3.5 percent butterfat, computed by combining the Class III price and the producer price differential.

[64 FR 47998, Sept. 1, 1999, as amended at 65 FR 82840, Dec. 28, 2000; 68 FR 7069, Feb. 12, 2003]

PAYMENTS FOR MILK

§1124.70 Producer-settlement fund.

See §1000.70.

§1124.71 Payments to the producer-settlement fund.

Each handler shall make payment to the producer-settlement fund in a manner that provides receipt of the funds by the market administrator no later than the 16th day after the end of the month (except as provided in §1000.90). Payment shall be the amount, if any, by which the amount specified in paragraph (a) of this section exceeds the amount specified in paragraph (b) of this section:

(a) The total value of milk to the handler for the month as determined pursuant to §1124.60.

(b) The sum of:

(1) An amount obtained by multiplying the total hundredweight of producer milk as determined pursuant to §1000.44(c) by the producer price differential as adjusted pursuant to §1124.75;

(2) An amount obtained by multiplying the total pounds of protein, other solids, and butterfat contained in producer milk by the protein, other solids, and butterfat prices respectively; and

(3) An amount obtained by multiplying the pounds of skim milk and butterfat for which a value was computed pursuant to §1124.60(h) by the producer price differential as adjusted pursuant to §1124.75 for the location of the plant from which received.

[64 FR 47998, Sept. 1, 1999, as amended at 65 FR 82840, Dec. 28, 2000; 68 FR 7069, Feb. 12, 2003]

§1124.72 Payments from the producer-settlement fund.

No later than the 18th day after the end of each month (except as provided in §1000.90), the market administrator shall pay to each handler the amount, if any, by which the amount computed pursuant to §1124.71(b) exceeds the

amount computed pursuant to §1124.71(a). If, at such time, the balance in the producer-settlement fund is insufficient to make all payments pursuant to this section, the market administrator shall reduce uniformly such payments and shall complete the payments as soon as the funds are available.

§1124.73 Payments to producers and to cooperative associations.

(a) Each handler shall pay each producer for producer milk for which payment is not made to a cooperative association pursuant to paragraph (b) of this section, as follows:

(1) *Partial payment.* For each producer who has not discontinued shipments as of the 18th day of the month, partial payment shall be made so that it is received by each producer on or before the last day of the month (except as provided in §1000.90) for milk received during the first 15 days of the month from the producer at not less than the lowest announced class price for the preceding month, less proper deductions authorized in writing by the producer.

(2) *Final payment.* For milk received during the month, payment shall be made so that it is received by each producer no later than the 19th day after the end of the month (except as provided in §1000.90) in an amount equal to not less than the sum of:

(i) The hundredweight of producer milk received times the producer price differential for the month as adjusted pursuant to §1124.75;

(ii) The pounds of butterfat received times the butterfat price for the month;

(iii) The pounds of protein received times the protein price for the month;

(iv) The pounds of other solids received times the other solids price for the month;

(v) Less any payment made pursuant to paragraph (a)(1) of this section;

(vi) Less proper deductions authorized in writing by such producer and plus or minus adjustments for errors in previous payments to such producer subject to approval by the market administrator; and

(vii) Less deductions for marketing services pursuant to §1000.86.

(b) *Payments for milk received from cooperative association members.* On or before the 2nd day prior to the dates specified in paragraphs (a)(1) and (a)(2) of this section (except as provided in § 1000.90), each handler shall pay to a cooperative association for milk from producers who market their milk through the cooperative association and who have authorized the cooperative to collect such payments on their behalf an amount equal to the sum of the individual payments otherwise payable for such producer milk pursuant to paragraphs (a)(1) and (a)(2) of this section.

(c) *Payment for milk received from cooperative association pool plants or from cooperatives as handlers pursuant to § 1000.9(c).* On or before the 2nd day prior to the dates specified in paragraphs (a)(1) and (a)(2) of this section (except as provided in § 1000.90), each handler who receives fluid milk products at its plant from a cooperative association in its capacity as the operator of a pool plant or who receives milk from a cooperative association in its capacity as a handler pursuant to § 1000.9(c), including the milk of producers who are not members of such association and who the market administrator determines have authorized the cooperative association to collect payment for their milk, shall pay the cooperative for such milk as follows:

(1) For bulk fluid milk products and bulk fluid cream products received from a cooperative association in its capacity as the operator of a pool plant and for milk received from a cooperative association in its capacity as a handler pursuant to § 1000.9(c) during the first 15 days of the month, at not less than the lowest announced class price per hundredweight for the preceding month.

(2) For the total quantity of bulk fluid milk products and bulk fluid cream products received from a cooperative association in its capacity as the operator of a pool plant, at not less than the total value of such products received from the association's pool plants, as determined by multiplying the respective quantities assigned to each class under § 1000.44, as follows:

(i) The hundredweight of Class I skim milk times the Class I skim milk price

for the month plus the pounds of Class I butterfat times the Class I butterfat price for the month. The Class I prices to be used shall be the prices effective at the location of the receiving plant;

(ii) The pounds of nonfat solids in Class II skim milk by the Class II nonfat solids price;

(iii) The pounds of butterfat in Class II times the Class II butterfat price;

(iv) The pounds of nonfat solids in Class IV times the nonfat solids price;

(v) The pounds of butterfat in Class III and Class IV milk times the butterfat price;

(vi) The pounds of protein in Class III milk times the protein price;

(vii) The pounds of other solids in Class III milk times the other solids price; and

(viii) Add together the amounts computed in paragraphs (c)(2)(i) through (vii) of this section and from that sum deduct any payment made pursuant to paragraph (c)(1) of this section; and

(3) For the total quantity of milk received during the month from a cooperative association in its capacity as a handler under § 1000.9(c) as follows:

(i) The hundredweight of producer milk received times the producer price differential as adjusted pursuant to § 1124.75;

(ii) The pounds of butterfat received times the butterfat price for the month;

(iii) The pounds of protein received times the protein price for the month;

(iv) The pounds of other solids received times the other solids price for the month; and

(v) Add together the amounts computed in paragraphs (c)(3)(i) through (iv) of this section and from that sum deduct any payment made pursuant to paragraph (c)(1) of this section.

(d) If a handler has not received full payment from the market administrator pursuant to § 1124.72 by the payment date specified in paragraph (a), (b) or (c) of this section, the handler may reduce pro rata its payments to producers or to the cooperative association (with respect to receipts described in paragraph (b) of this section, prorating the underpayment to the volume of milk received from the cooperative association in proportion to the total milk received from producers by the handler), but not by more than the amount of the underpayment. The payments shall be completed on the next scheduled payment date after receipt of the balance due from the market administrator.

(e) If a handler claims that a required payment to a producer cannot be made because the producer is deceased or cannot be located, or because the cooperative association or its lawful successor or assignee is no longer in existence, the payment shall be made to the producer-settlement fund, and in the event that the handler subsequently locates and pays the producer or a lawful claimant, or in the event that the handler no longer exists and a lawful claim is later established, the market administrator shall make the required payment from the producer-settlement fund to the handler or to the lawful claimant, as the case may be.

(f) In making payments to producers pursuant to this section, each handler shall furnish each producer, except a producer whose milk was received from a cooperative association handler described in § 1000.9(a) or (c), a supporting statement in a form that may be retained by the recipient which shall show:

(1) The name, address, Grade A identifier assigned by a duly constituted regulatory agency, and payroll number of the producer;

(2) The daily and total pounds, and the month and dates such milk was received from that producer;

(3) The total pounds of butterfat, protein, and other solids contained in the producer's milk;

(4) The minimum rate or rates at which payment to the producer is required pursuant to the order in this part;

(5) The rate used in making payment if the rate is other than the applicable minimum rate;

(6) The amount, or rate per hundredweight, or rate per pound of component, and the nature of each deduction claimed by the handler; and

(7) The net amount of payment to the producer or cooperative association.

[64 FR 47998, Sept. 1, 1999, as amended at 65 FR 82840, Dec. 28, 2000; 68 FR 7069, Feb. 12, 2003]

§ 1124.74 [Reserved]

§ 1124.75 Plant location adjustments for producer milk and nonpool milk.

For purposes of making payments for producer milk and nonpool milk, a plant location adjustment shall be determined by subtracting the Class I price specified in § 1124.51 from the Class I price at the plant's location. The difference, plus or minus as the case may be, shall be used to adjust the payments required pursuant to §§ 1124.73 and 1000.76.

§ 1124.76 Payments by a handler operating a partially regulated distributing plant.

See § 1000.76.

§ 1124.77 Adjustment of accounts.

See § 1000.77.

§ 1124.78 Charges on overdue accounts.

See § 1000.78.

ADMINISTRATIVE ASSESSMENT AND MARKETING SERVICE DEDUCTION

§ 1124.85 Assessment for order administration.

See § 1000.85.

§ 1124.86 Deduction for marketing services.

See § 1000.86.

PART 1125 [RESERVED]

PART 1126—MILK IN THE SOUTHWEST MARKETING AREA

Subpart—Order Regulating Handling

GENERAL PROVISIONS

AUTHORITY: 7 U.S.C. 601–674, and 7253.

SOURCE: 64 FR 48004, Sept. 1, 1999, unless otherwise noted.

Subpart—Order Regulating Handling

GENERAL PROVISIONS

§ 1126.1 General provisions.

The terms, definitions, and provisions in part 1000 of this chapter apply to this part 1126. In this part 1126, all references to sections in part 1000 refer to part 1000 of this chapter.

DEFINITIONS

§ 1126.2 Southwest marketing area.

The marketing area means all territory within the bounds of the following states and political subdivisions, including all piers, docks and wharves connected therewith and all craft moored thereat, and all territory occupied by government (municipal, State or Federal) reservations, installations, institutions, or other similar establishments if any part thereof is within any of the listed states or political subdivisions:

COLORADO COUNTIES

Archuleta, LaPlata, and Montezuma.

NEW MEXICO AND TEXAS

All of the States of New Mexico and Texas.

§ 1126.3 Route disposition.

See § 1000.3.

§ 1126.4 Plant.

See § 1000.4.

§ 1126.5 Distributing plant.

See § 1000.5.

§ 1126.6 Supply plant.

See § 1000.6.

§ 1126.7 Pool plant.

Pool plant means a plant specified in paragraphs (a) through (d) of this section, a unit of plants as specified in paragraph (e) of this section, or a plant specified in paragraph (h) of this section, but excluding a plant specified in paragraph (g) of this section. The pooling standards described in paragraphs (c) and (d) of this section are subject to modification pursuant to paragraph (f) of this section:

(a) A distributing plant, other than a plant qualified as a pool plant pursuant to paragraph (b) of this section or §_____.7(b) of any other Federal milk order, from which during the month 25 percent or more of the total quantity of fluid milk products physically received at the plant (excluding concentrated milk received from another plant by agreement for other than Class I use) are disposed of as route disposition or are transferred in the form of packaged fluid milk products to other distributing plants. At least 25 percent of such route disposition and transfers must be to outlets in the marketing area.

(b) Any distributing plant located in the marketing area which during the month processed at least 25 percent of the total quantity of fluid milk products physically received at the plant (excluding concentrated milk received from another plant by agreement for other than Class I use) into ultra-pasteurized or aseptically-processed fluid milk products.

(c) A supply plant from which 50 percent or more of the total quantity of milk that is physically received during the month from dairy farmers and handlers described in §1000.9(c), including milk that is diverted as producer milk to other plants, is transferred to pool distributing plants. Concentrated milk transferred from the supply plant to a distributing plant for an agreed-upon use other than Class I shall be excluded from the supply plant's shipments in computing the plant's shipping percentage.

(d) A plant located within the marketing area that is operated by a cooperative association if pool plant status under this paragraph is requested for such plant by the cooperative association and during the month at least 30 percent of the producer milk of members of such cooperative association is delivered directly from farms to pool distributing plants or is transferred to such plants as a fluid milk product (excluding concentrated milk transferred to a distributing plant for an agreed-upon use other than Class I) from the cooperative's plant.

(e) Two or more plants operated by the same handler and located within the marketing area may qualify for pool status as a unit by meeting the total and in-area route disposition requirements specified in paragraph (a) of this section and the following additional requirements:

(1) At least one of the plants in the unit must qualify as a pool plant pursuant to paragraph (a) of this section;

(2) Other plants in the unit must process only Class I or Class II products and must be located in a pricing zone providing the same or a lower Class I price than the price applicable at the distributing plant included in the unit pursuant to paragraph (e)(1) of this section; and

(3) A written request to form a unit, or to add or remove plants from a unit, must be filed with the market administrator prior to the first day of the month for which it is to be effective.

(f) The applicable shipping percentages of paragraphs (c) and (d) of this section may be increased or decreased by the market administrator if the market administrator finds that such adjustment is necessary to encourage needed shipments or to prevent uneconomic shipments. Before making such a finding, the market administrator shall investigate the need for adjustment either on the market administrator's own initiative or at the request of interested parties if the request is made in writing at least 15 days prior to the month for which the requested revision is desired effective. If the investigation shows that an adjustment of the shipping percentages might be appropriate, the market administrator shall issue a notice stating that an adjustment is being considered and invite data, views and arguments. Any decision to revise an applicable shipping percentage must be issued in writing at least one day before the effective date.

(g) The term pool plant shall not apply to the following plants:

(1) A producer-handler plant;

(2) An exempt plant as defined in §1000.8(e);

(3) A plant qualified pursuant to paragraph (a) of this section that is located within the marketing area if the plant also meets the pooling requirements of another Federal order, and

177

more than 50 percent of its route distribution has been in such other Federal order marketing area for 3 consecutive months;

(4) A plant qualified pursuant to paragraph (a) of this section which is not located within any Federal order marketing area that meets the pooling requirements of another Federal order and has had greater route disposition in such other Federal order's marketing area for 3 consecutive months;

(5) A plant qualified pursuant to paragraph (a) of this section that is located in another Federal order marketing area if the plant meets the pooling requirements of such other Federal order and does not have a majority of its route distribution in this marketing area for 3 consecutive months or if the plant is required to be regulated under such other Federal order without regard to its route disposition in any other Federal order marketing area;

(6) A plant qualified pursuant to paragraph (c) or (d) of this section which also meets the pooling requirements of another Federal order and from which greater qualifying shipments are made to plants regulated under the other Federal order than are made to plants regulated under the order in this part, or the plant has automatic pooling status under the other Federal order; and

(7) That portion of a pool plant designated as a nonpool plant that is physically separate and operated separately from the pool portion of such plant. The designation of a portion of a regulated plant as a nonpool plant must be requested in writing by the handler and must be approved by the market administrator.

(h) Any distributing plant, located within the marketing area as described on May 1, 2006, in § 1126.2;

(1) From which there is route disposition and/or transfers of packaged fluid milk products in any non-federally regulated marketing area(s) located within one or more States that require handlers to pay minimum prices for raw milk provided that 25 percent or more of the total quantity of fluid milk products physically received at such plant (excluding concentrated milk received from another plant by agreement for other than Class I use) is disposed of as route disposition and/or is transferred in the form of packaged fluid milk products to other plants. At least 25 percent of such route disposition and/or transfers, in aggregate, are in any non-federally regulated marketing area(s) located within one or more States that require handlers to pay minimum prices for raw milk. Subject to the following exclusions:

(i) The plant is described in § 1126.7(a), (b), or (e);

(ii) The plant is subject to the pricing provisions of a State-operated milk pricing plan which provides for the payment of minimum class prices for raw milk;

(iii) The plant is described in § 1000.8(a) or (e); or

(iv) A producer-handler described in § 1126.10 with less than three million pounds during the month of route disposition and/or transfers of packaged fluid milk products to other plants.

(2) [Reserved]

[64 FR 48004, Sept. 1, 1999, as amended at 71 FR 25501, May 1, 2006; 71 FR 28249, May 16, 2006]

§ 1126.8   Nonpool plant.

See § 1000.8.

§ 1126.9   Handler.

See § 1000.9.

§ 1126.10   Producer-handler.

Producer-handler means a person who:

(a) Operates a dairy farm and a distributing plant from which there is route disposition in the marketing area, and from which total route disposition and packaged sales of fluid milk products to other plants during the month does not exceed 3 million pounds;

(b) Receives fluid milk products from own farm production or milk that is fully subject to the pricing and pooling provisions of the order in this part or another Federal order;

(c) Receives no more than 150,000 pounds of fluid milk products from handlers fully regulated under any Federal order, including such products received at a location other than the producer-handler's processing plant for distribution on routes. This limitation shall not apply if the producer-handler's own farm production is less than 150,000 pounds during the month;

(d) Disposes of no other source milk as Class I milk except by increasing the nonfat milk solids content of the fluid milk products; and

(e) Provides proof satisfactory to the market administrator that the care and management of the dairy animals and other resources necessary to produce all Class I milk handled (excluding receipts from handlers fully regulated under any Federal order) and the processing and packaging operations are the producer-handler's own enterprise and at its own risk.

(f) Any producer-handler with Class I route dispositions and/or transfers of packaged fluid milk products in the marketing area described in §1131.2 of this chapter shall be subject to payments into the Order 1131 producer settlement fund on such dispositions pursuant to §1000.76(a) and payments into the Order 1131 administrative fund provided such dispositions are less than three million pounds in the current month and such producer-handler had total Class I route dispositions and/or transfers of packaged fluid milk products from own farm production of three million pounds or more the previous month. If the producer-handler has Class I route dispositions and/or transfers of packaged fluid milk products into the marketing area described in §1131.2 of this chapter of three million pounds or more during the current month, such producer-handler shall be subject to the provisions described in §1131.7 of this chapter or §1000.76(a).

[64 FR 48004, Sept. 1, 1999, as amended at 71 FR 25501, May 1, 2006; 75 FR 21161, Apr. 23, 2010]

§1126.11 [Reserved]

§1126.12 Producer.

(a) Except as provided in paragraph (b) of this section, *producer* means any person who produces milk approved by a duly constituted regulatory agency for fluid consumption as Grade A milk and whose milk (or components of milk) is:

(1) Received at a pool plant directly from the producer or diverted by the plant operator in accordance with §1126.13; or

(2) Received by a handler described in §1000.9(c).

(b) Producer shall not include:

(1) A producer-handler as defined in any Federal order;

(2) A dairy farmer whose milk is received at an exempt plant, excluding producer milk diverted to the exempt plant pursuant to §1126.13(d);

(3) A dairy farmer whose milk is received by diversion at a pool plant from a handler regulated under another Federal order if the other Federal order designates the dairy farmer as a producer under that order and the milk is allocated by request to a utilization other than Class I; and

(4) A dairy farmer whose milk is reported as diverted to a plant fully regulated under another Federal order with respect to that portion of the milk so diverted that is assigned to Class I under the provisions of such other order.

§1126.13 Producer milk.

Producer milk means the skim milk (or the skim equivalent of components of skim milk), including nonfat components, and butterfat contained in milk of a producer that is:

(a) Received by the operator of a pool plant directly from a producer or a handler described in §1000.9(c). All milk received pursuant to this paragraph shall be priced at the location of the plant where it is first physically received;

(b) Received by a handler described in §1000.9(c) in excess of the quantity delivered to pool plants;

(c) Diverted by a pool plant operator for the account of the handler operating such plant to another pool plant. Milk so diverted shall be priced at the location of the plant to which diverted; or

(d) Diverted by the operator of a pool plant or a handler described in §1000.9(c) to a nonpool plant, subject to the following conditions:

(1) Milk of a dairy farmer shall not be eligible for diversion unless a delivery of at least 40,000 pounds or one day's milk production, whichever is less, of such dairy farmer has been physically received as producer milk at a pool

plant and the dairy farmer has continuously retained producer status since that time;

(2) The total quantity of milk diverted during the month by a cooperative association shall not exceed 50 percent of the total quantity of producer milk that the cooperative association caused to be received at pool plants and diverted;

(3) The operator of a pool plant that is not a cooperative association may divert any milk that is not under the control of a cooperative association that diverts milk during the month pursuant to this paragraph. The total quantity of milk so diverted during the month shall not exceed 50 percent of the total quantity of the producer milk physically received at such plant (or such unit of plants in the case of plants that pool as a unit pursuant to §1126.7(e)) and diverted;

(4) Any milk diverted in excess of the limits prescribed in paragraphs (d)(2) and (3) of this section shall not be producer milk. If the diverting handler or cooperative association fails to designate the dairy farmers' deliveries that will not be producer milk, no milk diverted by the handler or cooperative association shall be producer milk;

(5) Diverted milk shall be priced at the location of the plant to which diverted; and

(6) The delivery requirement in paragraph (d)(1) and the diversion percentages in paragraphs (d)(2) and (3) of this section may be increased or decreased by the market administrator if there is a finding that such revision is necessary to assure orderly marketing and efficient handling of milk in the marketing area. Before making such a finding, the market administrator shall investigate the need for the revision either on the market administrator's own initiative or at the request of interested persons if the request is made in writing at least 15 days prior to the month for which the requested revision is desired effective. If the investigation shows that a revision might be appropriate, the market administrator shall issue a notice stating that the revision is being considered and inviting written data, views, and arguments. Any decision to revise the delivery day requirement or any diversion percentage must be issued in writing at least one day before the effective date.

§1126.14 Other source milk.

See §1000.14.

§1126.15 Fluid milk product.

See §1000.15.

§1126.16 Fluid cream product.

See §1000.16.

§1126.17 [Reserved]

§1126.18 Cooperative association.

See §1000.18.

§1126.19 Commercial food processing establishment.

See §1000.19.

HANDLER REPORTS

§1126.30 Reports of receipts and utilization.

Each handler shall report monthly so that the market administrator's office receives the report on or before the 8th day after the end of the month, in the detail and on prescribed forms, as follows:

(a) Each pool plant operator shall report for each of its operations the following information:

(1) Product pounds, pounds of butterfat, pounds of protein, pounds of nonfat solids other than protein (other solids), and the value of the somatic cell adjustment pursuant to §1000.50(p) contained in or represented by:

(i) Receipts of producer milk, including producer milk diverted by the reporting handler, from sources other than handlers described in §1000.9(c); and

(ii) Receipts of milk from handlers described in §1000.9(c);

(2) Product pounds and pounds of butterfat contained in:

(i) Receipts of fluid milk products and bulk fluid cream products from other pool plants;

(ii) Receipts of other source milk; and

(iii) Inventories at the beginning and end of the month of fluid milk products and bulk fluid cream products;

(3) The utilization or disposition of all milk and milk products required to be reported pursuant to this paragraph; and

(4) Such other information with respect to the receipts and utilization of skim milk, butterfat, milk protein, other nonfat solids, and somatic cell information, as the market administrator may prescribe.

(b) Each handler operating a partially regulated distributing plant shall report with respect to such plant in the same manner as prescribed for reports required by paragraph (a) of this section. Receipts of milk that would have been producer milk if the plant had been fully regulated shall be reported in lieu of producer milk. The report shall show also the quantity of any reconstituted skim milk in route disposition in the marketing area.

(c) Each handler described in §1000.9(c) shall report:

(1) The product pounds, pounds of butterfat, pounds of protein, pounds of solids-not-fat other than protein (other solids), and the value of the somatic cell adjustment pursuant to §1000.50(p), contained in receipts of milk from producers; and

(2) The utilization or disposition of such receipts.

(d) Each handler not specified in paragraphs (a) through (c) of this section shall report with respect to its receipts and utilization of milk and milk products in such manner as the market administrator may prescribe.

§1126.31 Payroll reports.

(a) On or before the 20th day after the end of each month, each handler that operates a pool plant pursuant to §1126.7 and each handler described in §1000.9(c) shall report to the market administrator its producer payroll for the month, in the detail prescribed by the market administrator, showing for each producer the information specified in §1126.73(e).

(b) Each handler operating a partially regulated distributing plant who elects to make payment pursuant to §1000.76(b) shall report for each dairy farmer who would have been a producer if the plant had been fully regulated in the same manner as prescribed for re-

ports required by paragraph (a) of this section.

§1126.32 Other reports.

In addition to the reports required pursuant to §§1126.30 and 1126.31, each handler shall report any information the market administrator deems necessary to verify or establish each handler's obligation under the order.

CLASSIFICATION OF MILK

§1126.40 Classes of utilization.

See §1000.40.

§1126.41 [Reserved]

§1126.42 Classification of transfers and diversions.

See §1000.42.

§1126.43 General classification rules.

See §1000.43.

§1126.44 Classification of producer milk.

See §1000.44.

§1126.45 Market administrator's reports and announcements concerning classification.

See §1000.45.

CLASS PRICES

§1126.50 Class prices, component prices, and advanced pricing factors.

See §1000.50.

§1126.51 Class I differential and price.

The Class I differential shall be the differential established for Dallas County, Texas, which is reported in §1000.52. The Class I price shall be the price computed pursuant to §1000.50(a) for Dallas County, Texas.

§1126.52 Adjusted Class I differentials.

See §1000.52.

§1126.53 Announcement of class prices, component prices, and advanced pricing factors.

See §1000.53.

§1126.54 Equivalent price.

See §1000.54.

PRODUCER PRICE DIFFERENTIAL

§ 1126.60 Handler's value of milk.

For the purpose of computing a handler's obligation for producer milk, the market administrator shall determine for each month the value of milk of each handler with respect to each of the handler's pool plants and of each handler described in § 1000.9(c) with respect to milk that was not received at a pool plant by adding the amounts computed in paragraphs (a) through (i) of this section and subtracting from that total amount the value computed in paragraph (j) of this section. Unless otherwise specified, the skim milk, butterfat, and the combined pounds of skim milk and butterfat referred to in this section shall result from the steps set forth in § 1000.44(a), (b), and (c), respectively, and the nonfat components of producer milk in each class shall be based upon the proportion of such components in producer skim milk. Receipts of nonfluid milk products that are distributed as labeled reconstituted milk for which payments are made to the producer-settlement fund of another Federal order under § 1000.76(a)(4) or (d) shall be excluded from pricing under this section.

(a) Class I value.

(1) Multiply the pounds of skim milk in Class I by the Class I skim milk price; and

(2) Add an amount obtained by multiplying the pounds of butterfat in Class I by the Class I butterfat price.

(b) Class II value.

(1) Multiply the pounds of nonfat solids in Class II skim milk by the Class II nonfat solids price; and

(2) Add an amount obtained by multiplying the pounds of butterfat in Class II times the Class II butterfat price.

(c) Class III value.

(1) Multiply the pounds of protein in Class III skim milk by the protein price;

(2) Add an amount obtained by multiplying the pounds of other solids in Class III skim milk by the other solids price; and

(3) Add an amount obtained by multiplying the pounds of butterfat in Class III by the butterfat price.

(d) Class IV value.

(1) Multiply the pounds of nonfat solids in Class IV skim milk by the nonfat solids price; and

(2) Add an amount obtained by multiplying the pounds of butterfat in Class IV by the butterfat price.

(e) Compute an adjustment for the somatic cell content of producer milk by multiplying the values reported pursuant to § 1126.30(a)(1) and (c)(1) by the percentage of total producer milk allocated to Class II, Class III, and Class IV pursuant to § 1000.44(c);

(f) Multiply the pounds of skim milk and butterfat overage assigned to each class pursuant to § 1000.44(a)(11) and the corresponding step of § 1000.44(b) by the skim milk prices and butterfat prices applicable to each class.

(g) Multiply the difference between the current month's Class I, II, or III price, as the case may be, and the Class IV price for the preceding month by the hundredweight of skim milk and butterfat subtracted from Class I, II, or III, respectively, pursuant to § 1000.44(a)(7) and the corresponding step of § 1000.44(b);

(h) Multiply the difference between the Class I price applicable at the location of the pool plant and the Class IV price by the hundredweight of skim milk and butterfat assigned to Class I pursuant to § 1000.43(d) and the hundredweight of skim milk and butterfat subtracted from Class I pursuant to § 1000.44(a)(3)(i) through (vi) and the corresponding step of § 1000.44(b), excluding receipts of bulk fluid cream products from plants regulated under other Federal orders and bulk concentrated fluid milk products from pool plants, plants regulated under other Federal orders, and unregulated supply plants.

(i) Multiply the difference between the Class I price applicable at the location of the nearest unregulated supply plants from which an equivalent volume was received and the Class III price by the pounds of skim milk and butterfat in receipts of concentrated fluid milk products assigned to Class I pursuant to § 1000.43(d) and § 1000.44(a)(3)(i) and the corresponding step of § 1000.44(b) and the pounds of skim milk and butterfat subtracted from Class I pursuant to § 1000.44(a)(8) and the corresponding step of

§ 1000.44(b), excluding such skim milk and butterfat in receipts of fluid milk products from an unregulated supply plant to the extent that an equivalent amount of skim milk or butterfat disposed of to such plant by handlers fully regulated under any Federal milk order is classified and priced as Class I milk and is not used as an offset for any other payment obligation under any order.

(j) For reconstituted milk made from receipts of nonfluid milk products, multiply $1.00 (but not more than the difference between the Class I price applicable at the location of the pool plant and the Class IV price) by the hundredweight of skim milk and butterfat contained in receipts of nonfluid milk products that are allocated to Class I use pursuant to § 1000.43(d).

[64 FR 478004, Sept. 1, 1999, as amended at 65 FR 82840, Dec. 28, 2000; 68 FR 7069, Feb. 12, 2003]

§ 1126.61 Computation of producer price differential.

For each month the market administrator shall compute a producer price differential per hundredweight. The report of any handler who has not made payments required pursuant to § 1126.71 for the preceding month shall not be included in the computation of the producer price differential, and such handler's report shall not be included in the computation for succeeding months until the handler has made full payment of outstanding monthly obligations. Subject to the conditions of this paragraph, the market administrator shall compute the producer price differential in the following manner:

(a) Combine into one total the values computed pursuant to § 1126.60 for all handlers required to file reports prescribed in § 1126.30;

(b) Subtract the total of the values obtained by multiplying each handler's total pounds of protein, other solids, and butterfat contained in the milk for which an obligation was computed pursuant to § 1126.60 by the protein price, other solids price, and the butterfat price, respectively, and the total value of the somatic cell adjustment pursuant to § 1126.30(a)(1) and (c)(1);

(c) Add an amount equal to the minus location adjustments and sub-

tract an amount equal to the plus location adjustments computed pursuant to § 1126.75;

(d) Add an amount equal to not less than one-half of the unobligated balance in the producer-settlement fund;

(e) Divide the resulting amount by the sum of the following for all handlers included in these computations:

(1) The total hundredweight of producer milk; and

(2) The total hundredweight for which a value is computed pursuant to § 1126.60(i); and

(f) Subtract not less than 4 cents nor more than 5 cents from the price computed pursuant to paragraph (e) of this section. The result shall be known as the producer price differential for the month.

[68 FR 7069, Feb. 12, 2003]

§ 1126.62 Announcement of producer prices.

On or before the 13th day after the end of each month, the market administrator shall announce the following prices and information:

(a) The producer price differential;

(b) The protein price;

(c) The nonfat solids price;

(d) The other solids price;

(e) The butterfat price;

(f) The somatic cell adjustment rate;

(g) The average butterfat, protein, nonfat solids, and other solids content of producer milk; and

(h) The statistical uniform price for milk containing 3.5 percent butterfat, computed by combining the Class III price and the producer price differential.

[64 FR 478004, Sept. 1, 1999, as amended at 65 FR 82841, Dec. 28, 2000; 68 FR 7069, Feb. 12, 2003]

PAYMENTS FOR MILK

§ 1126.70 Producer-settlement fund.

See § 1000.70.

§ 1126.71 Payments to the producer-settlement fund.

Each handler shall make payment to the producer-settlement fund in a manner that provides receipt of the funds by the market administrator no later than the 16th day after the end of the month (except as provided in § 1000.90).

Payment shall be the amount, if any, by which the amount specified in paragraph (a) of this section exceeds the amount specified in paragraph (b) of this section:

(a) The total value of milk to the handler for the month as determined pursuant to § 1126.60.

(b) The sum of:

(1) An amount obtained by multiplying the total hundredweight of producer milk as determined pursuant to § 1000.44(c) by the producer price differential as adjusted pursuant to § 1126.75;

(2) An amount obtained by multiplying the total pounds of protein, other solids, and butterfat contained in producer milk by the protein, other solids, and butterfat prices respectively;

(3) The total value of the somatic cell adjustment to producer milk; and

(4) An amount obtained by multiplying the pounds of skim milk and butterfat for which a value was computed pursuant to § 1126.60(i) by the producer price differential as adjusted pursuant to § 1126.75 for the location of the plant from which received.

[64 FR 48004, Sept. 1, 1999, as amended at 65 FR 82841, Dec. 28, 2000; 68 FR 7069, Feb. 12, 2003]

§ 1126.72 Payments from the producer-settlement fund.

No later than the 17th day after the end of each month (except as provided in § 1000.90), the market administrator shall pay to each handler the amount, if any, by which the amount computed pursuant to § 1126.71(b) exceeds the amount computed pursuant to § 1126.71(a). If, at such time, the balance in the producer-settlement fund is insufficient to make all payments pursuant to this section, the market administrator shall reduce uniformly such payments and shall complete the payments as soon as the funds are available.

§ 1126.73 Payments to producers and to cooperative associations.

(a) Each handler shall pay each producer for producer milk for which payment is not made to a cooperative association pursuant to paragraph (b) of this section, as follows:

(1) *Partial payment.* For each producer who has not discontinued shipments as of the 23rd day of the month, payment shall be made so that it is received by the producer on or before the 26th day of the month (except as provided in § 1000.90) for milk received during the first 15 days of the month at not less than the lowest announced class price for the preceding month, less proper deductions authorized in writing by the producer.

(2) *Final payment.* For milk received during the month, payment shall be made so that it is received by each producer no later than the 18th day after the end of the month (except as provided in § 1000.90) in an amount computed as follows:

(i) Multiply the hundredweight of producer milk received times the producer price differential for the month as adjusted pursuant to § 1126.75;

(ii) Multiply the pounds of butterfat received times the butterfat price for the month;

(iii) Multiply the pounds of protein received times the protein price for the month;

(iv) Multiply the pounds of other solids received times the other solids price for the month;

(v) Multiply the hundredweight of milk received times the somatic cell adjustment for the month;

(vi) Add the amounts computed in paragraphs (a)(2)(i) through (v) of this section, and from that sum:

(A) Subtract the partial payment made pursuant to paragraph (a)(1) of this section;

(B) Subtract the deduction for marketing services pursuant to § 1000.86;

(C) Add or subtract for errors made in previous payments to the producer subject to approval by the market administrator; and

(D) Subtract proper deductions authorized in writing by the producer.

(b) On or before the day prior to the dates specified for partial and final payments pursuant to paragraph (a) of this section (except as provided in § 1000.90), each handler shall pay a cooperative association for milk received as follows:

(1) *Partial payment to a cooperative association for bulk milk received directly from producers' farms.* For bulk milk

(including the milk of producers who are not members of such association and who the market administrator determines have authorized the cooperative association to collect payment for their milk) received during the first 15 days of the month from a cooperative association in any capacity, except as the operator of a pool plant, the payment shall be equal to the hundredweight of milk received multiplied by the lowest announced class price for the preceding month.

(2) *Partial payment to a cooperative association for milk transferred from its pool plant.* For bulk milk/skimmed milk products received during the first 15 days of the month from a cooperative association in its capacity as the operator of a pool plant, the partial payment shall be at the pool plant operator's estimated use value of the milk using the most recent class prices available at the receiving plant's location.

(3) *Final payment to a cooperative association for milk transferred from its pool plant.* Following the classification of bulk fluid milk products and bulk fluid cream products received during the month from a cooperative association in its capacity as the operator of a pool plant, the final payment for such receipts shall be determined as follows:

(i) The hundredweight of Class I skim milk times the Class I skim milk price for the month plus the pounds of Class I butterfat times the Class I butterfat price for the month. The Class I prices to be used shall be the prices effective at the location of the receiving plant;

(ii) The pounds of nonfat solids in Class II skim milk by the Class II nonfat solids price;

(iii) The pounds of butterfat in Class II times the Class II butterfat price;

(iv) The pounds of nonfat solids in Class IV times the nonfat solids price;

(v) The pounds of butterfat in Class III and Class IV milk times the butterfat price;

(vi) The pounds of protein in Class III milk times the protein price;

(vii) The pounds of other solids in Class III milk times the other solids price;

(viii) The hundredweight of Class II, Class III, and Class IV milk times the somatic cell adjustment; and

(ix) Add together the amounts computed in paragraphs (b)(3)(i) through (viii) of this section and from that sum deduct any payments made pursuant to paragraph (b)(2) of this section.

(4) *Final payment to a cooperative association for bulk milk received directly from producers' farms.* For bulk milk received from a cooperative association during the month, including the milk of producers who are not members of such association and who the market administrator determines have authorized the cooperative association to collect payment for their milk, the final payment for such milk shall be an amount equal to the sum of the individual payments otherwise payable for such milk pursuant to paragraph (a)(2) of this section.

(c) If a handler has not received full payment from the market administrator pursuant to §1126.72 by the payment date specified in paragraph (a) or (b) of this section, the handler may reduce pro rata its payments to producers or to cooperative associations pursuant to paragraphs (a) and (b) of this section, but by not more than the amount of the underpayment. The payments shall be completed on the next scheduled payment date after receipt of the balance due from the market administrator.

(d) If a handler claims that a required payment to a producer cannot be made because the producer is deceased or cannot be located, or because the cooperative association or its lawful successor or assignee is no longer in existence, the payment shall be made to the producer-settlement fund, and in the event that the handler subsequently locates and pays the producer or a lawful claimant, or in the event that the handler no longer exists and a lawful claim is later established, the market administrator shall make the required payment from the producer-settlement fund to the handler or to the lawful claimant as the case may be.

(e) In making payments to producers pursuant to this section, each pool plant operator shall furnish each producer, except a producer whose milk was received from a cooperative association handler described in §1000.9(a) or (c), a supporting statement in a form that may be retained by the recipient which shall show:

(1) The name, address, Grade A identifier assigned by a duly constituted regulatory agency, and the payroll number of the producer;

(2) The month and dates that milk was received from the producer, including the daily and total pounds of milk received;

(3) The total pounds of butterfat, protein, and other solids contained in the producer's milk;

(4) The somatic cell count of the producer's milk;

(5) The minimum rate or rates at which payment to the producer is required pursuant to the order in this part;

(6) The rate used in making payment if the rate is other than the applicable minimum rate;

(7) The amount, or rate per hundredweight, or rate per pound of component, and the nature of each deduction claimed by the handler; and

(8) The net amount of payment to the producer or cooperative association.

[64 FR 48004, Sept. 1, 1999, as amended at 65 FR 32010, May 22, 2000; 65 FR 82841, Dec. 28, 2000; 68 FR 7069, Feb. 12, 2003]

§ 1126.74 [Reserved]

§ 1126.75 Plant location adjustments for producer milk and nonpool milk.

For purposes of making payments for producer milk and nonpool milk, a plant location adjustment shall be determined by subtracting the Class I price specified in § 1126.51 from the Class I price at the plant's location. The difference, plus or minus as the case may be, shall be used to adjust the payments required pursuant to §§ 1126.73 and 1000.76.

§ 1126.76 Payments by a handler operating a partially regulated distributing plant.

See § 1000.76.

§ 1126.77 Adjustment of accounts.

See § 1000.77.

§ 1126.78 Charges on overdue accounts.

See § 1000.78.

PART 1131—MILK IN THE ARIZONA MARKETING AREA

Subpart—Order Regulating Handling

1131.53 Announcement of class prices, component prices, and advanced pricing factors.
1131.54 Equivalent price.

UNIFORM PRICES

1131.60 Handler's value of milk.
1131.61 Computation of uniform prices.
1131.62 Announcement of uniform prices.

PAYMENTS FOR MILK

1131.70 Producer-settlement fund.
1131.71 Payments to the producer-settlement fund.
1131.72 Payments from the producer-settlement fund.
1131.73 Payments to producers and to cooperative associations.
1131.74 [Reserved]
1131.75 Plant location adjustments for producers and nonpool milk.
1131.76 Payments by a handler operating a partially regulated distributing plant.
1131.77 Adjustment of accounts.
1131.78 Charges on overdue accounts.

ADMINISTRATIVE ASSESSMENT AND MARKETING SERVICE DEDUCTION

1131.85 Assessment for order administration.
1131.86 Deduction for marketing services.

AUTHORITY: 7 U.S.C. 601–674, and 7253.

SOURCE: 64 FR 48010, Sept. 1, 1999, unless otherwise noted.

Subpart—Order Regulating Handling

GENERAL PROVISIONS

§ 1131.1 General provisions.

The terms, definitions, and provisions in part 1000 of this chapter apply to this part 1131. In this part 1131, all references to sections in part 1000 refer to part 1000 of this chapter.

DEFINITIONS

§ 1131.2 Arizona marketing area.

The marketing area means all territory within the bounds of the following states and political subdivisions, including all piers, docks and wharves connected therewith and all craft moored thereat, and all territory occupied by government (municipal, State or Federal) reservations, installations, institutions, or other similar establishments if any part thereof is within any of the listed states or political subdivisions:

Arizona

All of the State of Arizona.

[71 FR 25502, May 1, 2006]

§ 1131.3 Route disposition.

See § 1000.3.

§ 1131.4 Plant.

See § 1000.4.

§ 1131.5 Distributing plant.

See § 1000.5.

§ 1131.6 Supply plant.

See § 1000.6.

§ 1131.7 Pool plant.

Pool Plant means a plant or unit of plants specified in paragraphs (a) through (e) of this section, but excluding a plant specified in paragraph (g) of this section. The pooling standards described in paragraphs (c) and (d) of this section are subject to modification pursuant to paragraph (f) of this section.

(a) A distributing plant, other than a plant qualified as a pool plant pursuant to paragraph (b) of this § _____ . 7(b) of any other Federal milk order, from which during the month 25 percent or more of the total quantity of fluid milk products physically received at the plant (excluding concentrated milk received from another plant by agreement for other than Class I use) are disposed of as route disposition or are transferred in the form of packaged fluid milk products to other distributing plants. At least 25 percent of such route disposition and transfers must be to outlets in the marketing area.

(b) Any distributing plant located in the marketing area which during the month processed at least 25 percent of the total quantity of fluid milk products physically received at the plant (excluding concentrated milk received from another plant by agreement for other than Class I use) into ultra-pasteurized or aseptically-processed fluid milk products.

(c) A supply plant from which 50 percent or more of the total quantity of milk that is physically received at such plant from dairy farmers and handlers described in § 1000.9(c), including milk that is diverted as producer milk to other plants, is transferred to pool distributing plants. Concentrated milk transferred from the supply plant to a distributing plant for an agreed-upon use other than Class I shall be excluded from the supply plant's shipments in computing the plant's shipping percentage.

(d) A plant located within the marketing area and operated by a cooperative association if, during the month, or the immediately preceding 12-month period ending with the current month, 35 percent or more of the producer milk of members of the association (and any producer milk of nonmembers and members of another cooperative association which may be marketed by the cooperative association) is physically received in the form of bulk fluid milk products (excluding concentrated milk transferred to a distributing plant for an agreed-upon use other that Class I) at plants specified in paragraph (a), (b), or (h) of this section either directly from farms or by transfer from supply plants operated by the cooperative association and from plants of the cooperative association for which pool plant status has been requested under this paragraph subject to the following conditions:

(1) The plant does not qualify as a pool plant under paragraph (a), (b), (c), or (h) of this section or under comparable provisions of another Federal order; and

(2) The plant is approved by a duly constituted regulatory agency for the handling of milk approved for fluid consumption in the marketing area.

(e) Two or more plants operated by the same handler and located in the marketing area may qualify for pool plant status as a unit by together meeting the requirements specified in paragraph (a) of this section and subject to all of the following additional requirements:

(1) At least one of the plants in the unit must qualify as a pool plant pursuant to paragraph (a) of this section;

(2) Other plants in the unit must process Class I or Class II products, using 50 percent or more of the total Grade A fluid milk products received in bulk form at such plant or diverted therefrom by the plant operator in Class I or Class II products, and must be located in a pricing zone providing the same or lower Class I price than the price applicable at the distributing plant included in the unit pursuant to paragraph (e)(1) of this section; and

(3) A written request to form a unit must be filed by the handler with the market administrator prior to the first day of the month for which such status is desired to be effective. The unit shall continue from month to month thereafter without further notification. The handler shall notify the market administrator in writing prior to the first day of any month for which termination or any change of the unit is desired.

(f) The applicable shipping percentages of paragraphs (c) and (d) of this section may be increased or decreased by the market administrator if the market administrator finds that such adjustment is necessary to encourage needed shipments or to prevent uneconomic shipments. Before making such a finding, the market administrator shall investigate the need for adjustment either on the market administrator's own initiative or at the request of interested parties if the request is made in writing at least 15 days prior to the month for which the requested revision is desired effective. If the investigation shows that an adjustment of the shipping percentages might be appropriate, the market administrator shall issue a notice stating that an adjustment is being considered and invite data, views and arguments. Any decision to revise an applicable shipping percentage must be issued in writing at least one day before the effective date.

(g) The term pool plant shall not apply to the following plants:

(1) A producer-handler as defined under any Federal order;

(2) An exempt plant as defined in § 1000.8(e);

(3) A plant located within the marketing area and qualified pursuant to paragraph (a) of this section which meets the pooling requirements of another Federal order, and from which more than 50 percent of its route disposition has been in the other Federal order marketing area for 3 consecutive months;

(4) A plant located outside any Federal order marketing area and qualified pursuant to paragraph (a) of this section that meets the pooling requirements of another Federal order and has had greater route disposition in such other Federal order's marketing area for 3 consecutive months;

(5) A plant located in another Federal order marketing area and qualified pursuant to paragraph (a) of this section that meets the pooling requirements of such other Federal order and does not have a majority of its route distribution in this marketing area for 3 consecutive months or if the plant is required to be regulated under such other Federal order without regard to its route disposition in any other Federal order marketing area;

(6) A plant qualified pursuant to paragraph (c) of this section which also meets the pooling requirements of another Federal order and from which greater qualifying shipments are made to plants regulated under the other Federal order than are made to plants regulated under the order in this part, or the plant has automatic pooling status under the other Federal order; and

(7) That portion of a regulated plant designated as a nonpool plant that is physically separate and operated separately from the pool portion of such plant. The designation of a portion of a regulated plant as a nonpool plant must be requested in advance and in writing by the handler and must be approved by the market administrator.

(h) Any distributing plant, located within the marketing area as described on May 1, 2006, in §1131.2;

(1) From which there is route disposition and/or transfers of packaged fluid milk products in any non-Federally regulated marketing area(s) located within one or more States that require handlers to pay minimum prices for raw milk provided that 25 percent or more of the total quantity of fluid milk products physically received at such plant (excluding concentrated milk received from another plant by agreement for other than Class I use) is disposed of as route disposition and/or is transferred in the form of packaged fluid milk products to other plants. At least 25 percent of such route disposition and/or transfers, in aggregate, are in any non-Federally regulated marketing area(s) located within one or more States that require handlers to pay minimum prices for raw milk. Subject to the following exclusions:

(i) The plant is described in §1131.7(a), (b), or (e);

(ii) The plant is subject to the pricing provisions of a State-operated milk pricing plan which provides for the payment of minimum class prices for raw milk;

(iii) The plant is described in §1000.8(a) or (e); or

(iv) A producer-handler described in §1131.10 with less than three million pounds during the month of route dispositions and/or transfers of packaged fluid milk products to other plants.

(2) [Reserved]

[64 FR 48010, Sept. 1, 1999, as amended at 71 FR 25502, May 1, 2006; 71 FR 28249, May 16, 2006]

§1131.8 Nonpool plant.

See §1000.8.

§1131.9 Handler.

See §1000.9.

§1131.10 Producer-handler.

Producer-handler means a person who operates a dairy farm and a distributing plant from which there is route disposition in the marketing area, from which total route disposition and packaged sales of fluid milk products to other plants during the month does not exceed 3 million pounds, and who the market administrator has designated a producer-handler after determining that all of the requirements of this section have been met.

(a) *Requirements for designation.* Designation of any person as a producer-handler by the market administrator shall be contingent upon meeting the conditions set forth in paragraphs (a)(1) through (5) of this section. Following the cancellation of a previous producer-handler designation, a person seeking to have their producer-handler designation reinstated must demonstrate that these conditions have been met for the preceding month.

(1) The care and management of the dairy animals and the other resources and facilities designated in paragraph (b)(1) of this section necessary to

produce all Class I milk handled (excluding receipts from handlers fully regulated under any Federal order) are under the complete and exclusive control, ownership and management of the producer-handler and are operated as the producer-handler's own enterprise and its own risk.

(2) The plant operation designated in paragraph (b)(2) of this section at which the producer-handler processes and packages, and from which it distributes, its own milk production is under the complete and exclusive control, ownership and management of the producer-handler and is operated as the producer-handler's own enterprise and at its sole risk.

(3) The producer-handler neither receives at its designated milk production resources and facilities nor receives, handles, processes, or distributes at or through any of its designated milk handling, processing, or distributing resources and facilities other source milk products for reconstitution into fluid milk products or fluid milk products derived from any source other than:

(i) Its designated milk production resources and facilities (own farm production);

(ii) Pool handlers and plants regulated under any Federal order within the limitation specified in paragraph (c)(2) of this section; or

(iii) Nonfat milk solids which are used to fortify fluid milk products.

(4) The producer-handler is neither directly nor indirectly associated with the business control or management of, nor has a financial interest in, another handler's operation; nor is any other handler so associated with the producer-handler's operation.

(5) No milk produced by the herd(s) or on the farm(s) that supply milk to the producer-handler's plant operation is:

(i) Subject to inclusion and participation in a marketwide equalization pool under a milk classification and pricing program under the authority of a State government maintaining marketwide pooling of returns, or

(ii) Marketed in any part as Class I milk to the non-pool distributing plant of any other handler.

(6) The producer-handler does not distribute fluid milk products to a wholesale customer who is served by a plant described in §1131.7(a), (b), or (e), or a handler described in §1000.8(c) that supplied the same product in the same-sized package with a similar label to a wholesale customer during the month.

(b) *Designation of resources and facilities.* Designation of a person as a producer-handler shall include the determination of what shall constitute milk production, handling, processing, and distribution resources and facilities, all of which shall be considered an integrated operation, under the sole and exclusive ownership of the producer-handler.

(1) Milk production resources and facilities shall include all resources and facilities (milking herd(s), buildings housing such herd(s), and the land on which such buildings are located) used for the production of milk which are solely owned, operated, and which the producer-handler has designated as a source of milk supply for the producer-handler's plant operation. However, for purposes of this paragraph, any such milk production resources and facilities which do not constitute an actual or potential source of milk supply for the producer-handler's operation shall not be considered a part of the producer-handler's milk production resources and facilities.

(2) Milk handling, processing, and distribution resources and facilities shall include all resources and facilities (including store outlets) used for handling, processing, and distributing fluid milk products which are solely owned by, and directly operated or controlled by the producer-handler or in which the producer-handler in any way has an interest, including any contractual arrangement, or over which the producer-handler directly or indirectly exercises any degree of management control.

(3) All designations shall remain in effect until canceled pursuant to paragraph (c) of this section.

(c) *Cancellation.* The designation as a producer-handler shall be canceled upon determination by the market administrator that any of the requirements of paragraph (a)(1) through (5) of this section are not continuing to be met, or under any of the conditions described in paragraphs (c)(1), (2) or (3) of this section. Cancellation of a producer-handler's status pursuant to this paragraph shall be effective on the first day of the month following the month in which the requirements were not met or the conditions for cancellation occurred.

(1) Milk from the milk production resources and facilities of the producer-handler, designated in paragraph (b)(1) of this section, is delivered in the name of another person as producer milk to another handler.

(2) The producer-handler handles fluid milk products derived from sources other than the milk production facilities and resources designated in paragraph (b)(1) of this section, except that it may receive at its plant, or acquire for route disposition, fluid milk products from fully regulated plants and handlers under any Federal order if such receipts do not exceed 150,000 pounds monthly. This limitation shall not apply if the producer-handler's own-farm production is less than 150,000 pounds during the month.

(3) Milk from the milk production resources and facilities of the producer-handler is subject to inclusion and participation in a marketwide equalization pool under a milk classification and pricing plan operating under the authority of a State government.

(d) *Public announcement.* The market administrator shall publicly announce:

(1) The name, plant location(s), and farm location(s) of persons designated as producer-handlers;

(2) The names of those persons whose designations have been cancelled; and

(3) The effective dates of producer-handler status or loss of producer-handler status for each. Such announcements shall be controlling with respect to the accounting at plants of other handlers for fluid milk products received from any producer-handler.

(e) *Burden of establishing and maintaining producer-handler status.* The burden rests upon the handler who is designated as a producer-handler to establish through records required pursuant to §1000.27 that the requirements set forth in paragraph (a) of this section have been and are continuing to be met, and that the conditions set forth

in paragraph (c) of this section for cancellation of the designation do not exist.

(f) Any producer-handler with Class I route dispositions and/or transfers of packaged fluid milk products in the marketing area described in §1131.2 shall be subject to payments into the Order 1131 producer settlement fund on such dispositions pursuant to §1000.76(a) and payments into the Order 1131 administrative fund provided such dispositions are less than three million pounds in the current month and such producer-handler had total Class I route dispositions and/or transfers of packaged fluid milk products from own farm production of three million pounds or more the previous month. If the producer-handler has Class I route dispositions and/or transfers of packaged fluid milk products into the marketing area described in §1131.2 of three million pounds or more during the current month, such producer-handler shall be subject to the provisions described in §1131.7 or §1000.76(a).

[71 FR 9433, Feb. 24, 2006, as amended at 71 FR 25502, May 1, 2006; 75 FR 21161, Apr. 23, 2010]

§1131.11 [Reserved]

§1131.12 Producer.

(a) Except as provided in paragraph (b) of this section, *producer* means any person who produces milk approved by a duly constituted regulatory agency for fluid consumption as Grade A milk and whose milk (or components of milk) is:

(1) Received at a pool plant directly from the producer or diverted by the plant operator in accordance with §1131.13; or

(2) Received by a handler described in §1000.9(c).

(b) Producer shall not include:

(1) A producer-handler as defined in any Federal order;

(2) A dairy farmer whose milk is received at an exempt plant, excluding producer milk diverted to the exempt plant pursuant to §1131.13(d);

(3) A dairy farmer whose milk is received by diversion at a pool plant from a handler regulated under another Federal order if the other Federal order

designates the dairy farmer as a producer under that order and that milk is allocated by request to a utilization other than Class I;

(4) A dairy farmer whose milk is reported as diverted to a plant fully regulated under another Federal order with respect to that portion of the milk so diverted that is assigned to Class I under the provisions of such other order; and

(5) A dairy farmer whose milk is received at a pool plant if during the month milk from the same farm is received at a nonpool plant (except a nonpool plant that has no utilization of milk products in any class other than Class III or Class IV) other than as producer milk under the order in this part or some other Federal order. Such a dairy farmer shall be known as a *dairy farmer for other markets*.

§ 1131.13 Producer milk.

Producer milk means the skim milk (or the skim equivalent of components of skim milk) and butterfat in milk of a producer that is:

(a) Received by the operator of a pool plant directly from a producer or a handler described in § 1000.9(c). All milk received pursuant to this paragraph shall be priced at the location of the plant where it is first physically received;

(b) Received by a handler described in § 1000.9(c) in excess of the quantity delivered to pool plants;

(c) Diverted by a pool plant operator to another pool plant. Milk so diverted shall be priced at the location of the plant to which diverted; or

(d) Diverted by the operator of a pool plant or a cooperative association described in § 1000.9(c) to a nonpool plant, subject to the following conditions:

(1) Milk of a dairy farmer shall not be eligible for diversion unless at least one day's production of such dairy farmer is physically received at a pool plant during the month;

(2) The total quantity of milk diverted by a handler in any month shall not exceed 50 percent of the total producer milk caused by the handler to be received at pool plants and diverted;

(3) Diverted milk shall be priced at the location of the plant to which diverted;

(4) Any milk diverted in excess of the limits prescribed in paragraph (d)(2) of this section shall not be producer milk. If the diverting handler or cooperative association fails to designate the dairy farmers' deliveries that are not to be producer milk, no milk diverted by the handler or cooperative association during the month to a nonpool plant shall be producer milk. In the event some of the milk of any producer is determined not to be producer milk pursuant to this paragraph, other milk delivered by such producer as producer milk during the month will not be subject to § 1131.12(b)(5); and

(5) The delivery day requirement in paragraph (d)(1) of this section and diversion percentage in paragraph (d)(2) of this section may be increased or decreased by the market administrator if the market administrator finds that such revision is necessary to assure orderly marketing and efficient handling of milk in the marketing area. Before making such a finding, the market administrator shall investigate the need for the revision either on the market administrator's own initiative or at the request of interested persons if the request is made in writing at least 15 days prior to the month for which the requested revision is desired effective. If the investigation shows that a revision might be appropriate, the market administrator shall issue a notice stating that the revision is being considered and inviting written data, views, and arguments. Any decision to revise the delivery day requirement or the diversion percentage must be issued in writing at least one day before the effective date.

(e) Producer milk shall not include milk of a producer that is subject to a marketwide equalization pool under a milk classification and pricing plan under the authority of a State government.

[64 FR 48010, Sept. 1, 1999, as amended at 70 FR 9848, Mar. 1, 2005]

§ 1131.14 Other source milk.

See § 1000.14.

§ 1131.15 Fluid milk product.

See § 1000.15.

§ 1131.16 Fluid cream product.

See § 1000.16.

§ 1131.17 [Reserved]

§ 1131.18 Cooperative association.

See § 1000.18.

§ 1131.19 Commercial food processing establishment.

See § 1000.19.

HANDLER REPORTS

§ 1131.30 Reports of receipts and utilization.

Each handler shall report monthly so that the market administrator's office receives the report on or before the 7th day after the end of the month, in the detail and on the forms prescribed by the market administrator, as follows:

(a) With respect to each of its pool plants, the quantities of skim milk and butterfat contained in or represented by:

(1) Receipts of producer milk, including producer milk diverted by the reporting handler, from sources other than handlers described in § 1000.9(c);

(2) Receipts of milk from handlers described in § 1000.9(c);

(3) Receipts of fluid milk products and bulk fluid cream products from other pool plants;

(4) Receipts of other source milk;

(5) Inventories at the beginning and end of the month of fluid milk products and bulk fluid cream products; and

(6) The utilization or disposition of all milk and milk products required to be reported pursuant to this paragraph.

(b) Each handler operating a partially regulated distributing plant shall report with respect to such plant in the same manner as prescribed for reports required by paragraph (a) of this section. Receipts of milk that would have been producer milk if the plant had been fully regulated shall be reported in lieu of producer milk. Such report shall show also the quantity of any reconstituted skim milk in route disposition in the marketing area.

(c) Each handler described in § 1000.9(c) shall report:

(1) The quantities of all skim milk and butterfat contained in receipts of milk from producers; and

(2) The utilization or disposition of all such receipts.

(d) Each handler described in § 1131.10 shall report:

(1) The pounds of milk received from each of the handler's own-farm production units, showing separately the production of each farm unit and the number of dairy cows in production at each farm unit;

(2) Fluid milk products and bulk fluid cream products received at its plant or acquired for route disposition from pool plants, other order plants, and handlers described in § 1000.9(c);

(3) Receipts of other source milk not reported pursuant to paragraph (d)(2) of this section;

(4) Inventories at the beginning and end of the month of fluid milk products and fluid cream products; and

(5) The utilization or disposition of all milk and milk products required to be reported pursuant to this paragraph.

(e) Each handler not specified in paragraphs (a) through (d) of this section shall report with respect to its receipts and utilization of milk and milk products in such manner as the market administrator may prescribe.

§ 1131.31 Payroll reports.

(a) On or before the 20th day after the end of each month, each handler that operates a pool plant pursuant to § 1131.7 and each handler described in § 1000.9(c) shall report to the market administrator its producer payroll for such month, in the detail prescribed by the market administrator, showing for each producer:

(1) The month;

(2) The producer's name and address;

(3) The daily and total pounds of milk received from the producer;

(4) The total butterfat content of such milk; and

(5) The price per hundredweight, the gross amount due, the amount and nature of any deductions, and the net amount paid.

(b) Each handler operating a partially regulated distributing plant who elects to make payment pursuant to § 1000.76(b) shall report for each dairy farmer who would have been a producer if the plant had been fully regulated in the same manner as prescribed for reports required by paragraph (a) of this section.

§ 1131.32 Other reports.

In addition to the reports required pursuant to § 1131.30 and § 1131.31, each handler shall report any information the market administrator deems necessary to verify or establish each handler's obligation under the order.

CLASSIFICATION OF MILK

§ 1131.40 Classes of utilization.

See § 1000.40.

§ 1131.41 [Reserved]

§ 1131.42 Classification of transfers and diversions.

See § 1000.42.

§ 1131.43 General classification rules.

See § 1000.43.

§ 1131.44 Classification of producer milk.

See § 1000.44.

§ 1131.45 Market administrator's reports and announcements concerning classification.

See § 1000.45.

CLASS PRICES

§ 1131.50 Class prices, component prices, and advanced pricing factors.

See § 1000.50.

§ 1131.51 Class I differential and price.

The Class I differential shall be the differential established for Maricopa County, Arizona, which is reported in § 1000.52. The Class I price shall be the price computed pursuant to § 1000.50(a) for Maricopa County, Arizona.

§ 1131.52 Adjusted Class I differentials.

See § 1000.52.

§ 1131.53 Announcement of class prices, component prices, and advanced pricing factors.

See § 1000.53.

§ 1131.54 Equivalent price.

See § 1000.54.

UNIFORM PRICES

§ 1131.60 Handler's value of milk.

For the purpose of computing a handler's obligation for producer milk, the market administrator shall determine for each month the value of milk of each handler with respect to each of the handler's pool plants and of each handler described in § 1000.9(c) with respect to milk that was not received at a pool plant by adding the amounts computed in paragraphs (a) through (e) of this section and subtracting from that total amount the value computed in paragraph (f) of this section. Receipts of nonfluid milk products that are distributed as labeled reconstituted milk for which payments are made to the producer-settlement fund of another Federal order under § 1000.76(a)(4) or (d) shall be excluded from pricing under this section.

(a) Multiply the pounds of skim milk and butterfat in producer milk that were classified in each class pursuant to § 1000.44(c) by the applicable skim milk and butterfat prices, and add the resulting amounts;

(b) Multiply the pounds of skim milk and butterfat overage assigned to each class pursuant to § 1000.44(a)(11) and the corresponding steps of § 1000.44(b) by the respective skim milk and butterfat prices applicable at the location of the pool plant;

(c) Multiply the difference between the current month's Class I, II, or III price, as the case may be, and the Class IV price for the preceding month by the hundredweight of skim milk and butterfat subtracted from Class I, II, or III, respectively, pursuant to § 1000.44(a)(7) and the corresponding step of § 1000.44(b);

(d) Multiply the difference between the Class I price applicable at the location of the pool plant and the Class IV price by the hundredweight of skim milk and butterfat assigned to Class I pursuant to § 1000.43(d) and the hundredweight of skim milk and butterfat subtracted from Class I pursuant to § 1000.44(a)(3)(i) through (vi) and the corresponding step of § 1000.44(b), excluding receipts of bulk fluid cream products from plants regulated under other Federal orders and bulk concentrated fluid milk products from pool plants, plants regulated under other Federal orders, and unregulated supply plants;

(e) Multiply the Class I skim milk and Class I butterfat prices applicable at the location of the nearest unregulated supply plants from which an equivalent volume was received by the pounds of skim milk and butterfat in receipts of concentrated fluid milk products assigned to Class I pursuant to §1000.43(d) and §1000.44(a)(3)(i) and the corresponding step of §1000.44(b) and the pounds of skim milk and butterfat subtracted from Class I pursuant to §1000.44(a)(8) and the corresponding step of §1000.44(b), excluding such skim milk and butterfat in receipts of fluid milk products from an unregulated supply plant to the extent that an equivalent amount of skim milk or butterfat disposed of to such plant by handlers fully regulated under any Federal milk order is classified and priced as Class I milk and is not used as an offset for any other payment obligation under any order.

(f) For reconstituted milk made from receipts of nonfluid milk products, multiply $1.00 (but not more than the difference between the Class I price applicable at the location of the pool plant and the Class IV price) by the hundredweight of skim milk and butterfat contained in receipts of nonfluid milk products that are allocated to Class I use pursuant to §1000.43(d).

[64 FR 48010, Sept. 1, 1999, as amended at 65 FR 82841, Dec. 28, 2000]

§1131.61 Computation of uniform prices.

On or before the 11th day of each month, the market administrator shall compute a uniform butterfat price, a uniform skim milk price, and a uniform price for producer milk receipts reported for the prior month. The report of any handler who has not made payments required pursuant to §1131.71 for the preceding month shall not be included in the computation of these prices, and such handler's report shall not be included in the computation for succeeding months until the handler has made full payment of outstanding monthly obligations.

(a) *Uniform butterfat price.* The uniform butterfat price per pound, round-

ed to the nearest one-hundredth cent, shall be computed by:

(1) Multiplying the pounds of butterfat in producer milk allocated to each class pursuant to §1000.44(b) by the respective class butterfat prices;

(2) Adding the butterfat value calculated in §1131.60(e) for other source milk allocated to Class I pursuant to §1000.43(d) and the steps of §1000.44(b) that correspond to §1000.44(a)(3)(i) and §1000.44(a)(8) by the Class I price; and

(3) Dividing the sum of paragraphs (a)(1) and (a)(2) of this section by the sum of the pounds of butterfat in producer milk and other source milk used to calculate the values in paragraphs (a)(1) and (a)(2) of this section.

(b) *Uniform skim milk price.* The uniform skim milk price per hundredweight, rounded to the nearest cent, shall be computed as follows:

(1) Combine into one total the values computed pursuant to §1131.60 for all handlers;

(2) Add an amount equal to the minus location adjustments and subtract an amount equal to the plus location adjustments computed pursuant to §1131.75;

(3) Add an amount equal to not less than one-half of the unobligated balance in the producer-settlement fund;

(4) Subtract the value of the total pounds of butterfat for all handlers. The butterfat value shall be computed by multiplying the sum of the pounds of butterfat in producer milk and other source milk used to calculate the values in paragraphs (a)(1) and (a)(2) of this section by the butterfat price computed in paragraph (a) of this section;

(5) Divide the resulting amount by the sum of the following for all handlers included in these computations:

(i) The total skim pounds of producer milk; and

(ii) The total skim pounds for which a value is computed pursuant to §1131.60(e); and

(6) Subtract not less than 4 cents and not more than 5 cents.

(c) *Uniform price.* The uniform price per hundredweight, rounded to the nearest cent, shall be the sum of the following:

(1) Multiply the uniform butterfat price for the month pursuant to paragraph (a) of this section times 3.5 pounds of butterfat; and

(2) Multiply the uniform skim milk price for the month pursuant to paragraph (b) of this section times .965.

[64 FR 48010, Sept. 1, 1999, as amended at 65 FR 82841, Dec. 28, 2000]

§ 1131.62 Announcement of uniform prices.

On or before the 11th day after the end of the month, the market administrator shall announce the uniform prices for the month computed pursuant to § 1131.61.

PAYMENTS FOR MILK

§ 1131.70 Producer-settlement fund.

See § 1000.70.

§ 1131.71 Payments to the producer-settlement fund.

Each handler shall make payment to the producer-settlement fund in a manner that provides receipt of the funds by the market administrator no later than the 13th day after the end of the month (except as provided in § 1000.90). Payments due the market administrator shall be deemed not to have been made until the money owed has been received at the market administrator's office, or deposited into the market administrator's bank account. Payment shall be the amount, if any, by which the amount specified in paragraph (a) of this section exceeds the amount specified in paragraph (b) of this section:

(a) The total value of milk to the handler for the month as determined pursuant to § 1131.60.

(b) The sum of:

(1) The value at the uniform prices for skim milk and butterfat, adjusted for plant location, of the handler's receipts of producer milk; and

(2) The value at the uniform price as adjusted pursuant to § 1131.75 applicable at the location of the plant from which received of other source milk for which a value is computed pursuant to § 1131.60(e).

§ 1131.72 Payments from the producer-settlement fund.

No later than the 14th day after the end of each month (except as provided in § 1000.90), the market administrator shall pay to each handler the amount, if any, by which the amount computed pursuant to § 1131.71(b) exceeds the amount computed pursuant to § 1131.71(a). If, at such time, the balance in the producer-settlement fund is insufficient to make all payments pursuant to this section, the market administrator shall reduce uniformly such payments and shall complete the payments as soon as the funds are available.

§ 1131.73 Payments to producers and to cooperative associations.

(a) Except as provided in paragraphs (b) and (c) of this section, each handler shall make payment to each producer from whom milk is received during the month as follows:

(1) *Partial Payment.* For each producer who has not discontinued shipments as of the 25th day of the month, payment shall be made so that it is received by the producer on or before the 27th day of each month (except as provided in § 1000.90) for milk received from such producer during the first 15 days of the month at not less than 1.3 times the lowest class price for the preceding month less proper deductions authorized in writing by the producer.

(2) *Final payment.* For milk received during the month, a payment computed as follows shall be made so that it is received by each producer one day after the payment date required in § 1131.72:

(i) Multiply the hundredweight of producer skim milk received times the uniform skim milk price for the month;

(ii) Multiply the pounds of producer butterfat received times the uniform butterfat price for the month;

(iii) Multiply the hundredweight of producer milk received times the plant location adjustment pursuant to § 1131.75; and

(iv) Add the amounts computed in paragraph (a)(2)(i), (ii), and (iii) of this section, and from that sum:

(A) Subtract the partial payment made pursuant to paragraph (a)(1) of this section;

(B) Subtract the deduction for marketing services pursuant to §1000.86;

(C) Add or subtract for errors made in previous payments to the producer, subject to approval by the market administrator; and

(D) Subtract proper deductions authorized in writing by the producer.

(b) Two days prior to the dates on which partial and final payments are due pursuant to paragraph (a) of this section, each handler shall pay a cooperative association for milk received as follows:

(1) *Partial payment to a cooperative association for bulk milk received directly from producers' farms.* For bulk milk (including the milk of producers who are not members of such association and who the market administrator determines have authorized the cooperative association to collect payment for their milk) received during the first 15 days of the month from a cooperative association in any capacity except as the operator of a pool plant, the payment shall be an amount not less than 1.3 times the lowest class price for the preceding month multiplied by the hundredweight of milk.

(2) *Partial payment to a cooperative association for milk transferred from its pool plant.* For bulk fluid milk products and bulk fluid cream products received during the first 15 days of the month from a cooperative association in its capacity as the operator of a pool plant, the partial payment shall be at the pool plant operator's estimated use value of the milk using the most recent class prices available for skim milk and butterfat at the receiving plant's location.

(3) *Final payment to a cooperative association for milk transferred from its pool plant.* For bulk fluid milk products and bulk fluid cream products received during the month from a cooperative association in its capacity as the operator of a pool plant, the final payment shall be the classified value of such milk as determined by multiplying the pounds of skim milk and butterfat assigned to each class pursuant to §1000.44 by the class prices for the month at the receiving plant's location, and subtracting from this sum the partial payment made pursuant to paragraph (b)(2) of this section.

(4) *Final payment to a cooperative association for bulk milk received directly from producers' farms.* For bulk milk received from a cooperative association during the month, including the milk of producers who are not members of such association and who the market administrator determines have authorized the cooperative association to collect payment for their milk, the final payment for such milk shall be an amount equal to the sum of the individual payments otherwise payable for such milk pursuant to paragraph (a)(2) of this section.

(c) If a handler has not received full payment from the market administrator pursuant to §1131.72 by the payment date specified in paragraph (a) or (b) of this section, the handler may reduce pro rata his payments pursuant to such paragraphs, but by not more than the amount of such underpayment. Payments to producers shall be completed on the next scheduled payment date after receipt of the balance due from the market administrator.

(d) If a handler claims that a required payment to a producer cannot be made because the producer is deceased or cannot be located, or because the cooperative association or its lawful successor or assignee is no longer in existence, the payment shall be made to the producer-settlement fund. In the event the handler subsequently locates and pays the producer or a lawful claimant, or in the event that the handler no longer exists and a lawful claim is later established, the market administrator shall make the required payment from the producer-settlement fund to the handler or the lawful claimant, as the case may be.

(e) In making payments to producers pursuant to this section, each pool plant operator shall furnish each producer, except a producer whose milk was received from a cooperative association described in §1000.9(a) or (c), a supporting statement in such form that it may be retained by the recipient which shall show:

(1) The month, and identity of the producer;

(2) The daily and total pounds and the total pounds of butterfat content of producer milk;

(3) The minimum rate at which payment to the producer is required pursuant to the order in this part;

(4) The rate used in making payments if the rate is other than the applicable minimum rate;

(5) The amount, rate per hundredweight, and nature of each deduction claimed by the handler; and

(6) The net amount of payment to the producer or cooperative association.

[64 FR 48010, Sept. 1, 1999, as amended at 65 FR 32010, May 22, 2000]

§ 1131.74 [Reserved]

§ 1131.75 Plant location adjustments for producers and nonpool milk.

For purposes of making payments for producer milk and nonpool milk, a plant location adjustment shall be determined by subtracting the Class I price specified in § 1131.51 from the Class I price at the plant's location. The difference, plus or minus as the case may be, shall be used to adjust the payments required pursuant to §§ 1131.73 and 1000.76.

§ 1131.76 Payments by handler operating a partially regulated distributing plant.

See § 1000.76.

§ 1131.77 Adjustment of accounts.

See § 1000.77.

§ 1131.78 Charges on overdue accounts.

See § 1000.78.

ADMINISTRATIVE ASSESSMENT AND MARKETING SERVICE DEDUCTION

§ 1131.85 Assessment for order administration.

See § 1000.85.

§ 1131.86 Deduction for marketing services.

See § 1000.86.

PARTS 1132–1134 [RESERVED]

PART 1135—MILK IN THE WESTERN MARKETING AREA

AUTHORITY: 7 U.S.C. 601–674, and 7253.

Subpart—Order Regulating Handling

§ 1135.1 General provisions.

The terms, definitions, and provisions in part 1000 of this chapter apply to this part 1135. In this part 1135, all references to sections in part 1000 refer to part 1000 of this chapter.

[64 FR 48015, Sept. 1, 1999]

PARTS 1136–1140 [RESERVED]

PART 1145—DAIRY FORWARD PRICING PROGRAM

Subpart A—Definitions

Sec.
1145.1 Definitions.

Subpart B—Program Rules

1145.2 Program.

Subpart C—Enforcement

1145.3 Enforcement.

AUTHORITY: 7 U.S.C. 8772.

SOURCE: 73 FR 64871, Oct. 31, 2008, unless otherwise noted.

Subpart A—Definitions

§ 1145.1 Definitions.

(a) *Program* means the dairy forward pricing program as established by Section 1502 of Public Law No. 110–246.

(b) *Eligible milk* means the quantity of milk equal to the contracting handler's Class II, III and IV utilization of producer milk, in product pounds, during the month, combining all plants of a single handler regulated under the same Federal milk marketing order.

(c) *Forward contract* means an agreement covering the terms and conditions for the sale of Class II, III or IV milk from a producer defined in 7 CFR 1001.12, 1005.12, 1006.12, 1007.12, 1030.12, 1032.12, 1033.12, 1124.12, 1126.12, 1131.12 or a cooperative association of producers defined in 7 CFR 1000.18, and a handler defined in 7 CFR 1000.9.

(d) *Contract milk* means the producer milk regulated under a Federal milk marketing order covered by a forward contract.

(e) *Disclosure statement* means the following statement which must be signed by each producer or cooperative representative entering into a forward contract with a handler before the Federal milk marketing order administrator will recognize the contract as satisfying the provisions of this program.

Attachment to §1145.1, paragraph (e):

DISCLOSURE STATEMENT

I am voluntarily entering into a forward contract with [insert handler's name]. I have been given a copy of the contract. By signing this form, I understand that I am forfeiting my right to receive the Federal milk marketing order's minimum prices for that portion of the milk which is under contract for the duration of the contract. I also understand that this contract milk will be priced in accordance with the terms and conditions of the contract.

Printed Name: _____
Signature: _____
Date: _____
Address: _____
Producer Number: _____

(f) *Other definitions.* The definition of any term in parts 1000–1131 of this chapter apply to, and are hereby made a part of this part, as appropriate.

Subpart B—Program Rules

§1145.2 Program.

(a) Any handler defined in 7 CFR 1000.9 may enter into forward contracts with producers or cooperative associations of producers for the handler's eligible volume of milk. Milk under forward contract in compliance with the provisions of this part will be exempt from the minimum payment provisions that would apply to such milk pursuant to 7 CFR 1001.73, 1005.73, 1006.73, 1007.73, 1030.73, 1032.73, 1033.73, 1051.73, 1124.73, 1126.73 and 1131.73 for the period of time covered by the contract.

(b) No forward price contract may be entered into under the program after September 30, 2023, and no forward contract entered into under the program may extend beyond September 30, 2026.

(c) Forward contracts must be signed and dated by the contracting handler and producer (or cooperative association) prior to the 1st day of the 1st month for which they are to be effective and must be received by the Federal milk market administrator by the 15th day of that month. The disclosure statement must be signed on the same date as the contract by each producer entering into a forward contract, and this signed disclosure statement must be attached to or otherwise included in each contract submitted to the market administrator.

(d) In the event that a handler's contract milk exceeds the handler's eligible milk for any month in which the specified contract price(s) are below the order's minimum prices, the handler must designate which producer milk shall not be contract milk. If the handler does not designate the suppliers of the over-contracted milk, the market administrator shall prorate the over-contracted milk to each producer and cooperative association having a forward contract with the handler.

(e) Payments for milk covered by a forward contract must be made on or before the dates applicable to payments for milk that are not under forward contract under the respective Federal milk marketing order.

(f) Nothing in this part shall impede the contractual arrangements that exist between a cooperative association and its members.

[73 FR 64871, Oct. 31, 2008, as amended at 79 FR 15636, Mar. 21, 2014; 84 FR 6962, Mar. 1, 2019]

Subpart C—Enforcement

§1145.3 Enforcement.

A handler may not require participation in a forward pricing contract as a condition of the handler receiving milk from a producer or cooperative association of producers. USDA will investigate all complaints made by producers or cooperative associations alleging coercion by handlers to enter into forward contracts and based on the results of the investigation will take appropriate action.

PART 1146—MILK DONATION REIMBURSEMENT PROGRAM

Subpart A—General Provisions

Sec.

Subpart B—Program Participation

Subpart C—Administrative Provisions

AUTHORITY: Sec. 1431, Pub. L. 113-79, 128 Stat. 695, as amended.

SOURCE: 84 FR 46658, Sept. 5, 2019, unless otherwise noted.

Subpart A—General Provisions

§ 1146.1   Definitions.

AMS means the Agricultural Marketing Service of the United States Department of Agriculture.

Eligible dairy organization means a dairy farmer cooperative or a dairy processor that

(1) Is regulated under a Federal milk marketing order (FMMO);

(2) Accounts to the FMMO marketwide pool; and

(3) Incurs qualified expenses described in § 1146.108.

Eligible distributor means a public or private non-profit organization that distributes donated eligible milk.

Eligible milk means Class I fluid milk products produced and processed in the United States that meet the specifications referenced in § 1146.3.

Eligible partnership means a partnership between an eligible dairy organization and an eligible distributor.

Fiscal year means the twelve-month period beginning October 1 of any year and ending September 30 of the following year.

Participating partnership means an eligible partnership for which AMS has approved a Milk Donation and Distribution Plan (Plan) for eligible milk under § 1146.104.

Program means the Milk Donation Reimbursement Program established in this part.

Secretary means the Secretary of the United States Department of Agriculture or a representative authorized to act in the Secretary's stead.

§ 1146.3   Commodity specifications.

(a) Eligible milk donations must meet the commodity specifications pursuant to (b) in effect on the date the milk products are shipped from the plant.

(b) AMS shall maintain on its website current commodity specifications for fluid milk products eligible for donation and reimbursement under the Milk Donation Reimbursement Program.

Subpart B—Program Participation

§ 1146.100   Program eligibility.

An eligible dairy organization must be a member of a participating partnership pursuant to § 1146.1 to be eligible to receive reimbursements for qualified expenses related to voluntary fluid milk donations, subject to the requirements and limitations specified in §§ 1146.102 and 1146.104.

§ 1146.102   Milk donation and distribution plans.

New and continuing program participants must submit completed Milk Donation and Distribution Plans to AMS in the form and manner established by AMS prior to the published deadline to be eligible for program consideration. The completed Milk Donation and Distribution Plans must –

(a) Include the physical location(s) of the eligible dairy organization's processing plant(s) and the eligible distributor's distribution site(s);

(b) Include an affirmation signed by both eligible partners regarding the partnership's ability to supply, transport, store, and distribute donated milk products consistent with the commodity specifications under § 1146.3;

(c) Include an estimate of the quantity of eligible milk that the eligible dairy organization plans to donate each year, based on –

(1) Preplanned donations and

(2) Contingency plans to address unanticipated donations; and

(d) Describe the rate at which the eligible dairy organization will be reimbursed, not to exceed 100 percent of qualified expenses pursuant to §1146.108.

§1146.104 Review and approval.

(a) *Program application and review.* Within 45 days of the announced application deadline, AMS will review all timely submitted applications and notify applicants regarding approval or disapproval for program participation during the applicable fiscal year. AMS's review will include the following considerations:

(1) Total annual funds available for program administration, including an appropriate reserve to cover costs related to increases in milk prices and emergencies including, but not limited to, natural disasters;

(2) The feasibility of the Milk Donation and Distribution Plan;

(3) The extent to which the Milk Donation and Distribution Plan –

(i) Promotes the donation of eligible milk,

(ii) Provides nutrition assistance to individuals in low-income groups, and

(iii) Reduces food waste; and

(4) The amount of funding and in-kind contributions the eligible dairy organization plans to provide to the eligible distributor in addition to the donations for which it will seek reimbursements.

(b) *Continued program participation.* Within 45 days of the announced application deadline, AMS will review and notify applicants regarding approval or disapproval of all timely submitted requests for continued program participation. AMS's review of requests for continued program participation will be based on consideration of the factors in paragraphs (a) and (b)(1) through (3) of this section:

(1) Eligible partnerships requesting continued program participation for a subsequent fiscal year can include information about the extent to which they provided funding and in-kind contributions in addition to eligible milk donations for which they were reimbursed through the program for the previous fiscal year.

(2) If there are no changes to the eligible partnership's approved Milk Donation and Distribution Plan from the previous fiscal year, the eligible partnership must request that AMS consider the partnership's previously approved Plan and provide the additional information described in paragraph (b)(1) of this sectoin, if applicable.

(3) If there are changes to the eligible partnership's approved Milk Donation and Distribution Plan from the previous fiscal year, the eligible partnership must submit a new Plan as described in paragraph (a) and provide the additional information described in paragraph (b)(1) of this section, if applicable.

(c) *Plan approval.* Subject to the provisions in paragraph (a) of this section, AMS will determine whether to approve new and continuing Milk Donation and Distribution Plans for all or a proportion of each Plan's proposed donations and reimbursements. For each approved Plan, AMS will determine:

(1) A reimbursement rate applicable to each claim for reimbursement during the fiscal year, and

(2) A total dollar amount available for reimbursement during the fiscal year.

(d) *Adjustments.* AMS will review the activity of approved Milk Donation and Distribution Plans during the fiscal year to determine whether adjustments should be made to the reimbursement amounts approved under paragraph (c) of this section.

(1) Determinations about adjustments will be based on –

(i) The participating partnership's performance,

(ii) Availability of program funds, and

(iii) Demand for eligible milk donations.

(2) AMS will provide 30 days' notice to participating partnerships prior to adjusting reimbursement amounts in their respective approved Milk Donation and Distribution Plans.

(e) *Request for increase.* Eligible partnerships with approved Milk Donation and Distribution Plans during any fiscal year may request an increase in the amount of reimbursement approved under paragraph (c) of this section based on changes in conditions.

(1) Requests for an increase must be submitted to AMS in the manner and form established by AMS, and must –

(i) Describe the change in conditions that would warrant an increase in reimbursement,

(ii) Indicate whether the requested increase is intended to be a long-term revision to the eligible partnership's approved Milk Donation and Distribution Plan or a short-term increase to respond to temporary conditions, and

(iii) Specify the amount of increased reimbursement requested.

(2) Within 30 days of receipt, AMS will review the request for an increase and will notify the requester regarding approval or disapproval of the request. AMS's determination about whether such an increase is feasible will be based on its evaluation of the factors described in paragraph (e)(1) of this section and the availability of funds.

(3) Based on the change in conditions identified by the requester, AMS will determine whether to provide interim approval of an increase requested under paragraph (e)(1) of this section and an incremental increase to the amount of reimbursement approved under paragraph (c) of this section prior to making a final determination regarding approval of the requested increase.

§ 1146.106 Reimbursement claims.

(a) In order for the eligible dairy organization partner to receive reimbursements for qualified expenses pursuant to § 1146.108, the participating partnership must submit a report and appropriate supporting documentation to AMS.

(1) *For each month* of the fiscal year pertaining to an approved Milk Donation and Distribution Plan (including the months prior to AMS's review and approval of the Plan), the report must include:

(i) The amount of eligible milk donated to the eligible distributor;

(ii) The location of the plant where the donated milk was processed;

(iii) The date the donated milk was shipped from the plant where the milk was processed;

(iv) The date the donated milk was received by the eligible distributor; and

(v) The applicable announced Federal milk marketing order prices for the month the milk was pooled:

(A) The Class I price at the plant location where the milk was processed; and

(B) The lowest classified price (either Class III or Class IV).

(2) Appropriate documentation to support the report required in paragraph (a)(1) of this section may include, but is not limited to, copies of processing records, shipping records, bills of lading, warehouse receipts, distribution records, or other documents that demonstrate the reported amount of eligible milk was processed, donated, and distributed in accordance with the approved Milk Donation and Distribution Plan and as reported in the eligible dairy organization's report.

(b) Reimbursement requests may be submitted to AMS at any time during the fiscal year and for up to 90 days after the close of the fiscal year.

(c) AMS will review and process reimbursement requests on a quarterly basis, including those submitted by the last day of the month following the end of each quarter of the fiscal year.

(d) Incomplete reimbursement requests will be returned to the submitter for revision or completion and resubmission as necessary.

§ 1146.108 Reimbursement calculation.

(a) For each reimbursement claim submitted by a participating partnership, the amount of reimbursement under § 1146.106 shall be the product of:

(1) The quantity of eligible milk donated by the eligible dairy organization to the eligible distributor member of the participating partnership;

(2) The rate described in the approved Milk Donation and Distribution Plan under § 1146.102(d); and

(3) The difference between the FMMO Class I price at the plant location and the lowest classified price (either Class III or Class IV), for the month in which the donation was pooled on a Federal Milk Marketing Order.

(b) Expenses eligible for reimbursement under § 1146.106 shall not exceed the value that an eligible dairy organization incurred by accounting to the Federal milk marketing order pool at the difference between the announced Class I milk price at the location of the plant where the milk was processed and the lower of the Class III or Class IV milk price for the applicable month.

(c) Claim reimbursements are subject to the limitations specified in paragraph (b) of this section.

(d) Total plan reimbursements are subject to the limitations specified in § 1146.104(c)(2).

Subpart C—Administrative Provisions

§ 1146.200 Opportunities to participate.

(a) AMS will announce opportunities to participate in the Milk Donation Reimbursement Program and the amount of program funding available for each fiscal year on the AMS website. The announcements will include invitations for interested parties to submit new or revised Milk Donation and Distribution Plans and will specify the manner and form in which program applications should be submitted.

(b) If, after making approval determinations for the fiscal year about each submitted program application, AMS determines that additional reimbursement funds are available, AMS will publish an announcement to that effect and invite further requests for Plan approvals pursuant to § 1146.104(a) through (c) or for increases in reimbursement amounts pursuant § 1146.104(e).

§ 1146.202 Rollover of fiscal year funds.

If reimbursement monies remain after all fiscal year reimbursement claims have been approved and distributed, the remaining monies will be available to fund reimbursement claims in subsequent fiscal years.

§ 1146.204 Prohibition on resale of products.

(a) *Prohibition in general.* An eligible distributor that receives eligible milk products donated under the Milk Dona-tion Reimbursement Program may not sell the donated milk products back into commercial markets.

(b) *Prohibition on future participation.* An eligible distributor that AMS determines has violated the prohibition in paragraph (a) of this section shall not be eligible for any future participation in the Milk Donation Reimbursement Program.

§ 1146.206 Enforcement.

AMS will verify the donated milk for which reimbursement is sought was pooled on a FMMO. AMS will also conduct spot checks, reviews, and audits of the reports and documentation submitted pursuant to § 1146.106(a) to verify their accuracy and to ensure the integrity of the Milk Donation Reimbursement Program.

§ 1146.208 Confidentiality.

AMS will collect only that information deemed necessary to administer the Milk Donation Reimbursement Program and will use the information only for that purpose. AMS will keep all proprietary business information collected under the program confidential.

§ 1146.210 Milk for other programs.

Milk sold or donated under other commodity or food assistance programs administered by the United States Department of Agriculture is not eligible for reimbursement under the Milk Donation Reimbursement Program in this part.

PART 1150—DAIRY PROMOTION PROGRAM

Subpart—Dairy Promotion and Research Order

DEFINITIONS

Subpart—Procedure for Certification of Milk Producer Organizations

AUTHORITY: 7 U.S.C. 4501–4514 and 7 U.S.C. 7401

SOURCE: 49 FR 11816, Mar. 28, 1984, unless otherwise noted.

Subpart—Dairy Promotion and Research Order

DEFINITIONS

§ 1150.101   Act.

Act means Title I, Subtitle B, of the Dairy and Tobacco Adjustment Act of 1983, Pub. L. 98–180, 97 Stat. 1128, as approved November 29, 1983, and any amendments thereto.

§ 1150.102   Department.

Department means the United States Department of Agriculture.

§ 1150.103   Secretary.

Secretary means the Secretary of Agriculture of the United States or any other officer or employee of the Department to whom authority has heretofore been delegated, or to whom authority may hereafter be delegated, to act in the Secretary's stead.

§ 1150.104   Board.

Board means the National Dairy Promotion and Research Board established pursuant to § 1150.131.

§ 1150.105   Person.

Person means any individual, group of individuals, partnership, corporation, association, cooperative or other entity.

§ 1150.106   United States.

United States means all of the States, the District of Columbia, and the Commonwealth of Puerto Rico.

[76 FR 14787, Mar. 18, 2011]

§ 1150.107   Fiscal period.

Fiscal period means the calendar year or such other annual period as the Board may determine.

§ 1150.108 Eligible organization.

Eligible organization means any organization which has been certified by the Secretary pursuant to §§ 1150.270 through 1150.278 of this part.

§ 1150.109 Qualified program.

Qualified program means any dairy product promotion, research or nutrition education program which is certified as a qualified program pursuant to § 1150.153.

[76 FR 14788, Mar. 18, 2011]

§ 1150.110 Producer.

Producer means any person engaged in the production of milk for commercial use.

§ 1150.111 Milk.

Milk means any class of cow's milk.

[76 FR 14788, Mar. 18, 2011]

§ 1150.112 Dairy products.

Dairy products means products manufactured for human consumption which are derived from the processing of milk, and includes fluid milk products.

§ 1150.113 Fluid milk products.

Fluid milk products means those milk products normally consumed in liquid form as a beverage.

§ 1150.114 Promotion.

Promotion means actions such as paid advertising, sales promotion, and publicity to advance the image and sales of, and demand for, dairy products generally.

§ 1150.115 Research.

Research means studies testing the effectiveness of market development and promotion efforts, studies relating to the nutritional value of milk and dairy products, and other related efforts to expand demand for dairy products.

§ 1150.116 Nutrition education.

Nutrition education means those activities intended to broaden the understanding of sound nutritional principles, including the role of milk and dairy products in a balanced diet.

§ 1150.117 Plans and projects.

Plans and projects means promotion, research and nutrition education plans, studies or projects pursuant to §§ 1150.139, 1150.140 and 1150.161.

§ 1150.118 Marketing.

Marketing means the sale or other disposition in commerce of dairy products.

§ 1150.119 Cooperative association.

Cooperative association means any cooperative marketing association of producers which is organized under the provisions of the Act of Congress of February 18, 1922, as amended, known as the "Capper-Volstead Act".

§ 1150.120 Imported dairy product.

Imported dairy product means any product that is imported into the United States under any of the Harmonized Tariff Schedule (HTS) classification numbers listed in § 1150.152(b)(1).

[76 FR 14788, Mar. 18, 2011]

§ 1150.121 Importer.

Importer means a person that imports imported dairy products into the United States as a principal or as an agent, broker, or consignee of any person who produces or handles dairy products outside of the United States for sale in the United States, and who is listed as the importer of record for such dairy products.

[76 FR 14788, Mar. 18, 2011]

§ 1150.122 CBP.

CBP means the United States Customs and Border Protection of the Department of Homeland Security.

[76 FR 14788, Mar. 18, 2011]

NATIONAL DAIRY PROMOTION AND
RESEARCH BOARD

§ 1150.131 Establishment and membership.

(a) There is hereby established a National Dairy Promotion and Research Board.

(b) Thirty-six members of the Board shall be United States producers. For purposes of nominating producers to the Board, the United States shall be divided into twelve geographic regions and the number of Board members from each region shall be as follows:

(1) Two members from region number one comprised of the following States: Alaska, Oregon and Washington.

(2) Seven members from region number two comprised of the following States: California and Hawaii.

(3) Two members from region number three comprised of the following States: Arizona, Colorado, Montana, Nevada, Utah and Wyoming.

(4) Four members from region number four comprised of the following States: Arkansas, Kansas, New Mexico, Oklahoma and Texas.

(5) Two members from region number five comprised of the following States: Minnesota, North Dakota and South Dakota.

(6) Five members from region number six comprised of the following State: Wisconsin.

(7) Two members from region number seven comprised of the following States: Illinois, Iowa, Missouri and Nebraska.

(8) Two members from region number eight comprised of the following State: Idaho.

(9) Three members from region number nine comprised of the following States: Indiana, Michigan, Ohio and West Virginia.

(10) Two members from region number ten comprised of the following States: Alabama, District of Columbia, Florida, Georgia, Kentucky, Louisiana, Mississippi, North Carolina, Commonwealth of Puerto Rico, South Carolina, Tennessee, and Virginia.

(11) Two members from region number eleven comprised of the following States: Delaware, Maryland, New Jersey and Pennsylvania.

(12) Three members from region number twelve comprised of the following States: Connecticut, Maine, Massachusetts, New Hampshire, New York, Rhode Island, and Vermont.

(c) One member of the board shall be an importer who is subject to assessments under §1150.152(b).

(d) The Board shall be composed of milk producers and importers appointed by the Secretary either from nominations submitted pursuant to

§1150.133 or in accordance with §1150.136. A milk producer may be nominated only to represent the region in which such producer's milk is produced.

(e) At least every five years, and not more than every three years, the Board shall review the geographic distribution of milk production volume throughout the United States and, if warranted, shall recommend to the Secretary a reapportionment of regions and/or a modification of the number of producer members from regions in order to best reflect the geographic distribution of milk production volume in the United States.

(f) At least once every three years, after the initial appointment of importer representatives on the Board, the Secretary shall review the average volume of domestic production of dairy products compared to the average volume of imports of dairy products into the United States during the previous three years and, on the basis of that review, if warranted, reapportion the importer representation on the Board to reflect the proportional shares of the United States market served by domestic production and imported dairy products. The basis for comparison of domestic production of dairy products to imported products shall be estimated total milk solids. The calculation of total milk solids of imported dairy products for reapportionment purposes shall be the same as the calculation of total milk solids of imported dairy products for assessment purposes.

(g) In determining the volume of milk produced and total milk solids of dairy products produced in the United States, the Board and Secretary shall utilize the information received by the Board pursuant to §1150.171(a) and data published by the Department.

[76 FR 14788, Mar. 18, 2011, as amended at 76 FR 80216, Dec. 23, 2011; 81 FR 53247, Aug. 12, 2016]

§1150.132   Term of office.

(a) The members of the Board shall serve for terms of three years, except that:

(1) The members appointed to the initial Board shall serve proportionately, for terms of one, two and three years.

(2) The 2 importer members initially appointed to the Board shall serve until October 31, 2013, and October 31, 2014.

(b) Each member of the Board shall serve until October 31 of the year in which his/her term expires, except that a retiring member may serve until a successor is appointed.

(c) No member shall serve more than two consecutive terms.

[49 FR 11816, Mar. 28, 1984, as amended at 60 FR 53253, Oct. 13, 1995; 76 FR 14788, Mar. 18, 2011]

§1150.133 Nominations.

Nominations for members of the Board shall be made in the following manner:

(a) The Secretary shall solicit nominations for producer representation on the Board from all eligible organizations. For nominations of producers, if the Secretary determines that a substantial number of producers are not members of, or their interests are not represented by, such eligible organizations, the Secretary shall also solicit nominations from such producers through general farmer organizations or by other means.

(b) After the appointment of the initial Board, the Secretary shall announce at least 120 days in advance when a Board member's term is expiring and shall solicit nominations for that position in the manner described in paragraph (a) of this section. Nominations for such position should be submitted to the Secretary not less than 60 days prior to the expiration of such term.

(c) An eligible producer organization may submit nominations only for positions on the Board that represent regions in which such eligible organization can establish that it represents a substantial number of producers. If there is more than one Board position for any such region, the organization may submit nominations for each position.

(d) Where there is more than one eligible organization representing producers in a specific geographic region, the organizations may caucus and jointly nominate producers for each position representing that region on the Board for which a member is to be appointed. If joint agreement is not reached with respect to any such nominations, or if no caucus is held, each eligible organization may submit to the Secretary nominations for each appointment to be made to represent that region.

(e) Nominations for representation of importers may be submitted by:

(1) Organizations that represent importers of dairy products, as approved by the Secretary. The primary considerations in determining if organizations adequately represent importers of dairy products shall be whether its membership consists primarily of importers of dairy products and whether a substantial interest of the organization is in the importation of dairy products and the promotion of the nutritional attributes of dairy products; and

(2) Individual importers of dairy products. Individual importers submitting nominations to represent importers on the Board must establish to the satisfaction of the Secretary that the persons submitting the nominations are importers of dairy products.

[49 FR 11816, Mar. 28, 1984, as amended at 76 FR 14788, Mar. 18, 2011]

§1150.134 Nominee's agreement to serve.

Any producer or importer nominated to serve on the Board shall file with the Secretary at the time of the nomination a written agreement to:

(a) Serve on the Board if appointed;

(b) Disclose any relationship with any organization that operates a qualified program or has a contractual relationship with the Board; and

(c) Withdraw from participation in deliberations, decision-making, or voting on matters where paragraph (b) applies.

[49 FR 11816, Mar. 28, 1984, as amended at 76 FR 14789, Mar. 18, 2011]

§1150.135 Appointments.

From the nominations made pursuant to §1150.133, the Secretary shall appoint the members of the Board on the bases of representation provided for in §§1150.131(b) and 1150.131(c).

[76 FR 14789, Mar. 18, 2011]

§ 1150.136 Vacancies.

To fill any vacancy occasioned by the death, removal, resignation, or disqualification of any member of the Board, the Secretary shall appoint a successor from the most recent list of nominations for the position or from nominations made by the Board.

§ 1150.137 Procedure.

(a) A majority of the members shall constitute a quorum at a properly convened meeting of the Board. Any action of the Board shall require the concurring votes of at least a majority of those present and voting. The Board shall establish rules concerning timely notice of meetings.

(b) The Board may take action upon the concurring votes of a majority of its members by mail, telephone, or telegraph when in the opinion of the chairman of the Board such action must be taken before a meeting can be called. Action taken by this emergency procedure is valid only if all members are notified and provided the opportunity to vote and any telephone vote is confirmed promptly in writing. Any action so taken shall have the same force and effect as though such action had been taken at a properly convened meeting of the Board.

[49 FR 11816, Mar. 28, 1984, as amended at 50 FR 9984, Mar. 13, 1985]

§ 1150.138 Compensation and reimbursement.

The members of the Board shall serve without compensation but shall be reimbursed for necessary and reasonable expenses, including a per diem allowance as recommended by the Board and approved by the Secretary, incurred by them in the performance of their duties under this subpart.

§ 1150.139 Powers of the Board.

The Board shall have the following powers:

(a) To receive and evaluate, or on its own initiative develop, and budget for plans or projects to promote the use of fluid milk and dairy products as well as projects for research and nutrition education and to make recommendations to the Secretary regarding such proposals;

(b) To administer the provisions of this subpart in accordance with its terms and provisions;

(c) To make rules and regulations to effectuate the terms and provisions of this subpart;

(d) To receive, investigate, and report to the Secretary complaints of violations of the provisions of this subpart;

(e) To disseminate information to producers, producer organizations, importers, and importer organizations through programs or by direct contact utilizing the public postage system or other systems;

(f) To select committees and subcommittees of Board members, and to adopt such rules for the conduct of its business as it may deem advisable;

(g) To establish advisory committees of persons other than Board members and pay the necessary and reasonable expenses and fees of the members of such committees;

(h) To recommend to the Secretary amendments to this subpart; and

(i) With the approval of the Secretary, to invest, pending disbursement pursuant to a plan or project, funds collected through assessments authorized under § 1150.152 in, and only in, obligations of the United States or any agency thereof, in general obligations of any State or any political subdivision thereof, in any interest-bearing account or certificate of deposit of a bank that is a member of the Federal Reserve System, or in obligations fully guaranteed as to principal and interest by the United States.

[49 FR 11816, Mar. 28, 1984, as amended at 76 FR 14789, Mar. 18, 2011]

§ 1150.140 Duties of the Board.

The Board shall have the following duties:

(a) To meet not less than annually, and to organize and select from among its members a chairman and such other officers as may be necessary;

(b) To appoint from its members an executive committee whose membership shall equally reflect each of the different geographic regions in the United States in which milk is produced and importer representation on the Board, and to delegate to the committee authority to administer the terms and provisions of this subpart under the direction of the Board and within the policies determined by the Board;

(c) To appoint or employ such persons as it may deem necessary and define the duties and determine the compensation of each;

(d) To review all programs that promote milk and dairy products on a brand or trade name basis that have requested certification pursuant to § 1150.153, and to recommend to the Secretary whether such request should be granted;

(e) To develop and submit to the Secretary for approval, promotion, research, and nutrition education plans or projects resulting from research or studies conducted either by the Board or others;

(f) To solicit, among other proposals, research proposals that would increase the use of fluid milk and dairy products by the military and by persons in developing nations, and that would demonstrate the feasibility of converting surplus nonfat dry milk to casein for domestic and export use;

(g) To prepare and submit to the Secretary for approval, budgets on a fiscal period basis of its anticipated expenses and disbursements in the administration of this subpart, including probable costs of promotion, research and nutrition education plans or projects, and also including a general description of the proposed promotion, research and nutrition education programs contemplated therein;

(h) To maintain such books and records, which shall be available to the Secretary for inspection and audit, and prepare and submit such reports from time to time to the Secretary as the Secretary may prescribe, and to make appropriate accounting with respect to the receipt and disbursement of all funds entrusted to it;

(i) With the approval of the Secretary, to enter into contracts or agreements with national, regional or State dairy promotion and research organizations or other organizations or entities for the development and conduct of activities authorized under §§ 1150.139 and 1150.161, and for the payment of the cost thereof with funds collected through assessments pursuant to § 1150.152. Any such contract or agreement shall provide that:

(1) The contractors shall develop and submit to the Board a plan or project together with a budgets or budget which shall show the estimated cost to be incurred for such plan or project;

(2) Any such plan or project shall become effective upon approval of the Secretary; and

(3) The contracting party shall keep accurate records of all of its transactions and make periodic reports to the Board of activities conducted and an accounting for funds received and expended, and such other reports as the Secretary or the Board may require. The Secretary or employees of the Board may audit periodically the records of the contracting party;

(j) To prepare and make public, at least annually, a report of its activities carried out and an accounting for funds received and expended;

(k) To have an audit of its financial statements conducted by a certified public accountant in accordance with generally accepted auditing standards, at least once each fiscal period and at such other times as the Secretary may request, and to submit a copy of each such audit report to the Secretary;

(l) To give the Secretary the same notice of meetings of the Board, committees of the Board and advisory committees as is given to such Board or committee members in order that the Secretary, or a representative of the Secretary, may attend such meetings;

(m) To submit to the Secretary such information pursuant to this subpart as may be requested; and

(n) To encourage the coordination of programs of promotion, research and nutrition education designed to strengthen the dairy industry's position in the marketplace and to maintain and expand:

(1) domestic markets and domestic uses for fluid milk and dairy products produced in the United States or imported into the United States; and

(2) foreign markets and foreign uses for fluid milk and dairy products produced in the United States.

[49 FR 11816, Mar. 28, 1984, as amended at 76 FR 14789, Mar. 18, 2011]

EXPENSES AND ASSESSMENTS

§ 1150.151 Expenses.

(a) The Board is authorized to incur such expenses (including provision for a reasonable reserve) as the Secretary finds are reasonable and likely to be incurred by the Board for its maintenance and functioning and to enable it to exercise its powers and perform its duties in accordance with the provisions of this subpart. However, after the first full year of operation of the order, administrative expenses incurred by the Board shall not exceed 5 percent of the projected revenue of that fiscal year. Such expenses shall be paid from assessments collected pursuant to § 1150.152.

(b) The Board shall reimburse the Secretary, from assessments collected pursuant to § 1150.152, for administrative costs incurred by the Department after May 1, 1984.

(c) The Board is authorized to expend up to the amount of the assessments collected from United States producers to promote dairy products produced in the United States in foreign markets.

[49 FR 11816, Mar. 28, 1984, as amended at 76 FR 14789, Mar. 18, 2011]

§ 1150.152 Assessments.

(a) *Domestic Assessments.* (1) Each person making payment to a producer for milk produced in the United States and marketed for commercial use shall collect an assessment on all such milk handled for the account of the producer at the rate of 15 cents per hundredweight of milk for commercial use, or the equivalent thereof, and shall remit the assessment to the Board.

(2) Any producer marketing milk of that producer's own production in the form of milk or dairy products to consumers, either directly or through retail or wholesale outlets, shall remit to the Board an assessment on such milk at the rate of 15 cents per hundredweight of milk for commercial use or the equivalent thereof.

(3) In determining the assessment due from each producer pursuant to § 1150.152(a)(1) and (a)(2), a producer who is participating in a qualified program(s) under § 1150.153 shall receive a credit for contributions to such program(s), but not to exceed 10 cents per hundredweight of milk marketed.

(4) In order for a producer described in § 1150.152(a)(1) to receive the credit authorized in § 1150.152(a)(3), either the producer or a cooperative association on behalf of the producer must establish to the person responsible for remitting the assessment to the Board that the producer is contributing to a qualified program under § 1150.153. Producers who contribute to a qualified program directly (other than through a payroll deduction) must establish with the person responsible for remitting the assessment to the Board, with validation by the qualified program, that they are making such contributions.

(5) In order for a producer described in § 1150.152(a)(2) to receive the credit authorized in § 1150.152(a)(3), the producer and the applicable qualified program must establish to the Board that the producer is contributing to the qualified program.

(6) The collection of assessments pursuant to § 1150.152(a)(1) and (a)(2) shall begin with respect to milk marketed on and after the effective date of this section and shall continue until terminated by the Secretary.

(7) Each person responsible for the remittance of the assessment pursuant to § 1150.152(a)(1) and (a)(2) shall remit the assessment to the Board not later than the last day of the month following the month in which the milk was marketed.

(8) Money remitted to the Board shall be in the form of a negotiable instrument made payable to "National Dairy Promotion and Research Board." Remittances and reports specified in § 1150.171(a) shall be mailed to the location designated by the Secretary or the Board.

(b) *Importer assessments.* (1) Each importer of dairy products identified in the following table, except for as provided for in § 1150.157, is responsible for paying an assessment of 7.5 cents per hundredweight of U.S. milk, or equivalent thereof. The importer shall use the assessment rate of $0.01327 per kilogram (kg) of milk solids to calculate and pay the assessment.

HTS Nos. for dairy import assessment	HTS Nos. for dairy import assessment
0401.10.0000	0404.10.9000
0401.20.2000	0404.90.1000
0401.20.4000	0404.90.3000
0401.30.0500	0404.90.5000
0401.30.2500	0404.90.7000
0401.30.5000	0405.10.1000
0401.30.7500	0405.10.2000
0402.10.1000	0405.20.2000
0402.10.5000	0405.20.3000
0402.21.0500	0405.20.4000
0402.21.2500	0405.20.6000
0402.21.3000	0405.20.7000
0402.21.5000	0405.20.8000
0402.21.7500	0405.90.1020
0402.21.9000	0405.90.1040
0402.29.1000	0405.90.2020
0402.29.5000	0405.90.2040
0402.91.1000	0406.10.0400
0402.91.3000	0406.10.0800
0402.91.7000	0406.10.1400
0402.91.9000	0406.10.1800
0402.99.1000	0406.10.2400
0402.99.3000	0406.10.2800
0402.99.4500	0406.10.3400
0402.99.5500	0406.10.3800
0402.99.7000	0406.10.4400
0402.99.9000	0406.10.4800
0403.10.1000	0406.10.5400
0403.10.5000	0406.10.5800
0403.10.9000	0406.10.6400
0403.90.0400	0406.10.6800
0403.90.1600	0406.10.7400
0403.90.2000	0406.10.7800
0403.90.4110	0406.10.8400
0403.90.4190	0406.10.8800
0403.90.4500	0406.20.1500
0403.90.5100	0406.20.2400
0403.90.5500	0406.20.2800
0403.90.6100	0406.20.3110
0403.90.6500	0406.20.3190
0403.90.7400	0406.20.3300
0403.90.7800	0406.20.3600
0403.90.8500	0406.20.3900
0403.90.9000	0406.20.4400
0403.90.9500	0406.20.4800
0404.10.0500	0406.20.5100
0404.10.1100	0406.20.5300
0404.10.1500	0406.20.6100
0404.10.2000	0406.20.6300
0404.10.5010	0406.20.6500
0404.10.5090	0406.20.6700

HTS Nos. for dairy import assessment	HTS Nos. for dairy import assessment
0406.20.6900	0406.90.3300
0406.20.7100	0406.90.3600
0406.20.7300	0406.90.3700
0406.20.7500	0406.90.4100
0406.20.7700	0406.90.4200
0406.20.7900	0406.90.4600
0406.20.8100	0406.90.4800
0406.20.8300	0406.90.4900
0406.20.8500	0406.90.5200
0406.20.8700	0406.90.5400
0406.20.8900	0406.90.6600
0406.20.9100	0406.90.6800
0406.30.0500	0406.90.7200
0406.30.1400	0406.90.7400
0406.30.1800	0406.90.7600
0406.30.2400	0406.90.7800
0406.30.2800	0406.90.8200
0406.30.3400	0406.90.8400
0406.30.3800	0406.90.8600
0406.30.4400	0406.90.8800
0406.30.4800	0406.90.9000
0406.30.5100	0406.90.9200
0406.30.5300	0406.90.9300
0406.30.6100	0406.90.9400
0406.30.6300	0406.90.9500
0406.30.6500	0406.90.9700
0406.30.6700	0406.90.9900
0406.30.6900	1517.90.5000
0406.30.7100	1517.90.6000
0406.30.7300	1702.11.0000
0406.30.7500	1702.19.0000
0406.30.7700	1704.90.5400
0406.30.7900	1704.90.5800
0406.30.8100	1806.20.2090
0406.30.8300	1806.20.2400
0406.30.8500	1806.20.2600
0406.30.8700	1806.20.2800
0406.30.8900	1806.20.3400
0406.30.9100	1806.20.3600
0406.40.4400	1806.20.3800
0406.40.4800	1806.20.8100
0406.40.5400	1806.20.8200
0406.40.5800	1806.20.8300
0406.40.7000	1806.20.8500
0406.90.0810	1806.20.8700
0406.90.0890	1806.20.8900
0406.90.1200	1806.32.0400
0406.90.1600	1806.32.0600
0406.90.1800	1806.32.0800
0406.90.3100	1806.32.1400
0406.90.3200	1806.32.1600

HTS Nos. for dairy import assessment	HTS Nos. for dairy import assessment
1806.32.1800	2106.90.7800
1806.32.6000	2106.90.8000
1806.32.7000	2106.90.8200
1806.32.8000	2202.90.1000
1806.90.0500	2202.90.2400
1806.90.0800	2202.90.2800
1806.90.1000	3501.10.1000
1806.90.1500	3501.10.5000
1806.90.1800	3501.90.6000
1806.90.2000	3502.20.0000
1806.90.2500	
1806.90.2800	
1806.90.3000	
1901.10.1500	
1901.10.3000	
1901.10.3500	
1901.10.4000	
1901.10.4500	
1901.20.0500	
1901.20.1500	
1901.20.2000	
1901.20.2500	
1901.20.3000	
1901.20.3500	
1901.20.4000	
1901.20.4500	
1901.20.5000	
1901.90.2800	
1901.90.3400	
1901.90.3600	
1901.90.4200	
1901.90.4300	
1901.90.7000	
2105.00.1000	
2105.00.2000	
2105.00.3000	
2105.00.4000	
2106.90.0600	
2106.90.0900	
2106.90.2400	
2106.90.2600	
2106.90.2800	
2106.90.3400	
2106.90.3600	
2106.90.3800	
2106.90.6400	
2106.90.6600	
2106.90.6800	
2106.90.7200	
2106.90.7400	
2106.90.7600	

(2) The assessment on imported dairy products shall be paid by the importer to CBP at the time of entry summary for any products identified in §1150.152(b)(1).

(3) The assessments collected by CBP pursuant to §1150.152(b)(2) of this section shall be transferred to the Board in compliance with an agreement between CBP and the Secretary.

(4) The Secretary, at his or her discretion, shall verify the information reported by importers to CBP to determine if additional money is due the Board or an amount is due to an importer based on the quantity imported and the milk solids content per unit. In the case of money due to an importer from the Board, the Board will issue payment promptly to the importer. In the case of money due from the importer to the Board, the Secretary will send an invoice for payment directly to the importer. The remittance will be due to the Secretary upon receipt of the invoice. The Secretary will promptly forward such payments received to the Board.

(5) If an importer elects to have funds remitted to a qualified program(s), the importer shall inform the Secretary of such designation by sending a letter to an address provided by the Secretary. Importer remittances for qualified program(s) shall not exceed 2.5 cents per hundredweight of milk, or equivalent thereof, of the 7.5 cents per hundredweight of milk, or equivalent thereof, paid by the importer pursuant to §1150.152(b)(1). The Secretary shall compute the funds due for each qualified program designated by importers and direct the Board to forward such funds to each qualified program.

(6) Assessments collected on imported dairy products shall not be used for foreign market promotion of United States dairy products.

(7) Any money received by the Board pursuant to § 1150.152(b)(1) before the Secretary appoints the initial importer representatives to the Board shall not be spent by the Board but shall be held in escrow until such appointment.

(8) The collection of assessments pursuant to § 1150.152(a) and (b) shall continue until terminated by the Secretary.

[76 FR 14789, Mar. 18, 2011]

§ 1150.153 Qualified dairy product promotion, research or nutrition education programs.

(a) Any producer organization that conducts a State or regional dairy product promotion, research or nutrition education program, authorized by Federal or State law; or has been an active and ongoing producer program before enactment of the Act; or is an importer organization that conducts a promotion, research, or nutrition education program may apply to the Secretary for certification of qualification so that:

(1) Producers may receive credit pursuant to § 1150.152(a)(3) for contributions to such program; and

(2) The Board may remit payments designated by importers pursuant to § 1150.152(b)(5).

(b) In order to be certified by the Secretary as a qualified program, the program must:

(1) Conduct activities as defined in §§ 1150.114, 1150.115, and 1150.116 that are intended to increase consumption of milk and dairy products generally;

(2) Except for producer programs operated under the laws of the United States or any State, and except for importer programs, have been active and ongoing before enactment of the Act;

(3) For producer organizations, be financed primarily by producers, either individually or through cooperative associations, or for importer organizations, be financed primarily by importers;

(4) Not use a private brand or trade name in its advertising and promotion of dairy products unless the Board recommends and the Secretary concurs that such preclusion should not apply;

(5) Certify to the Secretary that any requests from producers or importers for refunds under the program will be honored by forwarding to either the Board or a qualified program designated by the producer or importer that portion of such refunds equal to the amount that otherwise would be applicable to that program pursuant to § 1150.152(a)(3) or (b)(5); and

(6) Not use program funds for the purpose of influencing governmental policy or action.

(c) An application for certification of qualifications of any dairy product promotion, research or nutrition education program which does not satisfy the requirements specified in paragraph (b) of this section shall be denied. The certification of any qualified program which fails to satisfy the requirements specified in paragraph (b) of this section after certification shall be subject to suspension or termination.

(1) Prior to the denial of an application for certification of qualification, or the suspension or termination of an existing certification, the Director of the Dairy Division shall afford the applicant or the holder of an existing certification an opportunity to achieve compliance with the requirements for certification within a reasonable time, as determined by the Director.

(2) Any dairy product promotion, research or nutrition education program whose application for certification of qualification is to be denied, or whose certification of qualification is to be suspended or terminated shall be given written notice of such pending action and shall be afforded an opportunity to petition the Secretary for a review of the action. The petition shall be in writing and shall state the facts relevant to the matter for which the review is sought, and whether petitioner desires an informal hearing. If an informal hearing is not requested, the Director of the Dairy Division shall issue a final decision setting forth the action to be taken and the basis for such action. If petitioner requests a hearing, the Director of the Dairy Division, or a person designated by the Director,

shall hold an informal hearing in the following manner:

(i) Notice of a hearing shall be given in writing and shall be mailed to the last known address of the petitioner or of the program, or to an officer thereof, at least 20 days before the date set for the hearing. Such notice shall contain the time and place of the hearing and may contain a statement of the reason for calling the hearing and the nature of the questions upon which evidence is desired or upon which argument may be presented. The hearing place shall be as convenient to the program as can reasonably be arranged.

(ii) Hearings are not to be public and are to be attended only by representatives of the petitioner or the program and of the U.S. Government, and such other parties as either the program or the U.S. Government desires to have appear for purposes of submitting information or as counsel.

(iii) The Director of the Dairy Division, or a person designated by the Director, shall be the presiding officer at the hearing. The hearing shall be conducted in such manner as will be most conducive to the proper disposition of the matter. Written statements or briefs may be filed by the petitioner or the program, or other participating parties, within the time specified by the presiding officer.

(iv) The presiding officer shall prepare preliminary findings setting forth a recommendation as to what action should be taken and the basis for such action. A copy of such findings shall be served upon the petitioner or the program by mail or in person. Written exceptions to the findings may be filed within 10 days after service thereof.

(v) After due consideration of all the facts and the exceptions, if any, the Director of the Dairy Division shall issue a final decision setting forth the action to be taken and the basis for such action.

[49 FR 11816, Mar. 28, 1984, as amended at 56 FR 8258, Feb. 28, 1991; 76 FR 14791, Mar. 18, 2011]

§1150.154 Influencing governmental action.

No funds collected by the Board under this subpart shall in any manner be used for the purpose of influencing governmental policy or action, except to recommend to the Secretary amendments to this subpart.

§1150.155 Adjustment of accounts.

(a) Whenever the Board or the Department determines through an audit of a person's reports, records, books or accounts or through some other means that additional money is due the Board or that money is due such person from the Board in accordance with 1150.152(a), such person shall be notified of the amount due. The person shall then remit any amount due the Board by the next date for remitting assessments as provided in §1150.152(a). Overpayments shall be credited to the account of the person remitting the overpayment and shall be applied against amounts due in succeeding months.

(b) Any importer of dairy products against whose imports an assessment has been collected under §1150.152(b) who believes that such assessment or any portion of such assessment was made on milk solids of U.S. origin or milk solids other than cow's milk may apply to the Secretary for a reimbursement. The importer would be required to submit satisfactory proof to the Secretary that the importer paid the assessment for milk solids from milk produced from the U.S. or milk solids other than cow's milk solids. The Secretary will instruct the Board to send such reimbursement to the importer.

[76 FR 14791, Mar. 18, 2011]

§1150.156 Charges and penalties.

(a) *Late-payment charge.* Any unpaid assessments due to the Board pursuant to §1150.152 shall be increased 1.5 percent each month beginning with the day following the date such assessments were due. Any remaining amount due, which shall include any unpaid charges previously made pursuant to this section, shall be increased at the same rate on the corresponding day of each month thereafter until paid.

(1) For the purpose of this section, any assessment pursuant to §1150.152(a) that was determined at a date later than prescribed by this subpart because of a person's failure to submit a report to the Board when due shall be considered to have been payable by the date

it would have been due if the report had been filed when due. The timeliness of a payment to the Board shall be based on the applicable postmark date or the date actually received by the Board, whichever is earlier.

(2) For the purpose of this section, any assessment not collected by CBP at the time entry summary documents are filed by the importer is considered to be past due. If CBP does not collect an assessment from an importer, the importer shall be responsible for paying the assessment and any late charges to the Secretary in the form of a negotiable instrument made payable to "USDA." The payment shall be mailed to a location designated by the Secretary or sent in an electronic form approved by the Secretary.

(b) *Penalties.* Any person who willfully violates any provision of this subpart shall be assessed a civil penalty by the Secretary of not more than the amount specified in § 3.91(b)(1)(xx) of this title for each such violation and, in the case of a willful failure to pay, collect, or remit the assessment as required by this subpart, in addition to the amount due, a penalty equal to the amount of the assessment on the quantity of milk as to which the failure applies. The amount of any such penalty shall accrue to the United States and may be recovered in a civil suit brought by the United States. The remedies provided in this section shall be in addition to, and not exclusive of, other remedies that may be available by law or in equity.

[49 FR 11816, Mar. 28, 1984, as amended at 70 FR 29579, May 24, 2005; 76 FR 14791, Mar. 18, 2011]

§ 1150.157 Assessment exemption.

(a) A producer described in § 1150.152(a)(1) and (2) who operates under an approved National Organic Program (7 CFR part 205) (NOP) organic production system plan may be exempt from the payment of assessments under this part, provided that:

(1) Only agricultural products certified as "organic" or "100 percent organic" (as defined in the NOP) are eligible for exemption;

(2) The exemption shall apply to all certified "organic" or "100 percent organic" (as defined in the NOP) products

of the producer regardless of whether the agricultural commodity subject to the exemption is produced by a person that also produces conventional or nonorganic agricultural products of the same agricultural commodity as that for which the exemption is claimed;

(3) The producer maintains a valid certificate of organic operation as issued under the Organic Foods Production Act of 1990 (7 U.S.C. 6501–6522) (OFPA) and the NOP regulations issued under OFPA (7 CFR part 205); and

(4) Any producer so exempted shall continue to be obligated to pay assessments under this part that are associated with any agricultural products that do not qualify for an exemption under this section.

(b) To apply for exemption under this section, a producer subject to assessments pursuant to § 1150.152(a)(1) and (2) shall submit a request to the Board on an *Organic Exemption Request Form* (Form AMS–15) at any time during the year initially, and annually thereafter on or before July 1, for as long as the producer continues to be eligible for the exemption.

(c) A producer request for exemption shall include the following:

(1) The applicant's full name, company name, address, telephone and fax numbers, and email address;

(2) Certification that the applicant maintains a valid organic certificate issued under the OFPA and the NOP;

(3) Certification that the applicant produces organic products eligible to be labeled "organic" or "100 percent organic" under the NOP;

(4) A requirement that the applicant attach a copy of their certificate of organic operation issued by a USDA-accredited certifying agent under the OFPA and the NOP;

(5) Certification, as evidenced by signature and date, that all information provided by the applicant is true; and

(6) Such other information as may be required by the Board, with the approval of the Secretary.

(d) If a producer complies with the requirements of this section, the Board will grant an assessment exemption and issue a Certificate of Exemption to the producer within 30 days. If the application is disapproved, the Board will notify the applicant of the reason(s) for disapproval within the same time-frame.

(e) A producer approved for exemption under this section shall provide a copy of the Certificate of Exemption to each person responsible for remitting assessments to the Board on behalf of the producer pursuant to §1150.152(a).

(f) The person responsible for remitting assessments to the Board pursuant to §1150.152(a) shall maintain records showing the exempt producer's name and address and the exemption number assigned by the Board pursuant to §1150.172(a).

(g) An importer who imports products that are eligible to be labeled as "organic" or "100 percent organic" under the NOP, or certified as "organic" or "100 percent organic" under a U.S. equivalency arrangement established under the NOP, may be exempt from the payment of assessments on those products. Such importer may submit documentation to the Board and request an exemption from assessment on certified "organic" or "100 percent organic" dairy products on an *Organic Exemption Request Form* (Form AMS–15) at any time initially, and annually thereafter on or before July 1, as long as the importer continues to be eligible for the exemption. This documentation shall include the same information required of producers in paragraph (c) of this section. If the importer complies with the requirements of this section, the Board will grant the exemption and issue a Certificate of Exemption to the importer. The Board will also issue the importer an alphanumeric number valid for 1 year from the date of issue. This alphanumeric number should be entered by the importer on the CBP entry documentation. Any line item entry of "organic" or "100 percent organic" dairy products bearing this alphanumeric number assigned by the Board will not be subject to assessments. Any importer so exempted shall continue to be obligated to pay assessments under this part that are associated with any imported agricultural products that do not qualify for an exemption under this section.

(h) The exemption will apply not later than the last day of the month following the Certificate of Exemption issuance date.

(i) An importer who is exempt from payment of assessments under paragraph (g) of this section shall be eligible for reimbursement of assessments collected by the CBP on certified "organic" or "100 percent organic" dairy products and may apply to the Secretary for a reimbursement. The importer would be required to submit satisfactory proof to the Secretary that the importer paid the assessment on exempt organic products.

[76 FR 14792, Mar. 18, 2011, as amended at 80 FR 82021, Dec. 31, 2015]

PROMOTION, RESEARCH AND NUTRITION EDUCATION

§1150.161 Promotion, research and nutrition education.

(a) The Board shall receive and evaluate, or on its own initiative develop, and submit to the Secretary for approval any plans or projects authorized in §§1150.139, 1150.140 and this section. Such plans or projects shall provide for:

(1) The establishment, issuance, effectuation, and administration of appropriate plans or projects for promotion, research and nutrition education with respect to milk and dairy products; and

(2) The establishment and conduct of research and studies with respect to the sale, distribution, marketing and utilization of milk and dairy products and the creation of new products thereof, to the end that marketing and utilization of milk and dairy products may be encouraged, expanded, improved or made more acceptable. Included shall be research and studies of proposals intended to increase the use of fluid milk and dairy products by the military and by persons in developing nations and proposals intended to demonstrate the feasibility of converting nonfat dry milk to casein for domestic and export use.

(b) Each plan or project authorized under §1150.161(a) shall be periodically reviewed or evaluated by the Board to insure that the plan or project contributes to an effective program of promotion, research and nutrition education. If it is found by the Board that any such plan or project does not further the purposes of the Act, the Board shall terminate such plan or project.

(c) No plan or project authorized under § 1150.161(a) shall make use of unfair or deceptive acts or practices with respect to the quality, value or use of any competing product.

REPORTS, BOOKS AND RECORDS

§ 1150.171 Reports.

(a) Each producer marketing milk of that producer's own production directly to consumers and each person making payment to producers and responsible for the collection of the assessment under § 1150.152(a) shall be required to report at the time for remitting assessments to the Board such information as may be required by the Board or by the Secretary. Such information may include but not be limited to the following:

(1) The quantity of milk purchased, initially transferred or which, in any other manner, are subject to the collection of the assessment;

(2) The amount of assessment remitted;

(3) The basis, if necessary, to show why the remittance is less than the number of hundredweights of milk multiplied by 15 cents; and

(4) The date any assessment was paid.

(b) Importers of dairy products shall submit reports as requested by the Secretary as necessary to verify that provisions pursuant to § 1150.152(b) have been carried out correctly, including verification that correct amounts were paid based upon milk solids content of the imported dairy products pursuant to § 1150.152(b)(1).

[76 FR 14792, Mar. 18, 2011]

§ 1150.172 Books and records.

(a) Each producer who is subject to this subpart, and other persons subject to § 1150.171(a), shall maintain and make available for inspection by employees of the Board and the Secretary such books and records as are necessary to carry out the provisions of this subpart and the regulations issued hereunder, including such records as are necessary to verify any reports required. Such records shall be retained

for at least two years beyond the fiscal period of their applicability.

(b) Each importer of dairy products shall maintain and make available for inspection by the Secretary such books and records to verify that provisions pursuant to § 1150.152(b) have been carried out correctly, including verification that correct amounts were paid based upon milk solids content of the imported dairy products. Such records shall be retained for at least two years beyond the calendar period of their applicability. Such information may include but not be limited to invoices, packing slips, bills of lading, laboratory test results, and letters from the manufacturer on the manufacturer's letterhead stating the milk solids content of imported dairy products.

[76 FR 14792, Mar. 18, 2011]

§ 1150.173 Confidential treatment.

All information obtained from such books, records or reports under the Act and this subpart shall be kept confidential by all persons, including employees and former employees of the Board, all officers and employees and all former officers and employees of the Department, and by all officers and all employees and all former officers and employees of contracting agencies having access to such information, and shall not be available to Board members. Only those persons having a specific need for such information in order to effectively administer the provisions of this subpart shall have access to such information. In addition, only such information so furnished or acquired as the Secretary deems relevant shall be disclosed by them, and then only in a suit or administrative hearing brought at the discretion, or upon the request, of the Secretary, or to which the Secretary or any officer of the United States is a party, and involving this subpart. Nothing in this section shall be deemed to prohibit:

(a) The issuance of general statements based upon the reports of the number of persons subject to this subpart or statistical data collected therefrom, which statements do not identify the information furnished by any person; and

(b) The publication, by direction of the Secretary, of the name of any person who has been adjudged to have violated this subpart, together with a statement of the particular provisions of the subpart violated by such person.

MISCELLANEOUS

§ 1150.181 Proceedings after termination.

(a) Upon the termination of this subpart, the Board shall recommend not more than five of its members to the Secretary to serve as trustees for the purpose of liquidating the affairs of the Board. Such persons, upon designation by the Secretary, shall become trustees of all the funds and property owned, in the possession of, or under the control of the Board, including unpaid claims or property not delivered or any other claim existing at the time of such termination.

(b) The said trustees shall:

(1) Continue in such capacity until discharged by the Secretary;

(2) Carry out the obligations of the Board under any contract or agreements entered into by it pursuant to § 1150.140(i);

(3) From time to time account for all receipts and disbursements and deliver all property on hand, together with all books and records of the Board and of the trustees, to such persons as the Secretary may direct; and

(4) Upon the request of the Secretary, execute such assignments or other instruments necessary or appropriate to vest in such persons full title and right to all of the funds, property, and claims vested in the Board or the trustees pursuant to this subpart.

(c) Any person to whom funds, property, or claims have been transferred or delivered pursuant to this subpart shall be subject to the same obligation imposed upon the Board and upon the trustees.

(d) Any residual funds not required to defray the necessary expenses of liquidation shall be turned over to the Secretary to be used, to the extent practicable, in the interest of continuing one or more of the promotion, research or nutrition education plans or projects authorized pursuant to this subpart.

§ 1150.182 Effect of termination or amendment.

Unless otherwise expressly provided by the Secretary, the termination of this subpart or of any regulation issued pursuant hereto, or the issuance of any amendment to either thereof, shall not:

(a) Affect or waive any right, duty, obligation, or liability which shall have arisen or which may hereafter arise in connection with any provision of this subpart or any regulation issued thereunder;

(b) Release or extinguish any violation of this subpart or any regulation issued thereunder; or

(c) Affect or impair any rights or remedies of the United States, or of the Secretary, or of any person, with respect to any such violation.

§ 1150.183 Personal liability.

No member or employee of the Board shall be held personally responsible, either individually or jointly, in any way whatsoever to any person for errors in judgment, mistakes, or other acts of either commission or omission of such member or employee, except for acts of dishonesty or willful misconduct.

§ 1150.184 Patents, copyrights, inventions and publications.

Any patents, copyrights, trademarks, inventions or publications developed through the use of funds collected under the provisions of this subpart shall be the property of the U.S. Government as represented by the Board, and shall, along with any rents, royalties, residual payments, or other income from the rental, sale, leasing, franchising, or other uses of such patents, copyrights, inventions, or publications, inure to the benefit of the Board. Upon termination of this subpart, § 1150.181 shall apply to determine disposition of all such property.

§ 1150.185 Amendments.

The Secretary may from time to time amend provisions of this part. Any interested person or organization affected by the provisions of the Act may propose such amendments to the Secretary.

§ 1150.186 Separability.

If any provision of this subpart is declared invalid or the applicability thereof to any person or circumstances is held invalid, the validity of the remainder of this subpart or the applicability thereof to other persons or circumstances shall not be affected thereby.

§ 1150.187 Paperwork Reduction Act assigned number.

The information collection and recordkeeping requirements contained in §§ 1150.133, 1150.152, 1150.153, 1150.171, 1150.172, and 1150.273 of these regulations (7 CFR part 1150) have been approved by the Office of Management and Budget (OMB) under the provisions of 44 U.S.C. chapter 35 and have been assigned OMB Control Number 0581–0093 as appropriate.

[76 FR 14793, Mar. 18, 2011]

Subpart—Procedure for Certification of Milk Producer Organizations

§ 1150.270 General.

Organizations must be certified by the Secretary that they are eligible to represent milk producers and to participate in the making of nominations of milk producers to serve as members of the National Dairy Promotion and Research Board as provided in the Dairy and Tobacco Adjustment Act of 1983. Certifications of eligibility required of the Secretary shall be conducted in accordance with this subpart.

§ 1150.271 Definitions.

As used in this subpart:

(a) *Act* means Title I, Subtitle B, of the Dairy and Tobacco Adjustment Act of 1983, Pub. L. 98–180, 97 Stat. 1128, as approved November 29, 1983, and any amendments thereto;

(b) *Department* means the United States Department of Agriculture;

(c) *Secretary* means the Secretary of Agriculture of the United States, or any officer or employee of the Department to whom authority has heretofore been delegated, or to whom authority may hereafter be delegated to act in the Secretary's stead;

(d) *Dairy Division* means the Dairy Division of the Department's Agricultural Marketing Service;

(e) *Producer* means any person engaged in the production of milk for commercial use;

(f) *Dairy products* means products manufactured for human consumption which are derived from the processing of milk, and includes fluid milk products; and

(g) *Fluid milk products* means those milk products normally consumed in liquid form as a beverage.

§ 1150.272 Responsibility for administration of regulations.

The Dairy Division shall have the responsibility for administering the provisions of this subpart.

§ 1150.273 Application for certification.

Any organization whose membership consists primarily of milk producers may apply for certification. Applicant organizations should supply information for certification using as a guide "Application for Certification of Organizations," Form DA–26. Form DA–26 may be obtained from the Dairy Division, Agricultural Marketing Service, United States Department of Agriculture, Washington, DC 20250.

§ 1150.274 Certification standards.

(a) Certification of eligible organizations shall be based, in addition to other available information, on a factual report submitted by the organization, which shall contain information deemed relevant and specified by the Secretary for the making of such determination, including the following:

(1) Geographic territory covered by the organization's active membership;

(2) Nature and size of the organization's active membership including the total number of active milk producers represented by the organization;

(3) Evidence of stability and permanency of the organization;

(4) Sources from which the organization's operating funds are derived;

(5) Functions of the organization; and

(6) The organization's ability and willingness to further the aims and objectives of the Act.

(b) The primary considerations in determining the eligibility of an organization shall be whether its membership consists primarily of milk producers who produce a substantial volume of milk, and whether the primary or overriding interest of the organization is in the production or processing of fluid milk and dairy products and promotion of the nutritional attributes of fluid milk and dairy products.

(c) The Secretay shall certify any organization which he finds meets the criteria under this section and his determination as to eligibility shall be final.

§ 1150.275 Inspection and investigation.

The Secretary shall have the right, at any time after an application is received from an organization, to examine such books, documents, papers, records, files, and facilities of an organization as he deems necessary to verify the information submitted and to procure such other information as may be required to determine whether the organization is eligible for certification.

§ 1150.276 Review of certification.

Certifications issued pursuant to this subpart are subject to termination or suspension if the organization does not currently meet the certification standards. A certified organization may be requested at any time to supply the Dairy Division with such information as may be required to show that the organization continues to be eligible for certification. Any information submitted to satisfy a request pursuant to this section shall be subject to inspection and investigation as provided in § 1150.275.

§ 1150.277 Listing of certified organizations.

A copy of each certification shall be furnished by the Dairy Division to the respective organization. Copies also shall be filed in the Dairy Division where they will be available for public inspection.

§ 1150.278 Confidential treatment.

All documents and other information submitted by applicant organizations and otherwise obtained by the Department by investigation or examination of books, documents, papers, records, files, or facilities shall be kept confidential by all employees of the Department. Only such information so furnished or acquired as the Secretary deems relevant shall be disclosed by them, and then only in the issuance of general statements based upon the applications of a number of persons, which do not identify the information furnished by any one person.

PARTS 1151–1159 [RESERVED]

PART 1160—FLUID MILK PROMOTION PROGRAM

Subpart—Fluid Milk Promotion Order

DEFINITIONS

AUTHORITY: 7 U.S.C. 6401–6417 and 7 U.S.C. 7401.

SOURCE: 58 FR 46763, Sept. 3, 1993, unless otherwise noted.

Subpart—Fluid Milk Promotion Order

SOURCE: 58 FR 62503, Nov. 29, 1993, unless otherwise noted.

DEFINITIONS

§ 1160.101　Act.

Act means the Fluid Milk Promotion Act of 1990, Subtitle H of Title XIX of the Food, Agriculture, Conservation, and Trade Act of 1990, Public Law 101–624, 7 U.S.C. 6401–6417, and any amendments thereto.

§ 1160.102　Department.

Department means the United States Department of Agriculture.

§ 1160.103　Secretary.

Secretary means the Secretary of Agriculture of the United States or any officer or employee of the Department to whom authority has heretofore been delegated, or to whom authority may hereafter be delegated, to act in the Secretary's stead.

§ 1160.104　United States.

United States means the 48 contiguous states in the continental United States and the District of Columbia, except that United States means the 50 states of the United States of America and the District of Columbia under the following provisions: the petition and review under section 1999K of the Act, enforcement under section 1999L of the Act, and investigations and power to subpoena under section 1999M of the Act.

§ 1160.105　Board.

Board means the National Processor Advertising and Promotion Board established pursuant to 7 U.S.C. 6407(b)(1) and this subpart (hereinafter known as the National Fluid Milk Processor Promotion Board or Board).

§ 1160.106　Person.

Person means any individual, group of individuals, partnership, corporation, association, cooperative or other entity.

§ 1160.107　Fluid milk product.

Fluid milk product means any product that meets the definition provided in § 1000.15 for milk marketing orders issued pursuant to the Agricultural Marketing Agreement Act of 1937, as amended, 7 U.S.C. 601–674.

[67 FR 49858, Aug. 1, 2002]

§ 1160.108　Fluid milk processor.

(a) *Fluid milk processor* means any person who processes and markets commercially fluid milk products in consumer-type packages in the United States (excluding fluid milk products delivered directly to the place of residence of a consumer), except that the term fluid milk processor shall not include in each of the respective fiscal periods those persons who process and market not more than 3,000,000 pounds of such fluid milk products during the representative month, which shall be the first month of the fiscal period.

(b) Any person who did not qualify as a fluid milk processor for a fiscal period because of the 3,000,000-pound limitation shall not later qualify as a fluid milk processor during that fiscal period even though the monthly volume limitation is later exceeded during that period.

(c) Any person who qualified as a fluid milk processor for a fiscal period and whose monthly marketings of fluid milk products later become 3,000,000 pounds or less shall no longer qualify as a fluid milk processor during that fiscal period beginning with the month in which the marketings first dropped below the volume limitation.

(d) For the purpose of determining qualification as a fluid milk processor, each processor of fluid milk products shall report for the representative month of each fiscal period the hundredweight of fluid milk products processed and marketed by the processor.

[58 FR 62503, Nov. 29, 1993, as amended at 62 FR 3983, Jan. 28, 1997; 67 FR 49858, Aug. 1, 2002]

§ 1160.109 Milk.

Milk means any class of cow's milk produced in the United States.

§ 1160.110 Class I price.

Class I price is the price that is established for Class I milk in each marketing area under milk marketing orders authorized by the Agricultural Marketing Agreement Act of 1937, as amended, 7 U.S.C. 601–674.

§ 1160.111 Promotion.

Promotion means the following activities:

(a) *Consumer Education*, which means any program utilizing public relations, advertising or other means devoted to educating consumers about the desirable characteristics of fluid milk products and directed toward increasing the general demand for fluid milk products.

(b) *Advertising*, which means any advertising or promotion program involving only fluid milk products and directed toward educating consumers about the positive attributes of fluid milk and increasing the general demand for fluid milk products.

§ 1160.112 Research.

Research means market research to support advertising and promotion efforts, including educational activities, research directed to product characteristics, and product development, including new products or improved technology in production, manufacturing or processing of milk and the products of milk.

[62 FR 3983, Jan. 28, 1997]

§ 1160.113 Fiscal period.

Fiscal period means the initial period of up to 30 months that this subpart is effective. Thereafter, the fiscal period shall be such annual period as the Board may determine, except that the Board may provide for a lesser or greater period as it may find appropriate for the period immediately after the initial fiscal period to assure continuity of fiscal periods until the beginning of the first annual fiscal period.

[62 FR 3983, Jan. 28, 1997]

§ 1160.114 Eligible organization.

Eligible organization means an organization eligible to nominate members of the Board and which meets the following criteria:

(a) Is a nonprofit organization pursuant to section 501(c) (3), (5), or (6) of the Internal Revenue Code (26 U.S.C. 501(c) (3), (5), or (6));

(b) Is governed by a board comprised of a majority of fluid milk processors; and

(c) Represents fluid milk processors on a national basis whose members process more than 50 percent of the fluid milk products processed and marketed within the United States.

§ 1160.115 Milk marketing area.

Milk marketing area means each area within which milk being marketed is subject to a milk marketing order issued pursuant to the Agricultural Marketing Agreement Act of 1937, as amended, 7 U.S.C. 601–674, or applicable state laws.

§ 1160.116 [Reserved]

§ 1160.117 Continuation referendum.

Continuation referendum means that referendum among fluid milk processors that the Secretary shall conduct as provided in § 1160.501.

NATIONAL FLUID MILK PROCESSOR
PROMOTION BOARD

§ 1160.200 Establishment and membership.

(a) There is hereby established a National Fluid Milk Processor Promotion Board of 20 members, 15 of whom shall represent geographic regions and five of whom shall be at-large members of the Board. To the extent practicable, members representing geographic regions shall represent fluid milk processing operations of differing sizes. No fluid milk processor shall be represented on the Board by more than three members. The at-large members shall include at least three fluid milk processors and at least one member from the general public. Except for the non-processor member or members from the general public, nominees appointed to the Board must be active owners or employees of a fluid milk processor. The failure of such a member to own or work for such fluid milk processor shall disqualify that member for membership on the Board except that such member shall continue to serve on the Board for a period not to exceed 6 months following the disqualification or until appointment of a successor Board member to such position, whichever is sooner, provided that such person continues to meet the criteria for serving on the Board as a processor representative. Should a member representing the general public cease to be employed by the entity employing that member when appointed, gain employment with a new employer, or cease to own or operate the business which that member owned or operated at the date of appointment, such member shall be disqualified for membership on the Board, except that such member shall continue to serve on the Board for a period not to exceed 6 months, or until appointment of a successor Board member, whichever is sooner.

(b) In selecting the 15 Board members who represent geographic regions, one member shall be selected from each of the following regions:

Region 1. Connecticut, Maine, Massachusetts, New Hampshire, Rhode Island, and Vermont.
Region 2. New York and New Jersey.
Region 3. Delaware, Maryland, Pennsylvania, Virginia, and the District of Columbia.
Region 4. Georgia, North Carolina and South Carolina.
Region 5. Florida.
Region 6. Ohio and West Virginia.
Region 7. Michigan, Minnesota, North Dakota, South Dakota and Wisconsin.
Region 8. Illinois and Indiana.
Region 9. Alabama, Kentucky, Louisiana, Mississippi and Tennessee.
Region 10. Texas.
Region 11. Arkansas, Iowa, Kansas, Missouri, Nebraska and Oklahoma.
Region 12. Arizona, Colorado, New Mexico, Nevada, and Utah.
Region 13. Idaho, Montana, Oregon, Washington and Wyoming.
Region 14. Northern California which shall be composed of the Northern California Marketing Area and the South Valley Marketing Area as defined by the Stabilization and Marketing Plan, as amended, issued by the California Department of Food and Agriculture pursuant to the provisions of Chapter 2, Part 3, Division 21, of the California Food and Agriculture Code, effective February 3, 1992.
Region 15. Southern California which shall be composed of the Southern California Marketing Area as defined by the Stabilization and Marketing Plan, as amended, issued by the California Department of Food and Agriculture pursuant to the provisions of Chapter 2, Part 3, Division 21, of the California Food and Agriculture Code, effective February 3, 1992.

[58 FR 62503, Nov. 29, 1993, as amended at 62 FR 3983, Jan. 28, 1997; 63 FR 46639, Sept. 2, 1998; 65 FR 35810, June 6, 2000; 70 FR 14975, Mar. 24, 2005]

§ 1160.201 Term of office.

(a) The members of the Board shall serve for terms of three years, except that the members appointed to the initial Board shall serve proportionately, for terms of one year, two years, and three years, as determined by the Secretary. The terms of all Board members shall expire upon the suspension or termination of the order except as provided in § 1160.502.

(b) No member shall serve more than two consecutive terms, except that any member who is appointed to serve for an initial term of one or two years shall be eligible to be reappointed for two three-year terms. Appointment to another position on the Board is considered a consecutive term. Should a non-board member be appointed to fill a vacancy on the Board with a term of 18 months or less remaining, the appointee shall be entitled to serve two consecutive 3-year terms following the term of the vacant position to which the person was appointed.

[58 FR 62503, Nov. 29, 1993, as amended at 62 FR 3983, Jan. 28, 1997; 63 FR 46639, Sept. 2, 1998]

§ 1160.202 Nominations.

Nominations for members of the Board shall be made in the following manner:

(a) The Secretary shall solicit nominations for the initial Board from individual fluid milk processors and other interested parties, including eligible organizations. Fluid milk processors and other interested parties may submit nominations for positions on the Board for regions in which they are located or market fluid milk, and for at-large members. Eligible organizations may submit a slate of nominees for seats in all regions and for at-large members.

(b) After the appointment of the initial Board, the Secretary shall announce at least 180 days in advance of the expiration of members' terms that such terms are expiring, and shall solicit nominations for such positions in the manner described in paragraph (a) of this section. Nominations for such positions should be submitted to the Secretary not less than 120 days prior to the expiration of members' terms.

§ 1160.203 Nominee's agreement to serve.

Each nominee for Board membership must file with the Secretary at the time of nomination a written agreement to serve on the Board if appointed.

§ 1160.204 Appointment.

From the nominations made pursuant to § 1160.202, the Secretary shall appoint the members of the Board on the basis of representation provided for in §§ 1160.200 and 1160.201.

§ 1160.205 Vacancies.

To fill any vacancy occasioned by the death, removal, resignation, or disqualification of any member of the Board, the Secretary shall appoint a successor from the most recent list of nominations made by individual fluid milk processors and other interested parties, including eligible organizations, for the Board, or from nominations made by the Board.

§ 1160.206 Procedure.

(a) A majority of the members shall constitute a quorum at a properly convened meeting of the Board. Any action of the Board shall require the concurring votes of at least a majority of those present and voting. The Board shall establish rules concerning timely notice of meetings.

(b) The Board may take action upon the concurring votes of a majority of members by mail, telephone, telegraph, or other means of electronic communication when, in the opinion of the chairperson of the Board, such action must be taken before a meeting can be called. Action taken by this emergency procedure is valid only if all members are notified and provided the opportunity to vote and any telephone vote is confirmed promptly in writing. Any action so taken shall have the same force and effect as though such action had been taken at a properly convened meeting of the Board.

§ 1160.207 Compensation and reimbursement.

The members of the Board and trustees, if any, named under § 1160.502, shall serve without compensation but shall be reimbursed for necessary and reasonable expenses incurred by them in the performance of their duties under this subpart.

§ 1160.208 Powers of the Board.

The Board shall have the following powers:

(a) To receive and evaluate, or on its own initiative develop, and budget for plans or projects to educate consumers and promote the use of fluid milk products and to make recommendations to the Secretary regarding such proposals;

(b) To administer the provisions of this subpart in accordance with its terms and provisions;

(c) To make rules and regulations to effectuate the terms and provisions of this subpart;

(d) To receive, investigate, and report to the Secretary complaints of violations of the provisions of this subpart;

(e) To employ such persons as the Board deems necessary and determine the duties and compensation of such persons;

(f) To contract with eligible organizations or other persons to conduct activities authorized pursuant to this subpart;

(g) To select committees and subcommittees, to adopt bylaws, and to adopt such rules for the conduct of its business as it may deem advisable; the Board may establish working committees of persons other than Board members;

(h) To recommend to the Secretary amendments to this subpart; and

(i) With the approval of the Secretary, to invest, pending disbursement pursuant to a plan or project, funds collected through assessments authorized under § 1160.211 in, and only in, obligations of the United States or any agency thereof, in general obligations of any State or any political subdivision thereof, in any interest-bearing account or certificate of deposit of a bank that is a member of the Federal Reserve System, or in obligations fully guaranteed as to principal and interest by the United States.

[58 FR 62503, Nov. 29, 1993, as amended at 63 FR 46639, Sept. 2, 1998]

§ 1160.209   Duties of the Board.

The Board shall have the following duties:

(a) To meet not less than annually, and to organize and select from among its members a chairperson, who may serve for a term of a fiscal period pursuant to § 1160.113, and not more than two consecutive terms, and to select such other officers as may be necessary;

(b) To prepare and submit to the Secretary for approval a budget for each fiscal period of the anticipated expenses and disbursements in the administration of this subpart, including a description of and the probable costs of consumer education, promotion and research projects;

(c) To develop and submit to the Secretary for approval promotion and consumer education, and research plans or projects;

(d) To the extent practicable, carry out consumer education and promotion programs under § 1160.301 in such a manner as to ensure that advertising coverage in each of the regions defined in § 1160.200 is proportionate to funds collected from each such region;

(e) To disseminate information to fluid milk processors or eligible organizations;

(f) To maintain minutes, books and records that accurately reflect all of the acts and transactions of the Board, which shall be available to the Secretary for inspection and audit, and prepare and promptly report minutes of each Board meeting to the Secretary and submit such reports from time to time to the Secretary as the Secretary may prescribe, and to account with respect to the receipt and disbursement of all funds entrusted to it;

(g) To enter into contracts or agreements, with the approval of the Secretary, with such persons and organizations as the Board may approve for the development and conduct of activities authorized under this subpart and for the payment of the cost thereof with funds collected through assessments pursuant to § 1160.211 and income from such assessments. Any such contract or agreement shall provide that:

(1) The contractors shall develop and submit to the Board a plan or project together with a budget(s) showing the estimated cost of such plan or project;

(2) Any such plan or project shall be adopted upon approval of the Secretary; and

(3) The contracting party shall keep accurate records of all of its transactions and make periodic reports to the Board of all activities conducted pursuant to the contract or agreement, and provide accounts of all funds received and expended, and such other reports as the Secretary or the Board

may require. The Secretary or employees of the Board periodically may audit the records of the contracting parties;

(h) For the initial fiscal period, the Board shall contract, to the extent practicable and subject to the approval of the Secretary, with an eligible organization to carry out the provisions of this subpart;

(i) To prepare and make public, at least annually, a report of its activities and an accounting for funds received and expended;

(j) To have an audit of its financial statements conducted by a certified public accountant in accordance with generally accepted auditing standards, at the end of the first 15 months of the initial fiscal period, at the end of the initial fiscal period, and at least once each fiscal period thereafter as well as at such other times as the Secretary may request, and to submit a copy of each such audit report to the Secretary;

(k) To give the Secretary the same notice of meetings of the Board and committees of the Board, including actions conducted under §1160.206(b), as is given to such Board or committee members in order that the Secretary, or a representative of the Secretary, may attend such meetings;

(l) To submit to the Secretary such information pursuant to this subpart as may be requested;

(m) The Board shall take reasonable steps to coordinate the collection of assessments, and promotion, education, and research activities of the Board, with the National Dairy Promotion and Research Board established under section 113(b) of the Dairy Production Stabilization Act of 1983 (7 U.S.C. 4504(b)); and

(n) The Board shall conduct advertising using third parties only through contracts which shall prohibit the third party from selling, offering for sale, or otherwise making available advertising time or space to private industry members conducting brand-name advertising which immediately precedes, follows, appears in juxtaposition, or appears in the midst of Board-sponsored advertising.

[58 FR 62503, Nov. 29, 1993, as amended at 61 FR 27003, May 30, 1996; 62 FR 3983, Jan. 28, 1997]

§ 1160.210   **Expenses.**

(a) The Board is authorized to incur such expenses (including provision for a reasonable reserve) as the Secretary finds are reasonable and likely to be incurred by the Board for its administration, and to enable it to exercise its powers and perform its duties in accordance with the provisions of this subpart; except that, after the Board's first year, it shall not spend on its administration more than 5 percent of the assessments collected during any fiscal period subsequent to the initial fiscal period. Such administrative expenses shall be paid from assessments collected pursuant to §1160.211.

(b) The Board shall reimburse the Secretary for administrative costs incurred by the Department from assessments collected pursuant to §1160.211.

(c) Within 30 days after funds are remitted from Regions 14 and 15, the Board shall provide a grant of 80% of such funds to the entity authorized by the laws of the State of California to conduct an advertising program for fluid milk products in that State for the purpose of implementing a coordinated advertising program in the markets within those regions. Such grant shall be provided with the approval of the Secretary on the following conditions:

(1) The granted funds shall be utilized to implement a fluid milk promotion campaign within the markets within those regions. Verification of the implementation of this program shall be provided to the Board.

(2) The Board shall ensure that the recipients of these funds implement a research and evaluation program to determine the effect of such program on consumption of fluid milk within the region.

(3) The recipient of these funds must provide to the Board data from the research and evaluation programs so that the Board can determine the effect of the program on consumption of fluid milk.

§ 1160.211   **Assessments.**

(a)(1) Each fluid milk processor shall pay to the Board or its designated agent an assessment of $.20 per hundredweight of fluid milk products processed and marketed commercially in consumer-type packages in the United States by such fluid milk processor. Any fluid milk processor who markets milk of its own production directly to consumers as prescribed under section 113(g) of the Dairy Production Stabilization Act of 1983 (7 U.S.C. 4504(g)), and not exempt under §1160.108 or §1160.215, shall also pay the assessment under this subpart. The Secretary shall have the authority to receive assessments on behalf of the Board.

(2) The Secretary shall announce the establishment of the assessment each month in the Class I price announcement in each milk marketing area by adding it to the Class I price for the following month. In the event the assessment is suspended for a given month, the Secretary shall inform all fluid milk processors of the suspension in the Class I price announcement for that month. The Secretary shall also inform fluid milk processors marketing fluid milk in areas not subject to milk marketing orders administered by the Secretary of the establishment or suspension of the assessment.

(3) Each processor responsible for remitting an assessment shall remit it to the Board not later than the last day of the month following the month that the assessed milk was marketed.

(b) Such assessments shall not:

(1) Reduce the prices paid under the Federal milk marketing orders issued under section 8c of the Agricultural Adjustment Act (7 U.S.C. 608c), reenacted with amendments by the Agricultural Marketing Agreement Act of 1937;

(2) Otherwise be deducted from the amounts that handlers must pay to producers for fluid milk products sold to a processor; or

(3) Otherwise be deducted from the price of milk paid to a producer by a handler, as determined by the Secretary.

(c) Money remitted to the Board or the Board's designated agent shall be in the form of a negotiable instrument made payable to the Board or its agent, as the case may be. Processors must mail remittances and reports specified in §§1160.108, 1160.211(a)(1), 1160.213, 1160.214, and 1160.401 to the location designated by the Board or its agent.

[58 FR 62503, Nov. 29, 1993, as amended at 62 FR 3983, Jan. 28, 1997; 70 FR 2753, Jan. 14, 2005]

§ 1160.212 Influencing governmental action.

No funds collected by the Board under this subpart shall in any manner be used for the purpose of influencing governmental policy or action, except to recommend to the Secretary amendments to this subpart.

§ 1160.213 Adjustment of accounts.

Whenever the Board or the Secretary determines through an audit of a processor's reports, records, books or accounts or through some other means that additional money is due the Board or to such processor from the Board, the Board shall notify that person of the amount due or overpaid. If the processor owes money to the Board, it shall remit that amount by the next date for remitting assessments as provided in §1160.211. For the first two erroneous reports submitted by a processor in the preceding 12-month period, late-payment charges assessed pursuant to §1160.214 shall not begin to accrue until the day following such date. For all additional erroneous reports submitted by a processor during the 12-month period, late-payment charges shall accrue from the date the payment was due. If the processor has overpaid, that amount shall be credited to its account and applied against amounts due in succeeding months.

[73 FR 29390, May 21, 2008]

§ 1160.214 Charges and penalties.

(a) Late-payment charge. Any unpaid assessments shall be increased 1.5 percent each month beginning with the day following the date such assessments were due. Any remaining amount due, which shall include any unpaid charges previously made pursuant to this section, shall be increased at the same rate on the corresponding day of each month thereafter until paid. For the purpose of this section, any assessment determined at a date later than prescribed by this subpart because of the failure of a processor to submit a report to the Board when due shall be considered to have been payable by the date it would have been due if the report had been filed when due. The receipt of a payment by the Board will be based on the earlier of the postmark date or the actual date of receipt.

(b) *Penalties.* The Secretary may assess any person who violates any provision of this subpart a civil penalty of not less than nor more than the minimum and maximum amounts specified in §3.91(b)(1)(xxxv) of this title for each such violation. In the case of a willful failure to pay an assessment as required by this subpart, in addition to the amount due, the Secretary may assess an additional penalty of not less than nor more than the minimum and maximum amounts specified in §3.91(b)(1)(xxxv) of this title for each such violation. The amount of any such penalty shall accrue to the United States, which may recover such amount in a civil suit. The remedies provided in this section are in addition to, and not exclusive of, other remedies that may be available by law or in equity.

[58 FR 62503, Nov. 29, 1993, as amended at 70 FR 29579, May 24, 2005]

§1160.215 **Assessment exemption.**

(a) No assessment shall be required on fluid milk products exported from the United States.

(b) A fluid milk processor described in §1160.211(a) who operates under an approved National Organic Program (7 CFR part 205) (NOP) organic handling system plan may be exempt from the payment of assessments under this part, provided that:

(1) Only agricultural products certified as "organic" or "100 percent organic" (as defined in the NOP) are eligible for exemption;

(2) The exemption shall apply to all certified "organic" or "100 percent organic" (as defined in the NOP) products of a fluid milk processor regardless of whether the agricultural commodity subject to the exemption is processed by a person that also processes conventional or nonorganic agricultural products of the same agricultural commodity as that for which the exemption is claimed;

(3) The fluid milk processor maintains a valid certificate of organic operation as issued under the Organic Foods Production Act of 1990 (7 U.S.C. 6501–6522)(OFPA) and the NOP regulations issued under OFPA (7 CFR part 205); and

(4) Any fluid milk processor so exempted shall continue to be obligated to pay assessments under this part that are associated with any agricultural products that do not qualify for an exemption under this section.

(c) To apply for an assessment exemption, a fluid milk processor described in §1160.211(a) shall submit a request to the Board on an *Organic Exemption Request Form* (Form AMS–15) at any time during the year initially, and annually thereafter on or before July 1, for as long as the processor continues to be eligible for the exemption.

(d) A fluid milk processor request for exemption shall include the following information:

(1) The applicant's full name, company name, address, telephone and fax numbers, and email address;

(2) Certification that the applicant maintains a valid organic certificate issued under the OFPA and the NOP;

(3) Certification that the applicant processes organic products eligible to be labeled "organic" or "100 percent organic" under the NOP;

(4) A requirement that the applicant attach a copy of their certificate of organic operation issued by a USDA-accredited certifying agent under the OFPA and the NOP;

(5) Certification, as evidenced by signature and date, that all information provided by the applicant is true; and

(6) Such other information as may be required by the Board, with the approval of the Secretary.

(e) If a fluid milk processor complies with the requirements of this section, the Board will grant an assessment exemption and issue a Certificate of Exemption to the processor within 30 days. If the application is disapproved, the Board will notify the applicant of the reason(s) for disapproval within the same timeframe.

(f) The exemption will apply not later than the last day of the month following the Certificate of Exemption issuance date.

[70 FR 2754, Jan. 14, 2005, as amended at 80 FR 82022, Dec. 31, 2015]

PROMOTION, CONSUMER EDUCATION AND RESEARCH

§ 1160.301　**Promotion, consumer education and research.**

(a) The Board shall receive and evaluate, or on its own initiative develop, and submit to the Secretary for approval any plans or projects authorized in §§ 1160.208 and 1160.209. Such plans or projects shall provide for:

(1) The establishment, issuance, effectuation, and administration of consumer education, promotion and research activities with respect to fluid milk products; and

(2) The evaluation of consumer education, promotion and research activities implemented under the direction of the Board, and the communication of such evaluation to fluid milk processors and the public.

(b) The Board shall periodically review or evaluate each plan or project authorized under § 1160.301(a) to ensure that it contributes to an effective program of promotion, consumer education and research. If the Board finds that any such plan or project does not further the purposes of the Act, the Board shall terminate that plan or project.

(c) No plan or project authorized under § 1160.301(a) may employ unfair or deceptive acts or practices with respect to the quality, value or use of any competing product.

(d) No plan or project authorized under § 1160.301(a) may make use of a brand or trade name of a fluid milk product, except that this paragraph does not preclude the Board from offering program material to commercial parties to use under such terms and conditions as the Board may prescribe, subject to approval by the Secretary.

REPORTS, BOOKS AND RECORDS

§ 1160.401　**Reports.**

Each fluid milk processor marketing milk and paying an assessment under § 1160.211 shall be required to report upon the remittance of such assessments such information as the Board or the Secretary may require. Such information shall include but not be limited to the following:

(a) The quantity of fluid milk products marketed that is subject to the collection of the assessment;

(b) The amount of assessment remitted;

(c) The reason, if necessary, why the remittance is less than the number of hundredweights of milk multiplied by 20 cents; and

(d) The date any assessment was paid.

§ 1160.402　**Books and records.**

Each person subject to this subpart shall maintain and make available for inspection by agents of the Board and the Secretary such books and records as are necessary to carry out the provisions of this subpart and the regulations issued hereunder, including such records as are necessary to verify any reports required. Such books and records shall be retained for at least two years beyond the fiscal period of their applicability.

§ 1160.403　**Confidential treatment.**

(a) All persons, including agents and former agents of the Board, all officers and employees and all former officers and employees of the Department, and all officers and all employees and all former officers and employees of contracting agencies having access to commercial or financial information obtained from such books, records or reports under the Act and this subpart shall keep such information confidential, and not make it available to Board members. Only those persons, as determined by the Secretary, who have a specific need for such information in order to effectively administer the provisions of this subpart shall have access to such information. In addition, they shall disclose only that information the Secretary deems relevant, and then only in a suit or administrative hearing brought at the discretion, or upon the request, of the Secretary, or to which the Secretary or any officer of the United States is a party, and involving this subpart. Nothing in this section, however, shall be deemed to prohibit:

(1) The issuance of general statements based upon the reports of the number of processors, individuals, groups of individuals, partnerships, corporations, associations, cooperatives, or other entities subject to this subpart or statistical data collected from such sources, which statements do not identify the information furnished by any such parties, and

(2) The publication, at the direction of the Secretary, of the name of any processor, individuals, group of individuals, partnership, corporation, association, cooperative, or other entity that has been adjudged to have violated this subpart, together with a statement of the particular provisions of the subpart so violated.

(b) Except as otherwise provided in this subpart, information obtained under this subpart may be made available to another agency of the Federal Government for a civil or criminal law enforcement activity if the activity is authorized by law and if the head of the agency has made a written request to the Secretary specifying the particular information desired and the law enforcement activity for which the information is sought.

(c) Any person violating this section, on conviction, shall be subject to a fine of not more than $1,000 or to imprisonment for not more than 1 year, or both, and if such person is an agent of the Board or an officer or employee of the Department shall be removed from office.

(d) Nothing in this subsection authorizes the Secretary to withhold information from a duly authorized committee or subcommittee of Congress.

MISCELLANEOUS

§ 1160.501   Continuation referenda.

(a) The Secretary at any time may conduct a referendum among those persons who the Secretary determines were fluid milk processors during a representative period, as determined by the Secretary, on whether to suspend or terminate the order. The Secretary shall hold such a referendum at the request of the Board or of any group of such processors that marketed during a

representative period, as determined by the Secretary, 10 percent or more of the volume of fluid milk products marketed in the United States by fluid milk processors voting in the preceding referendum.

(b) Any suspension or termination of the order on the basis of a referendum conducted pursuant to this section must be favored:

(1) By at least 50 percent of the fluid milk processors voting in the referendum; and

(2) By fluid milk processors voting in the referendum that marketed during a representative period, as determined by the Secretary, 40 percent or more of the volume of fluid milk products marketed in the United States by fluid milk processors voting in the referendum.

(c) If the Secretary determines that the suspension or termination of the order is favored in the manner set forth in § 1160.501(b), the Secretary shall take such action within 6 months of such determination.

[58 FR 62503, Nov. 29, 1993, as amended at 62 FR 3983, Jan. 28, 1997]

§ 1160.502   Proceedings after suspension or termination.

(a) Upon the suspension or termination of this subpart, the Board shall recommend to the Secretary not more than five of its members to serve as trustees for the purpose of liquidating the affairs of the Board. Once the Secretary has designated such members as trustees, they shall become trustees of all the funds and property that the Board owns, possesses, or controls, including unpaid and undelivered property or any other unpaid claim existing at the time of such termination. The actions of such trustees shall be subject to approval by the Secretary.

(b) The said trustees shall:

(1) Serve as trustees until discharged by the Secretary;

(2) Carry out the obligations of the Board under any contract or agreements that it entered pursuant to §§ 1160.208 and 1160.209;

(3) Account for all receipts and disbursements and deliver to any person designated by the Secretary all property on hand, together with all books

and records of the Board and the trustees; and

(4) At the request of the Secretary, execute such assignments or other instruments necessary or appropriate to vest in the Secretary's designee full title and right to all of the funds, property, and claims of the Board or the trustees.

(c) The Secretary's designee shall be subject to the same obligations with respect to funds, property or claims transferred or delivered pursuant to this subpart as the Board and the trustees.

(d) The Board, the trustees or the Secretary's designee shall deliver to the Secretary any residual funds not required to pay liquidation expenses, which funds may be used, to the extent practicable, to continue one or more of the promotion, research or nutrition education plans or projects authorized pursuant to this subpart.

§ 1160.503 Effect of suspension, termination or amendment.

Unless otherwise expressly provided by the Secretary, the suspension or termination of this subpart or of any regulation issued pursuant hereto, or the issuance of any amendment to either thereof, shall not:

(a) Affect or waive any right, duty, obligation, or liability of the Board or its trustees which shall have arisen or which may hereafter arise in connection with any provision of this subpart or any regulation issued thereunder;

(b) Release or extinguish any violation of this subpart or any regulation issued thereunder; or

(c) Affect or impair any rights or remedies of the United States, the Secretary, or any person, with respect to any such violation.

§ 1160.504 Personal liability.

No member or employee of the Board shall be held personally responsible, either individually or jointly, in any way whatsoever to any person for errors in judgment, mistakes, or other acts of either commission or omission by such member or employee, except for acts of dishonesty or willful misconduct.

§ 1160.505 Patents, copyrights, inventions and publications.

(a) Any patents, copyrights, trademarks, inventions or publications developed through the use of funds collected under the provisions of this subpart are the property of the United States Government as represented by the Board, and shall, along with any rents, royalties, residual payments, or other income from the rental, sale, leasing, franchising, or other uses of such patents, copyrights, inventions, or publications, inure to the benefit of the Board. Section 1160.502 governs the disposition of all such property upon suspension or termination of this subpart.

(b) Should patents, copyrights, inventions, and publications be developed through the use of funds collected by the Board under this subpart, and funds contributed by another organization or person, ownership and related rights to such patents, copyrights, inventions, and publications shall be determined by the agreement between the Board and the party contributing funds towards the development of such patent, copyright, invention, and publication in a manner consistent with paragraph (a) of this section.

[58 FR 62503, Nov. 29, 1993, as amended at 63 FR 46639, Sept. 2, 1998]

§ 1160.506 Amendments.

The Secretary may from time to time amend provisions of this subpart. Any interested person or organization affected by the provisions of the Act may propose amendments to the Secretary.

§ 1160.507 Report.

The Secretary shall provide annually for an independent evaluation of the effectiveness of the fluid milk promotion program carried out under this subtitle during the previous fiscal year, in conjunction with the evaluation of the National Dairy Promotion and Research Board established under section 113(b) of the Dairy Production Stabilization Act of 1983 (7 U.S.C. 4504(b)).

§ 1160.508 Separability.

If any provision of this subpart is declared invalid or the applicability thereof to any person or any circumstances is held invalid, such declaration or holding shall not offset the validity of the remainder of this subpart or the applicability thereof to other persons or circumstances.

Subpart—Procedure for Conduct of Referenda in Connection with a Fluid Milk Promotion Order

§ 1160.600 General.

Referenda to determine whether eligible fluid milk processors favor the issuance, continuance, termination or suspension of a Fluid Milk Promotion Order authorized by the Fluid Milk Promotion Act of 1990 shall be conducted in accordance with this subpart.

§ 1160.601 Definitions.

As used in this subpart:

(a) *Act* means the Fluid Milk Promotion Act of 1990 (Subtitle H of Title XIX of the Food, Agriculture, Conservation, and Trade Act of 1990, Pub. L. 101–624, 7 U.S.C. 6401–6417) and any amendments thereto.

(b) *Department* means the United States Department of Agriculture.

(c) *Secretary* means the Secretary of Agriculture of the United States or any officer or employee of the Department to whom authority has heretofore been delegated, or to whom authority may hereafter be delegated, to act in the Secretary's stead.

(d) *Administrator* means the Administrator of the Agricultural Marketing Service, with power to redelegate, or any officer or employee of the Department to whom authority has been delegated or may hereafter be delegated to act in the Administrator's stead.

(e) *Order* means a Fluid Milk Promotion Order, and any amendments thereto, authorized by the Act.

(f) *Board* means the National Fluid Milk Processor Promotion Board established pursuant to the Act.

(g) *Assessment* means the monies that are collected and remitted to the Board pursuant to the Act.

(h) *Person* means any individual, group of individuals, partnership, corporation, association, cooperative association or other entity.

(i) *Fluid milk processor* means any person who is defined as a fluid milk processor under the order, or under the proposed order on which the initial referendum is held.

(j) *Referendum agent* means the person designated by the Secretary to conduct the referendum.

(k) *Representative period* means the period designated by the Secretary pursuant to Sections 1999N and 1999O of the Act.

§ 1160.602 Conduct of referendum.

(a) The referendum shall be conducted by mail in the manner prescribed in this subpart. The referendum agent may utilize such personnel or agencies of the Department as are deemed necessary by the Administrator. There shall be no voting except within the time specified by the referendum agent.

(b) The referendum agent shall mail to each fluid milk processor that has properly registered to participate in the referendum:

(1) A ballot containing a description of the question(s) upon which the referendum is being held;

(2) Instructions for completing the ballot; and

(3) A statement as to the time within which the ballot must be mailed to the referendum agent.

§ 1160.603 Who may vote.

(a) Each person who was a fluid milk processor during the representative period, as determined by the Secretary, and who at the time of voter registration and when voting is processing and marketing commercially fluid milk products in consumer-type packages in the United States shall be entitled to vote in a referendum, and no such person shall be refused a ballot. Any person casting more than one ballot with conflicting votes shall thereby invalidate all ballots cast by such person in such referendum. Each person voting shall have registered with the referendum agent prior to the voting period. Each ballot cast shall contain a certification by the person casting the ballot that such person is qualified to vote. All information required on the ballot pertinent to the identification of the person voting must be supplied and certified to as being correct in order for the ballot to be valid.

(b) Voting by proxy or agent will not be permitted. However, the ballot of a fluid milk processor who is other than an individual may be cast by a person who is duly authorized to do so, and such ballot shall contain a certification by such person that the entity on whose behalf the ballot is cast was a fluid milk processor during the representative period. All information required on the ballot pertinent to the identification of the fluid milk processor on whose behalf the ballot is cast must be supplied and certified to as being correct in order for the ballot to be valid.

§ 1160.604 Duties of the referendum agent.

The referendum agent, in addition to any other duties imposed by this subpart, shall:

(a) For the purpose of adjusting the rate of assessment, determine and publicly announce prior to the voting period the total volume of fluid milk products marketed by all processors of fluid milk in the United States during the representative period and the portion of such volume that must be represented by those fluid milk processors voting in favor of the question included on the ballot if the referendum question is to pass.

(b)(1) Within 12 days after the deadline for registering to vote in the referendum, the referendum agent shall make available upon request a list of those fluid milk processors that properly registered. Any challenge of a processor's eligibility to vote must be received by the referendum agent within 17 days of the deadline for voter registration.

(2) If the voting eligibility of any fluid milk processor is challenged within the timeframe specified in § 1160.604(b)(1), the referendum agent shall review the challenge and make a final determination regarding the processor's eligibility to vote.

(3) Prior to the time of mailing ballots to fluid milk processors, the referendum agent shall prepare a final list of eligible voters and make such list available upon request.

(c) Verify the eligibility of all persons voting in the referendum by reviewing all ballots cast to assure that each ballot:

(1) Was mailed within the prescribed time;

(2) Contains all certifications required attesting to the eligibility of the person to vote, and that the person voting filed with the referendum agent prior to the voting period the advance registration required pursuant to § 1160.606(a)(1); and

(3) Was completed with respect to all necessary information pertinent to the identification of the person voting so that additional verification can be conducted by the referendum agent to substantiate the eligibility of each such person to vote.

(d) Conduct further verification, as necessary, to determine the eligibility of each person to vote. Such verification may be completed by reviewing readily available sources of information, including the following:

(1) Records of the Department;

(2) Fluid milk processors' records; and

(3) Any other reliable sources of information which may be available to the referendum agent.

(e) Further verify ballots to avoid a duplication of votes. The following criteria shall serve as a guide:

(1) Each fluid milk processor that is other than an individual shall be regarded as one person for voting purposes;

(2) No more than one vote may be cast on behalf of any one fluid milk processor; and

(3) In the event that more than one individual claim the right to vote and cast a ballot for a fluid milk processor, concurring votes of such individuals shall be treated as one vote while any conflicting votes shall thereby invalidate all ballots cast by such individuals.

[58 FR 62503, Nov. 29, 1993, as amended at 62 FR 3983, Jan. 28, 1997]

§ 1160.605 Scheduling of referendum.

A referendum shall be held:

(a) Whenever prescribed by the order;

(b) For the purpose of adjusting the rate of assessment:

(1) At the direction of the Secretary; or

(2) Upon request of the Board or upon request of any group of fluid milk processors that marketed during a representative period, as determined by the Secretary, 10 percent or more of the volume of fluid milk products marketed by all processors of fluid milk in the United States during that period; or

(c) For the purpose of suspending or terminating the order:

(1) At the direction of the Secretary; or

(2) Upon request of the Board or upon request of any group of fluid milk processors that marketed during a representative period, as determined by the Secretary, 10 percent or more of the volume of fluid milk products marketed by fluid milk processors voting in the preceding referendum.

[62 FR 3984, Jan. 28, 1997]

§ 1160.606 Notice of referendum.

The referendum agent shall provide at least 30 days' notice of any referendum authorized by the Act by:

(a) Mailing to each known person processing fluid milk products a notice of referendum, which shall include:

(1) An advance registration form to be filed with the referendum agent prior to the voting period by any person choosing to vote in the referendum, with a statement as to the time within which the registration form must be mailed to the referendum agent;

(2) A copy of the final rule, when applicable;

(3) A sample ballot containing a description of the question(s) upon which the referendum is being held; and

(4) Rules for participating in the referendum, including a statement as to the time within which the ballot must be mailed to the referendum agent; and

(b) Giving public notice of the referendum:

(1) By furnishing press releases and other information to available media of public information (including but not limited to press, radio, and television facilities) announcing the time within which ballots must be completed and mailed to the referendum agent, eligibility requirements, required certifications to cast a valid ballot, where additional information, ballots and instructions may be obtained, and other pertinent information; and

(2) By such other means as the referendum agent may deem advisable.

§ 1160.607 Tabulation of ballots.

(a) The referendum agent shall verify the validity of all ballots cast in accordance with the instructions and requirements specified in §§ 1160.602 through 1160.606. Ballots that are not valid shall be marked "disqualified" with a notation on the ballot as to the reason for the disqualification.

(b) The total number of ballots cast, including the disqualified ballots, shall be ascertained. The number of ballots cast approving, the number of ballots cast disapproving, and the pounds of fluid milk products distributed during the representative period by the processors represented in each grouping of ballots, shall also be ascertained. The ballots marked "disqualified" shall not be considered as approving or disapproving, and the persons who cast such ballots shall not be regarded as participating in the referendum.

(c) The referendum agent shall notify the Administrator of the number of ballots cast, the count of the votes, the number of disqualified ballots, and the volume of fluid milk products associated with the ballots cast as prescribed in § 1160.607(b). The referendum agent shall seal the ballots and transmit to the Administrator a complete detailed report of all actions taken in connection with the referendum and all other information furnished to or compiled by the referendum agent.

(d) Announcement of the results of the referendum will be made only at the direction of the Secretary. The referendum agent or others who assist in the referendum shall not disclose the results of the referendum or the total number of ballots and votes cast.

§ 1160.608 Confidential information.

The ballots cast, the identity of any person who voted, or the manner in which any person voted and all information furnished to, compiled by, or in the possession of the referendum agent, except the list of eligible voters, shall be regarded as confidential.

§ 1160.609 Supplementary instructions.

The Administrator is authorized to issue instructions and to prescribe forms and ballots, not inconsistent with the provisions of this subpart, to govern the conduct of referenda by referendum agents.

PARTS 1161–1169 [RESERVED]

PART 1170—DAIRY PRODUCT MANDATORY REPORTING

Sec.
1170.1   Secretary.
1170.2   Act.
1170.3   Person.
1170.4   Dairy products.
1170.5   Manufacturer.
1170.6   Store.

AUTHORITY: 7 U.S.C. 1637–1637b, as amended by Pub. L. 106–532, 114 Stat. 2541; Pub. L. 107–171, 116 Stat. 207; and Pub. L. 111–239, 124 Stat. 2501.

SOURCE: 73 FR 34181, June 17, 2008, unless otherwise noted.

§ 1170.1 Secretary.

Secretary means the Secretary of Agriculture of the United States or any other officer or employee of USDA to whom authority has been delegated.

§ 1170.2 Act.

Act means the Agricultural Marketing Act of 1946, 7 U.S.C. 1621 et seq., as amended by the Dairy Market Enhancement Act of 2000, Pub. L. 106–532, 114 Stat. 2541; the Farm Security and Rural Investment Act of 2002, Pub. L. 107–171, 116 Stat. 207; and the Mandatory Price Reporting Act of 2010, Pub. L. 111–239, 124 Stat. 2501.

[77 FR 8721, Feb. 15, 2012]

§ 1170.3 Person.

Person means an individual, partnership, corporation, association, or any other business unit.

§ 1170.4 Dairy products.

Dairy Products means:
(a) Manufactured dairy products that are used by the Secretary to establish minimum prices for Class III and Class IV milk under a Federal milk marketing order issued under section 8c of the Agricultural Adjustment Act (7 U.S.C. 608c), reenacted with amendments by the Agricultural Marketing Agreement Act of 1937; and
(b) Substantially identical products designated by the Secretary in this part.

§ 1170.5 Manufacturer.

Manufacturer means any person engaged in the business of buying milk in commerce for the purpose of manufacturing dairy products in one or more locations.

§ 1170.6 Store.

(a) Store means to place cheese or butter in a warehouse or facility which is artificially cooled to a temperature of 50 degrees Fahrenheit or lower and hold these dairy products for 30 days or more; or
(b) Store means to place nonfat dry milk or dry whey in a manufacturing plant, packaging plant, distribution point, or shipment in transit.

DAIRY PRODUCT REPORTING PROGRAMS

§ 1170.7 Reporting requirements.

(a) All dairy product manufacturers, with the exception of those who are exempt as described in § 1170.9, shall submit a report weekly to the Agricultural Marketing Service (AMS) by Tuesday, 12 noon local time of reporting entities, on all products sold as specified in § 1170.8 during the 7 days ending 12 midnight of the previous Saturday, local time of the plant or storage facility where the sales are made. If a Federal holiday falls on Monday through Wednesday of a particular week, the due date for report submission may be adjusted. Prior to the beginning of each calendar year, AMS shall release, to manufacturers that are required to report, the times and dates that reports are due. For the applicable products, the report shall be submitted by electronic means specified by AMS and shall indicate the name, address, plant location(s), quantities sold, total sales dollars, dollars per pound, and the moisture content where applicable. Each sale shall be reported for the time period when the transaction is completed, i.e. the product is "shipped out" and title transfer occurs. Each sale shall be reported either f.o.b. plant if the product is "shipped out" from the plant or f.o.b. storage facility location if the product is "shipped out" from a storage facility. In calculating the total dollars received and dollars per pound, the reporting entity shall neither add transportation charges incurred at the time the product is "shipped out" or after the product is "shipped out" nor deduct transportation charges incurred before the product is "shipped out." In calculating the total dollars received and dollars per pound, the reporting entity shall not deduct brokerage fees or clearing charges paid by the manufacturer.

(b) Manufacturers or other persons storing dairy products are required to report, on a monthly basis, stocks of dairy products (as defined in § 1170.4) on hand, on the appropriate forms supplied by the National Agricultural Statistic Service. The report shall indicate the name, address, and stocks on hand at the end of the month for each storage location.

[77 FR 8721, Feb. 15, 2012]

§ 1170.8 Price reporting specifications.

The following are the reporting specifications for each dairy product:

(a) Specifications for Cheddar Cheese Prices:

(1) Variety: Cheddar cheese.

(2) Style: 40-pound blocks or 500-pound barrels.

(3) Moisture Content:

(i) 40-pound blocks: Moisture content is not reported. Exclude cheese that will be aged.

(ii) 500-pound barrels: Report weighted average moisture content of cheese sold. AMS will adjust price to a benchmark of 38.0 percent based on standard moisture adjustment formulas. Ex-

clude cheese with moisture content exceeding 37.7 percent.

(4) Age: Not less than 4 days or more than 30 days on date of sale.

(5) Grade:

(i) 40-pound blocks: Product meets Wisconsin State Brand or USDA Grade A or better standards.

(ii) 500-pound barrels: Product meets Wisconsin State Brand or USDA Extra Grade or better standards.

(6) Color:

(i) 40-pound blocks: colored and within the color range of 6–8 on the National Cheese Institute color chart.

(ii) 500-pound barrels: white.

(7) Packaging:

(i) 40-pound blocks: Price should reflect cheese wrapped in a sealed, airtight package in corrugated or solid fiberboard containers with a reinforcing inner liner or sleeve. Exclude all other packaging costs from the reported price.

(ii) 500-pound barrels: Exclude all packaging costs from the reported price.

(8) Exclude: Intra-company sales, resales of purchased cheese, forward pricing sales (sales in which the selling price was set [not adjusted] 30 or more days before the transaction was completed), cheese produced under faith-based close supervision and marketed at a higher price than the manufacturer's wholesale market price for the basic commodity (for example, kosher cheese produced with a rabbi on site who is actively involved in supervision of the production process), sales under the Dairy Export Incentive Program or other premium-assisted sales (for example, export assistance sales through the Cooperatives Working Together program), and cheese certified as organic by a USDA-accredited certifying agent.

(b) Specifications for Butter Prices:

(1) Variety: 80 percent butterfat, salted, fresh or storage.

(2) Grade: Product meets USDA Grade AA standards.

(3) Packaging: 25-kilogram and 68-pound box sales.

(4) Exclude: Unsalted and Grade A butter, intra-company sales, resales of purchased butter, forward pricing sales (sales in which the selling price was set [not adjusted] 30 or more days before

the transaction was completed), butter produced under faith-based close supervision and marketed at a higher price than the manufacturer's wholesale market price for the basic commodity (for example, kosher butter produced with a rabbi on site who is actively involved in supervision of the production process), sales under the Dairy Export Incentive Program or other premium-assisted sales (for example, export assistance sales through the CWT program), and butter certified as organic by a USDA-accredited certifying agent.

(c) Specifications for Dry Whey Prices:

(1) Variety: Edible nonhygroscopic.

(2) Age: No more than 180 days.

(3) Grade: Product meets USDA Extra Grade standards.

(4) Packaging or container: 25-kilogram bag, 50-pound bag, tote, or tanker.

(5) Exclude: Sales of Grade A dry whey, intra-company sales, resales of purchased dry whey, forward pricing sales (sales in which the selling price was set [not adjusted] 30 or more days before the transaction was completed), dry whey produced under faith-based close supervision and marketed at a higher price than the manufacturer's wholesale market price for the basic commodity (for example, kosher dry whey produced with a rabbi on site who is actively involved in supervision of the production process), premium-assisted sales, and dry whey certified as organic by a USDA-accredited certifying agent.

(d) Specifications for the Nonfat Dry Milk Prices:

(1) Variety: Non-fortified.

(2) Age: No more than 180 days.

(3) Grade: Product meets USDA Extra Grade or USPH[2] Grade A standards.

(4) Packaging or container: 25-kilogram bag, 50-pound bag, tote, or tanker.

(5) Exclude: Nonfat dry milk manufactured using high heat process, sales of instant nonfat dry milk, sales of dry buttermilk products, intra-company sales, resales of purchased nonfat dry

---

[2] USPH refers to the US Department of Health and Human Services—Public Health Service/Food and Drug Administration.

milk, forward pricing sales (sales in which the selling price was set [not adjusted] 30 or more days before the transaction was completed), nonfat dry milk produced under faith-based close supervision and marketed at a higher price than the manufacturer's wholesale market price for the basic commodity (for example, kosher nonfat dry milk produced with a rabbi on site who is actively involved in supervision of the production process), sales under the Dairy Export Incentive Program or other premium-assisted sales, and nonfat dry milk certified as organic by a USDA-accredited certifying agent.

[73 FR 34181, June 17, 2008, as amended at 77 FR 8721, Feb. 15, 2012]

§ 1170.9 Price reporting exemptions.

(a) Any manufacturer that processes and markets less than 1 million pounds of cheddar cheese per calendar year is exempt from reporting cheddar cheese sales as specified in § 1170.8(a).

(b) Any manufacturer that processes and markets less than 1 million pounds of butter per calendar year is exempt from reporting butter sales as specified in § 1170.8(b).

(c) Any manufacturer that processes and markets less than 1 million pounds of dry whey per calendar year is exempt from reporting dry whey sales as specified in § 1170.8(c).

(d) Any manufacturer that processes and markets less than 1 million pounds of nonfat dry milk per calendar year is exempt from reporting nonfat dry milk sales as specified in § 1170.8(d).

§ 1170.10 Storage reporting specifications.

(a) Cold Storage Report:

(1) Reporting universe: All warehouses or facilities, artificially cooled to a temperature of 50 degrees Fahrenheit or lower, where dairy products generally are placed and held for 30 days or more. Excluded are stocks in refrigerated space maintained by wholesalers, jobbers, distributors, and chain stores; locker plants containing individual lockers; and frozen food processors whose inventories are turned over more than once a month.

(2) Products required to be reported:

(i) Natural cheese, domestic and foreign made, including barrel and cheese to be processed; American type cheeses, (cheddar, Monterey, Colby, etc.), including government owned stocks; Swiss; other natural cheese types (brick, mozzarella, Muenster, Parmesan, etc.). Exclude processed cheese.

(ii) Salted and unsalted butter, anhydrous milkfat (AMF), butter oil, including government owned stocks.

. (b) Dairy Products Report:

(1) Reporting universe: All manufacturing plants.

(2) Products required to be reported:

(i) Nonfat dry milk.

(ii) Dry whey.

§1170.11 Records.

Each person required to report information to the Secretary shall maintain, and make available to the Secretary, on request, original contracts, agreements, receipts, and other records associated with the sale or storage of any dairy products during the 2-year period beginning on the date of the creation of the records.

§1170.12 Confidential information.

Except as otherwise directed by the Secretary or the Attorney General for enforcement purposes, no officer, employee, or agent of the United States shall make available to the public information, statistics, or documents obtained from or submitted by any person in compliance with the Dairy Product Mandatory Reporting program other than in a manner that ensures that confidentiality is preserved regarding the identity of person, including parties to a contract, and proprietary business information.

VERIFICATION AND ENFORCEMENT

§1170.13 Verification of reports.

For the purpose of assuring compliance and verification, records and reports required to be filed by manufacturers or other persons pursuant to section 273(b)(1)(A)(i) of the Act, the Agricultural Marketing Service, through its duly authorized agents, shall have access to any premises where applicable records are maintained, where dairy products are produced or stored, and at any time during reasonable business hours shall be permitted to in-

spect such manufacturer or person, and any original contracts, agreements, receipts, and other records associated with the sale of any dairy products.

§1170.14 Noncompliance procedures.

(a) When the Secretary becomes aware that a manufacturer or person may have willfully delayed reporting of, or failed or refused to provide, accurate information pursuant to section 273(b)(1)(A)(i) of the Act, the Secretary may issue a cease and desist order.

(b) Prior to the issuance of a cease and desist order, the Secretary shall provide notice and an opportunity for an informal hearing regarding the matter to the manufacturer or person involved.

(c) The notice shall contain the following information:

(1) That the issuance of a cease and desist order is being considered;

(2) That the reasons for the proposed cease and desist order in terms sufficient to put the person on notice of the conduct or lack thereof upon which the notice is based;

(3) That within 30 days after receipt of the notice, the manufacturer or person may submit, in person, in writing, or through a representative, information and argument in opposition to the proposed cease and desist order; and

(4) That if no response to the notice is received within the 30 days after receipt of the notice, that a cease and desist order may be issued immediately.

(d) If a manufacturer or person requests a hearing, the hearing should be held at a location and time that is convenient to the parties concerned, if possible. The hearing will be held before the Deputy Administrator, Dairy Programs, Agricultural Marketing Service, or a designee. The manufacturer or person may be represented. Witnesses may be called by either party.

(e) The Deputy Administrator, Dairy Programs, Agricultural Marketing Service, or a designee will make a decision on the basis of all the information in the administrative record, including any submission made by the manufacturer or person. The decision of whether a cease and desist order should be issued shall be made within 30 days after receipt of any information and argument submitted by the manufacturer or person. The cease and desist order shall be final unless the affected manufacturer or person requests a reconsideration of the order to the Administrator, Agricultural Marketing Service, within 30 days after the date of the issuance of the order.

§ 1170.15 Appeals.

If the cease and desist order is confirmed by the Administrator, Agricultural Marketing Service, the manufacturer or person may appeal the order in the appropriate United States District Court not later than 30 days after the date of the confirmation of the order.

§ 1170.16 Enforcement.

(a) If a person subject to the Dairy Product Mandatory Reporting program fails to obey a cease and desist order after the order has become final and unappealable, or after the appropriate United States district court has entered a final judgment in favor of the Administrator, Agricultural Marketing Service, the United States may apply to the appropriate United States district court for enforcement of the order.

(b) If the court determines that the cease and desist order was lawfully made and duly served and that the manufacturer or person violated the .order, the court shall enforce the order.

(c) If the court finds that the manufacturer or person violated the cease and desist order, the manufacturer or person shall be subject to a civil penalty of not more than the amount specified at § 3.91(b)(1)(liv) of this title for each offense.

[73 FR 34181, June 17, 2008, as amended at 75 FR 17561, Apr. 7, 2010]

§ 1170.17 Publication of statistical information.

Not later than 3 p.m. Eastern Time on the Wednesday of each week, AMS shall publish aggregated information obtained from manufacturers or other persons of all products sold as specified in § 1170.8. If a Federal holiday falls on Monday through Wednesday of a particular week, the due date for report publication may be adjusted. The public shall be notified of report times prior to the beginning of the calendar year.

[77 FR 8721, Feb. 15, 2012]

PARTS 1171–1199 [RESERVED]

FINDING AIDS

A list of CFR titles, subtitles, chapters, subchapters and parts and an alphabetical list of agencies publishing in the CFR are included in the CFR Index and Finding Aids volume to the Code of Federal Regulations which is published separately and revised annually.

Table of CFR Titles and Chapters
Alphabetical List of Agencies Appearing in the CFR
List of CFR Sections Affected

Table of CFR Titles and Chapters

(Revised as of January 1, 2020)

Title 1—General Provisions

Title 2—Grants and Agreements

Title 2—Grants and Agreements—Continued

Title 3—The President

Title 4—Accounts

Title 5—Administrative Personnel

246

Title 15—Commerce and Foreign Trade—Continued

Chap.

XIII   East-West Foreign Trade Board (Parts 1300—1399)

XIV   Minority Business Development Agency (Parts 1400—1499)

SUBTITLE C—REGULATIONS RELATING TO FOREIGN TRADE AGREE-
MENTS

XX   Office of the United States Trade Representative (Parts 2000—2099)

SUBTITLE D—REGULATIONS RELATING TO TELECOMMUNICATIONS
AND INFORMATION

XXIII   National Telecommunications and Information Administration,
Department of Commerce (Parts 2300—2399) [Reserved]

Title 16—Commercial Practices

I   Federal Trade Commission (Parts 0—999)

II   Consumer Product Safety Commission (Parts 1000—1799)

Title 17—Commodity and Securities Exchanges

I   Commodity Futures Trading Commission (Parts 1—199)

II   Securities and Exchange Commission (Parts 200—399)

IV   Department of the Treasury (Parts 400—499)

Title 18—Conservation of Power and Water Resources

I   Federal Energy Regulatory Commission, Department of Energy
(Parts 1—399)

III   Delaware River Basin Commission (Parts 400—499)

VI   Water Resources Council (Parts 700—799)

VIII   Susquehanna River Basin Commission (Parts 800—899)

XIII   Tennessee Valley Authority (Parts 1300—1399)

Title 19—Customs Duties

I   U.S. Customs and Border Protection, Department of Homeland
Security; Department of the Treasury (Parts 0—199)

II   United States International Trade Commission (Parts 200—299)

III   International Trade Administration, Department of Commerce
(Parts 300—399)

IV   U.S. Immigration and Customs Enforcement, Department of
Homeland Security (Parts 400—599) [Reserved]

Title 20—Employees' Benefits

I   Office of Workers' Compensation Programs, Department of
Labor (Parts 1—199)

II   Railroad Retirement Board (Parts 200—399)

III   Social Security Administration (Parts 400—499)

250

251

Title 23—Highways—Continued

Title 24—Housing and Urban Development

Title 25—Indians

Title 26—Internal Revenue

Title 27—Alcohol, Tobacco Products and Firearms

Title 28—Judicial Administration

Title 29—Labor

254

Title 31—Money and Finance: Treasury—Continued

Title 32—National Defense

Title 33—Navigation and Navigable Waters

Title 34—Education

Title 34—Education—Continued

Title 35 [Reserved]

Title 36—Parks, Forests, and Public Property

Title 37—Patents, Trademarks, and Copyrights

Title 38—Pensions, Bonuses, and Veterans' Relief

Title 39—Postal Service

Title 40—Protection of Environment

Title 41—Public Contracts and Property Management

Title 41—Public Contracts and Property Management—Continued

Title 42—Public Health

Title 43—Public Lands: Interior

Title 44—Emergency Management and Assistance

Title 45—Public Welfare

Title 48—Federal Acquisition Regulations System

54 Defense Logistics Agency, Department of Defense (Parts 5400—5499)

57 African Development Foundation (Parts 5700—5799)

61 Civilian Board of Contract Appeals, General Services Administration (Parts 6100—6199)

99 Cost Accounting Standards Board, Office of Federal Procurement Policy, Office of Management and Budget (Parts 9900—9999)

Title 49—Transportation

SUBTITLE A—OFFICE OF THE SECRETARY OF TRANSPORTATION (PARTS 1—99)

SUBTITLE B—OTHER REGULATIONS RELATING TO TRANSPORTATION

I Pipeline and Hazardous Materials Safety Administration, Department of Transportation (Parts 100—199)

II Federal Railroad Administration, Department of Transportation (Parts 200—299)

III Federal Motor Carrier Safety Administration, Department of Transportation (Parts 300—399)

IV Coast Guard, Department of Homeland Security (Parts 400—499)

V National Highway Traffic Safety Administration, Department of Transportation (Parts 500—599)

VI Federal Transit Administration, Department of Transportation (Parts 600—699)

VII National Railroad Passenger Corporation (AMTRAK) (Parts 700—799)

VIII National Transportation Safety Board (Parts 800—999)

X Surface Transportation Board (Parts 1000—1399)

XI Research and Innovative Technology Administration, Department of Transportation (Parts 1400—1499) [Reserved]

XII Transportation Security Administration, Department of Homeland Security (Parts 1500—1699)

Title 50—Wildlife and Fisheries

I United States Fish and Wildlife Service, Department of the Interior (Parts 1—199)

II National Marine Fisheries Service, National Oceanic and Atmospheric Administration, Department of Commerce (Parts 200—299)

III International Fishing and Related Activities (Parts 300—399)

IV Joint Regulations (United States Fish and Wildlife Service, Department of the Interior and National Marine Fisheries Service, National Oceanic and Atmospheric Administration, Department of Commerce); Endangered Species Committee Regulations (Parts 400—499)

V Marine Mammal Commission (Parts 500—599)

Alphabetical List of Agencies Appearing in the CFR

(Revised as of January 1, 2020)

Agency	CFR Title, Subtitle or Chapter
Administrative Conference of the United States	1, III
Advisory Council on Historic Preservation	36, VIII
Advocacy and Outreach, Office of	7, XXV
Afghanistan Reconstruction, Special Inspector General for	5, LXXXIII
African Development Foundation	22, XV
Federal Acquisition Regulation	48, 57
Agency for International Development	2, VII; 22, II
Federal Acquisition Regulation	48, 7
Agricultural Marketing Service	7, I, VIII, IX, X, XI; 9, II
Agricultural Research Service	7, V
Agriculture, Department of	2, IV; 5, LXXIII
Advocacy and Outreach, Office of	7, XXV
Agricultural Marketing Service	7, I, VIII, IX, X, XI; 9, II
Agricultural Research Service	7, V
Animal and Plant Health Inspection Service	7, III; 9, I
Chief Financial Officer, Office of	7, XXX
Commodity Credit Corporation	7, XIV
Economic Research Service	7, XXXVII
Energy Policy and New Uses, Office of	2, IX; 7, XXIX
Environmental Quality, Office of	7, XXXI
Farm Service Agency	7, VII, XVIII
Federal Acquisition Regulation	48, 4
Federal Crop Insurance Corporation	7, IV
Food and Nutrition Service	7, II
Food Safety and Inspection Service	9, III
Foreign Agricultural Service	7, XV
Forest Service	36, II
Information Resources Management, Office of	7, XXVII
Inspector General, Office of	7, XXVI
National Agricultural Library	7, XLI
National Agricultural Statistics Service	7, XXXVI
National Institute of Food and Agriculture	7, XXXIV
Natural Resources Conservation Service	7, VI
Operations, Office of	7, XXVIII
Procurement and Property Management, Office of	7, XXXII
Rural Business-Cooperative Service	7, XVIII, XLII
Rural Development Administration	7, XLII
Rural Housing Service	7, XVIII, XXXV
Rural Utilities Service	7, XVII, XVIII, XLII
Secretary of Agriculture, Office of	7, Subtitle A
Transportation, Office of	7, XXXIII
World Agricultural Outlook Board	7, XXXVIII
Air Force, Department of	32, VII
Federal Acquisition Regulation Supplement	48, 53
Air Transportation Stabilization Board	14, VI
Alcohol and Tobacco Tax and Trade Bureau	27, I
Alcohol, Tobacco, Firearms, and Explosives, Bureau of	27, II
AMTRAK	49, VII
American Battle Monuments Commission	36, IV
American Indians, Office of the Special Trustee	25, VII
Animal and Plant Health Inspection Service	7, III; 9, I
Appalachian Regional Commission	5, IX
Architectural and Transportation Barriers Compliance Board	36, XI

264

266

Agency	CFR Title, Subtitle or Chapter
Industry and Security, Bureau of	15, VII
Information Resources Management, Office of	7, XXVII
Information Security Oversight Office, National Archives and Records Administration	32, XX
Inspector General	
Agriculture Department	7, XXVI
Health and Human Services Department	42, V
Housing and Urban Development Department	24, XII, XV
Institute of Peace, United States	22, XVII
Inter-American Foundation	5, LXIII; 22, X
Interior, Department of	2, XIV
American Indians, Office of the Special Trustee	25, VII
Endangered Species Committee	50, IV
Federal Acquisition Regulation	48, 14
Federal Property Management Regulations System	41, 114
Fish and Wildlife Service, United States	50, I, IV
Geological Survey	30, IV
Indian Affairs, Bureau of	25, I, V
Indian Affairs, Office of the Assistant Secretary	25, VI
Indian Arts and Crafts Board	25, II
Land Management, Bureau of	43, II
National Indian Gaming Commission	25, III
National Park Service	36, I
Natural Resource Revenue, Office of	30, XII
Ocean Energy Management, Bureau of	30, V
Reclamation, Bureau of	43, I
Safety and Enforcement Bureau, Bureau of	30, II
Secretary of the Interior, Office of	2, XIV; 43, Subtitle A
Surface Mining Reclamation and Enforcement, Office of	30, VII
Internal Revenue Service	26, I
International Boundary and Water Commission, United States and Mexico, United States Section	22, XI
International Development, United States Agency for	22, II
Federal Acquisition Regulation	48, 7
International Development Cooperation Agency, United States	22, XII
International Development Finance Corporation, U.S.	5, XXXIII; 22, VII
International Joint Commission, United States and Canada	22, IV
International Organizations Employees Loyalty Board	5, V
International Trade Administration	15, III; 19, III
International Trade Commission, United States	19, II
Interstate Commerce Commission	5, XL
Investment Security, Office of	31, VIII
James Madison Memorial Fellowship Foundation	45, XXIV
Japan–United States Friendship Commission	22, XVI
Joint Board for the Enrollment of Actuaries	20, VIII
Justice, Department of	2, XXVIII; 5, XXVIII; 28, I, XI; 40, IV
Alcohol, Tobacco, Firearms, and Explosives, Bureau of	27, II
Drug Enforcement Administration	21, II
Federal Acquisition Regulation	48, 28
Federal Claims Collection Standards	31, IX
Federal Prison Industries, Inc.	28, III
Foreign Claims Settlement Commission of the United States	45, V
Immigration Review, Executive Office for	8, V
Independent Counsel, Offices of	28, VI
Prisons, Bureau of	28, V
Property Management Regulations	41, 128
Labor, Department of	2, XXIX; 5, XLII
Employee Benefits Security Administration	29, XXV
Employees' Compensation Appeals Board	20, IV
Employment and Training Administration	20, V
Employment Standards Administration	20, VI
Federal Acquisition Regulation	48, 29
Federal Contract Compliance Programs, Office of	41, 60
Federal Procurement Regulations System	41, 50

List of CFR Sections Affected

All changes in this volume of the Code of Federal Regulations (CFR) that were made by documents published in the FEDERAL REGISTER since January 1, 2015 are enumerated in the following list. Entries indicate the nature of the changes effected. Page numbers refer to FEDERAL REGISTER pages. The user should consult the entries for chapters, parts and subparts as well as sections for revisions.

For changes to this volume of the CFR prior to this listing, consult the annual edition of the monthly List of CFR Sections Affected (LSA). The LSA is available at *www.govinfo.gov*. For changes to this volume of the CFR prior to 2001, see the "List of CFR Sections Affected, 1949–1963, 1964–1972, 1973–1985, and 1986–2000" published in 11 separate volumes. The "List of CFR Sections Affected 1986–2000" is available at *www.govinfo.gov*.